CONTEMPORARY INSECT DIAGNOSTICS

Dedication

To Loni
my committed, charming, cute and compassionate wife.

Who doubles as my consultant, chum, center, cheerleader, copartner, chief of staff, crutch, comforter, co-conspirator, compass, celestial companion, comrade, confidant, constant, champion, cornerstone, counselor, and when need be, critic, cook and cuddler.

.... and those are just the 'C' words.

I love you.

You are beyond compare

CONTEMPORARY INSECT DIAGNOSTICS

THE ART AND SCIENCE OF PRACTICAL ENTOMOLOGY

Timothy Gibb

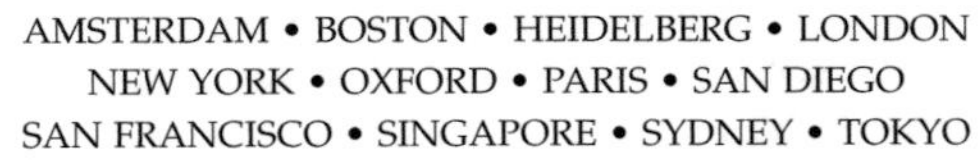

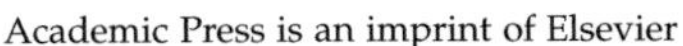

Academic Press is an imprint of Elsevier
The Boulevard, Langford Lane, Kidlington, Oxford, OX5 1GB, UK
225 Wyman Street, Waltham, MA 02451, USA

ISBN: 978-0-12-404623-8

British Library Cataloguing in Publication Data
A catalogue record for this book is available from the British Library

Library of Congress Cataloging in Publication Data
A catalog record for this book is available from the Library of Congress

For information on all Academic Press publications
visit our website at http://store.elsevier.com/

Typeset by TNQ Books and Journals
www.tnq.co.in

Printed and bound in the United States

Contents

6. Understanding the Client

7. Responding, Educating and Record-Keeping

8. Making Management Recommendations Using IPM

9. Networking

Acknowledgements

I express my sincere appreciation to all of my coworkers and friends in Cooperative Extension. Thanks to your support, encouragement and friendship over many years, this book finally has finally become a reality. One of the best decisions I have made in my life was choosing to work in Extension where the spirit of sharing, giving and helping others is the center of all that we do.

Photographic credits:

Thanks to John Obermeyer, who is the expert behind the camera for the majority of the photographs in this book - certainly all of the really good ones.

Preface

Diagnosticians include a widely diverse group of entomologists who are engaged in problem solving. Whether in academia or industry, every practicing entomologist can benefit with improved diagnostic skills.

This book is designed to provide a basic understanding of the workings of an insect diagnostic laboratory. It describes the required techniques and equipment as well as an update of new developments, tools, procedures, and advances in diagnostic technology, necessary for beginning or established laboratories. It describes the responsibilities of an insect diagnostician and provides case examples of practical, applied entomology.

Many entomology departments, particularly those housed in land grant universities, employ a specialist with primary responsibilities in insect diagnostics. In smaller departments, diagnostic responsibilities are divided among extension faculty.

Entomologists working in industry often have a minor but important responsibility in insect diagnostics. For example, pest auditors, consultants, and quality control managers in large food or pharmaceutical plants also have responsibilities for insect identification and making control recommendations.

Integrated pest managers rely on insect identification and the principles of pest management as a basis for what they do. Extension agents, professional urban pest managers and agricultural insect scouts all can be considered practicing insect diagnosticians.

As a resource text, this book is valuable in many college courses. It may be a primary reference text for courses in extension, insect IPM or insect diagnostics, and because of its practical nature, it provides valuable instruction for other college courses in insect classification and related pest management fields.

The principles taught herein are indeed the essence of practical entomology.

CHAPTER

1

Introduction

THE INTERACTIONS OF INSECTS AND PEOPLE

Insects have always been part of the environment in which humans live and regardless of our efforts to eradicate them, we will always live with insects in one way or another.

Insect pest management predates recorded history. The first time an early humanoid swatted at a mosquito could be considered the first incident of insect pest management. As time went on, the principles of pest management were built upon man first being able to recognize an insect as a pest insect and then study its behavior and biology sufficiently to devise a method of control. Whether it was physical (fly swatter), cultural (excluding them from a dwelling) or chemical (using smoke or plant-derived toxins to kill or repel), control methods were tested, improved upon and made available to others experiencing similar infestations.

Early pest management successes were tied very closely to medical, sanitation or other health-related advancements. Improved sanitation and hygiene not only prevented diseases but also insect pest problems. For example, the Romans built massive aqueducts that provided levels of cleanliness previously unavailable. Because bathing had the added benefit of suppressing lice and other personal pests, insects and diseases that tormented people decreased (Figure 1.1).

Unfortunately the road to better health and better pest management was not without setbacks. The Roman Empire eventually collapsed, and the medical and pest management advances of the Greeks and others were lost along with it. Books were destroyed, learned people were persecuted and science was lost. Civilization was thrust into a period called the Dark Ages where pests and pestilences ran rampant.

The bubonic plague was one of several disease epidemics vectored by insects that killed millions of people. Mankind no longer made the connection between pests and pestilences. Ignorance prevailed and death, famine and disease reigned for hundreds of years.

With time, however, came a transition from the Middle Ages to the Renaissance that gradually led to the rediscovery of science-based medicine and pest management. Progress continued during the 16th, 17th and 18th centuries. A scientific and systematic study of symptoms and causal agents advanced both medicine and pest management.

Subsequent centuries have seen unparalleled progress. We continue to learn better ways to manage insect pests, improving agriculture as well as human health.

Today, though we may boast of modern methods of insect control, on close inspection these techniques are primarily just improvements upon original methods. Once people understood the value of identifying the pest and then devising ways to control it through understanding its

Contemporary Insect Diagnostics
http://dx.doi.org/10.1016/B978-0-12-404623-8.00001-6

FIGURE 1.1 Our desire to manage pests has remained unchanged.

biology and behavior, pest management moved forward and human comfort and safety improved.

This is not to say that all cultures progressed at an even rate. Those that disregarded the importance of specific pest identification and associated biology, simply lumping all 'vermin' together, progressed more slowly.

The understanding of how each pest is unique and has its own biology and behavior was facilitated by the advent of scientific taxonomy. Effective pest management has always relied on an accurate identification of the pest and some understanding of its biology. These two go hand in hand.

Effective insect pest management strategies are built upon the foundation of accurate identification and an understanding of the insect's life history and habits.

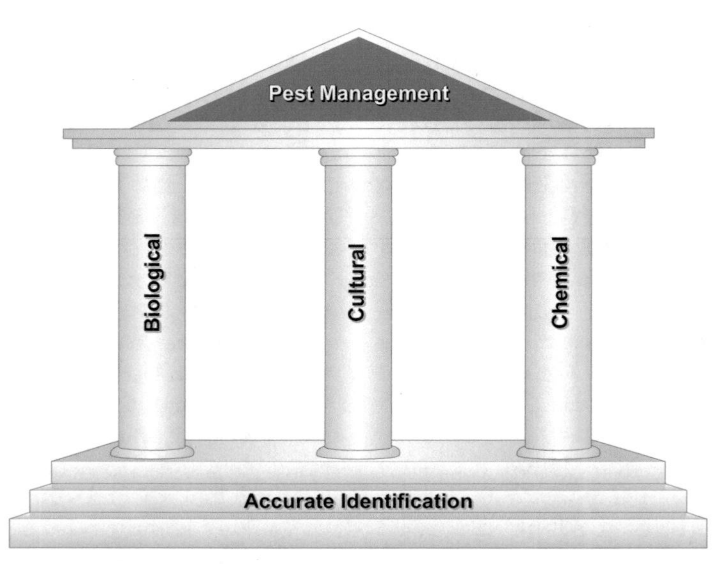

FIGURE 1.2 Effective insect pest management strategies are built upon the foundation of accurate identification and an understanding of the insect's life history and habits.

Even today, the first step in managing an insect pest is to accurately identify it, learn its biology and life history in an effort to determine how and why it became a pest, and then to use a commonsense and logical series of control techniques to not necessarily eradicate it but rather render it non-threatening. Shortcuts often lead to disaster (Figure 1.2).

THE VALUE OF INSECTS

Even a cursory study of our environment reveals that insects are nearly everywhere. Biologists clearly recognize insects as the most common animals on our planet. Today, more than 1.5 million species of insects have been named. This is three times the number of all other animals combined. Some scientists say that the list of insects that have been given names represents only a small fraction of all insects in nature. No doubt many are yet to be discovered.

Insects can be found in almost every conceivable habitat. Their size, shape, color, biology, and life history are diverse, making the study of insects challenging yet absolutely fascinating.

Insects are valuable to humans in many ways. In fact, without insects our society would not be the same. Insects pollinate many of our fruits, flowers and vegetables. We would not have many of the agricultural and horticultural products that we enjoy and rely upon without the pollinating services of bees. Insects not only improve agriculture, but their by-products (honey, beeswax, silk, dye, and other useful products), have become valuable commodities in our world.

Insects affect our world in other ways as well. They feed on a seemingly endless variety of foods. Many insects are omnivorous, meaning that they can eat a variety of foods, including living or dead plants and animals, fungus, decaying organic matter, and nearly anything else they encounter. Some can even digest wood. Still others are particular specialists and may totally rely on only one particular part of one species of plant to survive.

Many insects are predators or parasites, preying either on other insects or animals, including people. Such insects are valuable to people by helping to keep the population of pest insects or weeds tolerable.

Without insects to help break down and dispose of organic wastes such as dead animals and plants, our environment would be messy indeed. A great many insects and other arthropods assist by breaking down naturally occurring organic wastes. In so doing they also facilitate fertilization and aerate the soils they help create.

Insects are underappreciated for their role in the food web. They are the sole source of food for many amphibians, reptiles, birds and even mammals. They form an important part of the diet of many others, making them an essential component in nature.

As if these attributes were not enough, insects make one more valuable contribution: they make our world much more interesting. Naturalists and laypersons alike derive a great deal of satisfaction in watching ants work, bees pollinate, or dragonflies patrol. Can you imagine how dull life would become without having butterflies or lightning beetles to add interest to a landscape?

People benefit in many ways by sharing their world with insects.

INSECTS AS PESTS

In spite of all their many positive attributes, some small percentage of insects can become pests. We know from history that insects can feed on the blood of people or other vertebrate animals and can transmit diseases that create serious health concerns. Most parasitic or blood-sucking insects are not only an irritation, but they can be deadly because of their role in vectoring diseases. Even today, mosquitoes cause millions of deaths each year due to any of several diseases that they transmit, including malaria. Epidemics and even forecasted pandemics of various strains of influenza can be vectored by mosquitoes. Ticks, lice, other biting flies and bugs also transmit deadly diseases.

Nearly every kind of plant in nature is food to some insect or another. It follows that any agricultural crop or horticultural plant that is of value to people may also be consumed by insects and is at risk from the time the seed is planted until the crop is harvested, stored or consumed. This creates a conflict between people and insects. When insects destroy our food, we suffer.

A pest can generally be defined as any animal, plant, or other organism whose biology, behavior, or location places it in direct conflict with humans. Because some insects threaten human health, destroy food, damage structures or landscapes, or cause general annoyance or anxiety, they are considered pests.

Insects that harm people or animals, destroy foods, damage structures or products or that harm humans in any way are called pests. Insects that annoy or contaminate or that make life less pleasurable to people are also considered pests.

Interestingly, most people are much more aware of the relatively few insects that cause problems than they are of beneficial insects. As a result, most texts (including the one you are reading) tend to focus on the few insect pests that cause us harm.

Uninformed people may inaccurately generalize and consider all insects as 'bad or harmful' and in need of control. This is precisely the thinking of the Dark Ages. Enlightened societies recognize that the good done by the many beneficial insects far outweighs any bad caused by a few pest species. We must study each insect species and judge them based on their individual behavior and biology.

IDENTIFYING INSECTS

Common Names

Creating names is the human way of facilitating communication. By creating and using names we have a common reference point. Common insects almost always have a common name. This is a name given to them by laypeople and most often describes an insect's appearance, behavior or even more commonly, where it is found or what it is feeding on. Names such as 'beetles,' 'moths' or even 'bugs' refer to large groups or orders of insects. While these names are not incorrect, they offer little specificity in identifying just what insect is being discussed. Common names of families are also used. These names are much more detailed and offer a more specific idea. Still more specific are the common names of individual insect species such as 'stink bug,' 'tomato worm,' or 'burrowing beetle.' These names are valuable only to the extent that everyone associates these particular insects with the name. Even though common names are generally helpful, they can sometimes be ambiguous and confusing.

This creates serious confusion in communication. Ultimately, identifying or communicating about an insect by using common names alone becomes dubious and arbitrary. To avoid confusion, especially in science, communication must be based on a more exact naming system and reference to specific insect species.

Binomial Nomenclature

Insects, like all plants and animals, can be identified using scientifically accepted names. Science has a very discrete set of rules for reference to all organisms. Scientific names are usually based in Latin, and are composed of two parts: a genus name (always capitalized) and a species name (always lowercase). By rule, scientific names are always italicized when typed and underlined when hand-written.

The system for naming plants and animals was developed by a Swedish taxonomist by the name of Carl von Linne (1707–1778). He is known today as the father of taxonomic nomenclature and is famous for actually changing his own name Carl Linne to *Carolus linnaeus*, reflecting his dedication to Latinizing names of all organisms for classification.

This two-part (genus/species) naming system is called binomial nomenclature. It radically simplifies sharing research and discovery information in the scientific community. Whereas a single insect could go by two, three, or even more different common names depending on who is speaking and what geographical region they come from, a reference in binomial nomenclature can only correspond to one organism. Relying only on common names would make it difficult and frustrating for scientists to share research with one another around the world because it is never absolutely certain as to what organism one may be referring. It is less complicated for scientists to learn one Latin name for an organism rather than many common names in many languages and regions.

The scientific code of zoological nomenclature generally promotes dependability and reliability as well as accuracy and universality of an organism's scientific name. Having said that, stability of scientific names is not a given. Over time, taxonomic studies often lump, separate or reassign organisms to different taxon, making the science of taxonomy somewhat fluid. This makes it difficult for a generalist to know what the most current scientific name actually is. Entomologists specializing in the classification of insects are called insect taxonomists or systematists. They deal in heavily in scientific names and classification systems that have rigid and specific rules regarding their use. The public record (peer-reviewed published papers) is currently accepted as the official regulating process.

By comparison, common names, if established and agreed upon, are not subject to the same rules as are scientific names and thus can actually remain more stable than scientific taxa. In the early part of the 20th century, entomologists in America began discussing a method of overseeing a list of accepted common names, linked to their scientific (genus and species) name. Considerable effort has gone into creating and monitoring this list. Maintaining the list (Common Names of Insects and Related Organisms) (reference) is now a function of the Entomological Society of America. Many of our most common insects have become part of this list. To date, over 2000 common names have been approved. Great effort has gone into ensuring that the common name is complete and that it in some way describes the insect in question.

The list of common names is intended to contain insects of general concern or interest because of their economic or medical importance, striking appearance or abundant occurrence. It contains most of the insects that are common to homeowners, gardeners, farmers, health care providers, and pest managers in America.

Most laypersons prefer to use common rather than scientific names, as these are easier to employ in common speech and are easier to remember. For example, referring to an insect as a boxelder bug rather than *Boisea trivittata* (its scientific name) is usually preferred. In most cases, insect diagnosticians also prefer to use accepted common names over scientific names when possible. Keep in mind that not all insects have an accepted common name and furthermore, that many insects are similar in appearance. A person dedicated to the science of entomology is required to accurately and reliably identify them.

A more detailed examination of common and scientific names is found in Chapter 5 in the discussion of arthropod identification.

WHAT IS AN INSECT DIAGNOSTICIAN?

An insect diagnostician is an entomologist who not only identifies insects but also provides pertinent insect-related information to clients. An insect diagnostician functions much differently than an insect taxonomist, who constructs name classification systems based on relationships of form or morphology.

In the real world, as opposed to the scientific world, most laypeople are less interested in scientific names than in practical information. If a person finds an insect and seeks its identification, in most cases, a common name will suffice. However, a common name alone without added facts would be considered insufficient information. Most submitters want to know something about the insect in addition to its identification. Will it bite or harm me? Does it spread disease? Why is it in my home? Will it damage my tomatoes? Should I attempt to control it? When and how is the best method of managing it? What should I use to control it with? Will it persist? These are some of the many questions of submitters, both expressed and unexpressed.

An effective diagnostician can often anticipate questions even before they are asked. An adequate response to a sample submitted can be ascertained by whether or not a follow-up question is forthcoming. Responses take time and resources. Providing all of the information that a client asks for or should ask for in the first response obviates need for follow-up communication and is therefore the most appropriate response.

Responding properly to clients' questions is an art that diagnosticians must learn along the way. The information contained in the proper response is dictated by the ability to see a sample, note who submitted it, acknowledge what questions accompany the submission and also determine what questions should have been asked.

Professional insect diagnosticians almost always request that a submission form accompany an insect identification request.

Submission forms vary slightly but often have the same general components. They typically include information about the submitter, how and why the specimen was collected, what response is requested, and the urgency or priority of the request. Obviously, the more detailed the information is, the easier it may be for the diagnostician to make the proper diagnosis. Additional discussion about sample submission forms is found in Chapter 7.

VALUE OF INSECT DIAGNOSTICS

It is difficult to place a monetary value on insect diagnostics. In health care professions, it goes without saying that correctly diagnosing and treating patients saves lives. Properly diagnosing and recommending controls for insects and other arthropods also saves humans and animals from frustrations, pain and even death, but the major value of insects diagnostics lies in plant, crop, food, and structure protection. The value of crops, foods, plants and structures is enormous, thus the value of ensuring their protection is also enormous. Even so, placing a monetary value on diagnostic services is not easily done.

Perhaps the best measure of the value of insect diagnostics is made by assessing the satisfaction of the clients. Clients demand the services provided by insect diagnosticians. Most states support at least one public diagnostic laboratory, usually housing both insect and disease diagnosticians. These laboratories are located at land-grant universities. Based on informal surveys, satisfaction of clients is extremely high.

Private companies also employ their own diagnosticians, either full or part-time. Quality assurance personnel with skills in insect diagnostics are critical. Pest management firms also use diagnostic skills on a daily basis. County-based extension personnel are always more effective if they have developed skills in insect diagnostics.

In sum, insect diagnostics is important and would be sorely missed if it was lost in either the public or private domain.

INSECT IDENTIFICATION FORM

Submitter Information:

Name ______________
Address ______________
Company ______________
E-mail ______________ Tel# ______________ Fax ______________
Client name ______________ address ______________

Collection Information: *(Please provide as much information as possible. Circle options or add words as appropriate.)*

Date Collected ______________
Location (county, city, etc) ______________
Collector (home owner, pest manager, etc) ______________
Specific locality where found (on tree, in basement, etc) ______________
Description of infestation ______________
found, damage ______________
Behavior of insect at collection ______________
Biting people or pets ______________
Feeding on plant ______________
Crawling, flying, jumping ______________
Damage description ______________
Plant (boring, leaf feeding, etc) ______________
Wood ______________
Fabric textile ______________
Structural ______________
Food ______________
Biting, stinging, annoying ______________
Contamination ______________
Other ______________

Information Requested

Control information ______________
Identification only ______________
Problem potential ______________
Other (please explain) ______________

Priority: Urgent Routine ______________

CHAPTER

2

Equipping a Diagnostic Laboratory

INTRODUCTION

A functional insect diagnostic laboratory can range anywhere from a simple setup, consisting only of desk, microscope and reference book, to a multimillion-dollar state of the art facility which meets federal containment rules and has an in-house database, customized microscopes, rearing chambers, museum-quality reference collections, and interactive, real-time photography and communication equipment. Differences are primarily dictated by budget and purpose.

Equipment, tools and procedures required for sampling, preserving, and identifying insects will be discussed below. Inherent in any discussion of tools is an understanding of proper techniques for their use.

Most insect diagnosticians are experienced in the use of insect collection and handling tools through their formal training in entomology. While we will discuss some of these tools generally, we will devote more time in describing tools and techniques used for special cases and in unusual circumstances.

BASIC TOOLS AND SUPPLIES FOR HANDLING SAMPLES

Standard supplies and tools that are used in any biological laboratory are sure to find a place in a diagnostic laboratory. In addition to standard supplies, diagnosticians may find the following tools valuable. Fine, lightweight forceps are strongly recommended for collectors and diagnosticians alike. Specialized forceps may be selected depending upon individual need (Fig. 2.1).

Lightweight spring-steel forceps are designed to prevent the crushing of fragile and small insects. Extra-fine precision may be obtained with sharp-pointed 'watchmaker' forceps (Fig. 2.1A); however, care must be taken not to puncture specimens. When possible, grasp specimens with the part of the forceps slightly behind the points. Curved forceps often make this easier. When the forceps are not in use, their sharp tips should be protected. This can be accomplished by thrusting the tips into a small piece of Styrofoam or cork, or by using a small section of flexible tubing as a collar.

Other supplies and equipment used in a diagnostic laboratory include eye droppers, a selection of insect pins, straight and curved-tip teasing needles, fine point scissors, Syracuse watch glass, a soft polyethylene squeeze bottle for dispensing alcohol, a pocket knife, pinning boards, a microscope-adaptable insect pin observation stage, pinning blocks, insect trays, polyethylene foam blocks for holding pinned specimens, a card punch and a syringe (Gibb, 2006).

The temporary holding of pinned specimens is of prime importance. Blocks of Styrofoam or polyethylene suit this purpose nicely. Use of

Contemporary Insect Diagnostics
http://dx.doi.org/10.1016/B978-0-12-404623-8.00002-8

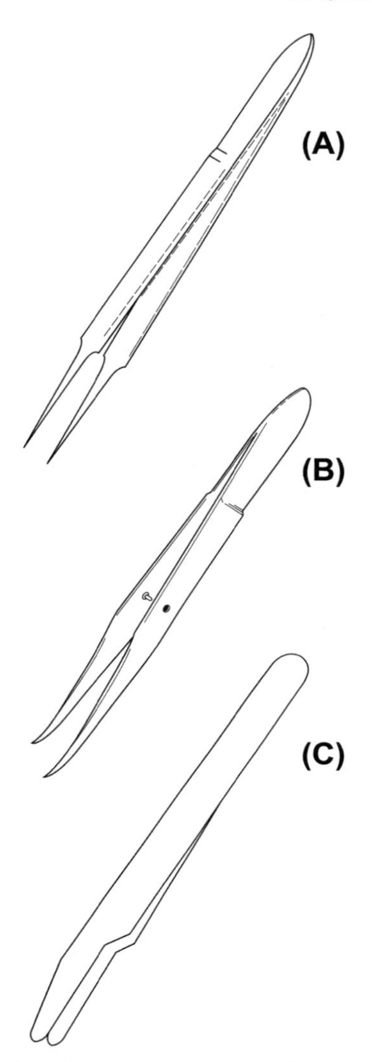

FIGURE 2.1 Forceps for insect collecting. (A) Fine watchmaker forceps (B); curved metal collecting forceps; (C) soft forceps.

unit trays with polyethylene or Plastazote bottoms is even better, as they will not tip over and have sides to protect the specimen as it is being handled and moved to and from the reference collection and the microscope.

Glass vials are indispensable in an insect diagnostic laboratory. They are often used by frequent submitters to send samples to the diagnostician. Diagnosticians, in turn, use them for temporarily holding specimens while they are being diagnosed, as well as for permanent storage in a reference collection. Sample vials of various sizes can be readily obtained from biological supply outlets. Screw top vials are most commonly used with specially designed leak-proof caps for collecting, shipping, processing and permanent storage of specimens. The utility of these vials is great in that they can be used to store many species, sizes, and life stages of arthropods, can be conveniently stored in storage racks, and offer transparent glass that allows for instant observation and label identification (Figs. 2.2 and 2.3).

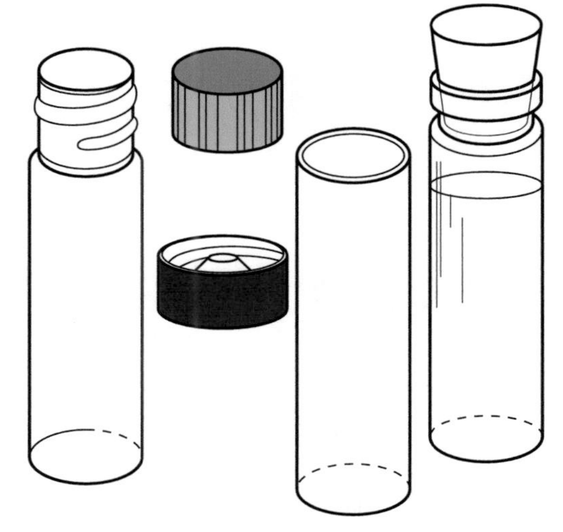

FIGURE 2.2 Sample vials.

A supply of ethyl alcohol (70–80%) is a staple in a diagnostic laboratory. The general procedure for preserving and storing arthropods in an insect diagnostic laboratory is to contain them in alcohol. Alcohol is also generally considered the

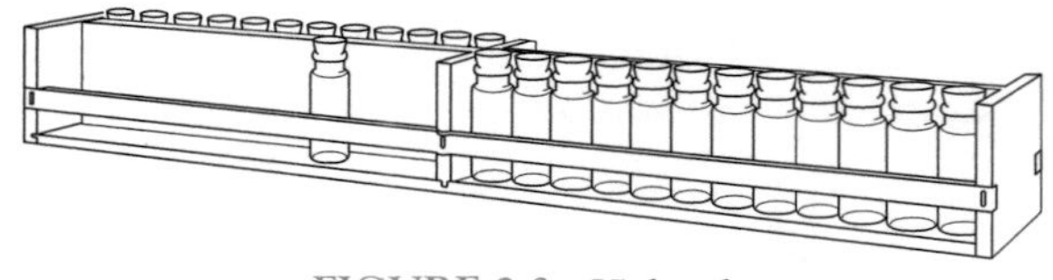

FIGURE 2.3 Vial rack.

best general killing agent. Dropping a live insect into a vial of alcohol simultaneously preserves and dispatches it.

Alcohol in a laboratory must be properly labeled and is easily dispensed using small plastic squirt bottles designed for this purpose.

COMMUNICATION DEVICES

In its most rudimentary form, an insect diagnostic laboratory is a place where samples can be received, stored, and examined by a diagnostician, who then responds to the submitter. If nothing else, a functioning diagnostic laboratory is a center for communication. Requests come in, responses go out. Like any functioning service laboratory it must have basic communication equipment such as a telephone, access to mail delivery, and a computer. These are all vital vehicles of communication that clients use to submit samples and diagnosticians use to return responses. Some clients actually hand deliver samples (walk-ins) and while this is a submission method that must be made available, it is far from the most common method for either receiving samples or responding to them.

We will not discuss computers and internet connections, other than to state the obvious: they are essential. Not only do they provide a means of instantaneous electronic communication, but they also provide access to internet resources and data storage. Much of what a diagnostician does on a day to day basis revolves around the use of a computer.

MICROSCOPES AND MICROSCROPY

Both compound microscopes and dissecting microscopes can be found in diagnostic laboratories; however, of the two, the dissecting scope is most valuable for viewing insects and their relatives. When compared, the dissecting microscope does not offer magnification equal to compound microscopes. Dissecting microscopes have an objective lens which often allows a continuous range of magnification (from 2–40×), controlled by a magnification knob.

A dissecting microscope is configured to allow magnification of three-dimensional objects, as well as objects much larger than the compound microscope can accommodate. Furthermore, the two separate lenses of the binocular dissecting microscope allow one to see objects in three dimensions; i.e., in stereo.

Another difference between dissecting and compound microscopes is that the stage of the dissecting scope is much farther from the objective lens, allowing a large object to be placed on the stage and manipulated so that it can be viewed from any of several angles (Fig. 2.4).

Dissecting microscopes utilize two types of light: incident light (direct illumination) or transmitted light. In most instances, however, diagnosticians rely only on direct lighting. In this case the illuminator can be free-standing, mounted in an opening in the arm of the microscope,

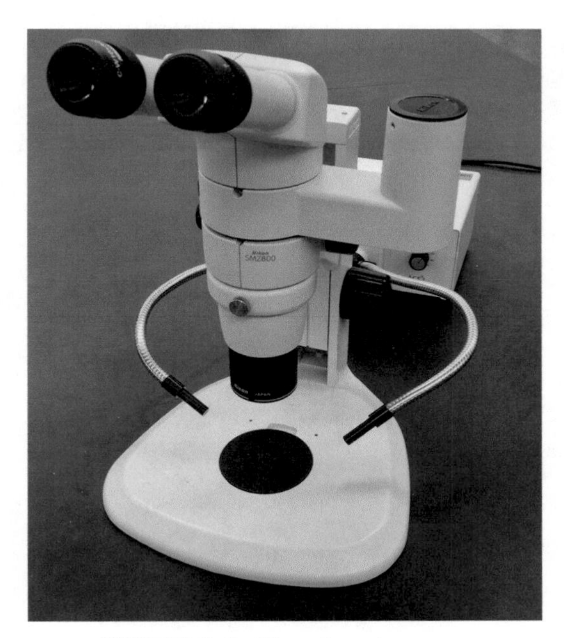

FIGURE 2.4 Dissection microscope.

or in an adapter ring attached to a separate illuminator base (transformer). Many illuminators have rheostats to allow adjustment of the intensity of the light. Light is directed onto the specimen. The specimen is placed on a stage insert which has a reversible white side and a black side for preference in observing the specimen.

Microscopic Examination Techniques

Insects are mounted on pins to allow them to be moved, sorted, or viewed under a microscope when necessary. Many of the external morphological features of insects, which are important for identification, can be seen under low-power microscopy, such as with a dissecting scope. Pinned specimens can be handled most easily and with the least amount of damage if they are transferred from their permanent storage box to a small observation base constructed of cork and a small amount of modeling clay or Plasticine® mounted on a wooden or cardboard base (Fig. 2.5).

The base should be small enough to sit directly on the stage of the dissecting microscope, allowing enough room for it to be turned as needed. The pinned insect can then be inserted into either the clay or the cork at nearly any angle such that observations can be made of any side of the insect. Commercially available observation stages (Fig. 2.6) can be purchased that offer even more flexibility under a microscope.

There are situations where a portion of an insect is key to making specific diagnoses. Legs, wings and genitalia are then separated from the body of the insect and preserved in such a manner that they are easier to manipulate under a microscope. Large, thick, or complex structures that must be examined from several angles may be best preserved in a liquid (such as glycerin) and held in microvials. These can be labeled and stored with the whole specimen or may be kept separately. In either case, proper labeling techniques are mandatory.

Small insects and mites (thrips, whiteflies, aphids, scale insects, leas, lice, and some parasitic wasps) often must be examined at relatively high magnification. A quality dissecting microscope (80–250×) or a compound microscope (320–970×) is required for this level of magnification. A microscope with this level

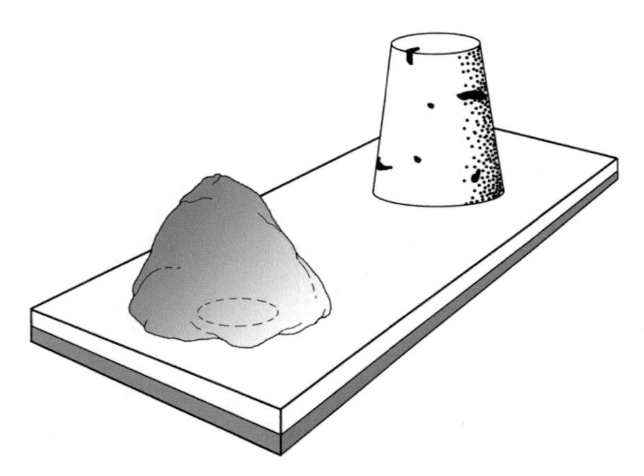

FIGURE 2.5 Observation block.

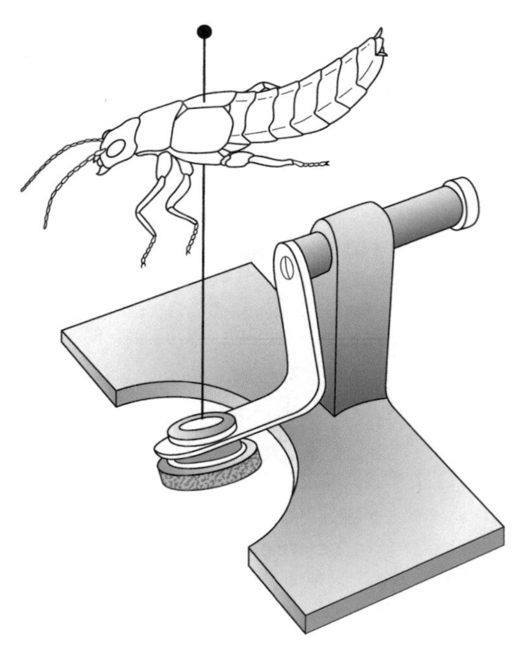

FIGURE 2.6 Commercially available observation stage.

of magnification is also necessary to clearly observe minute details of larger insects, such as the genitalia, hair patterns, sutures, and bristles.

Small specimens (ca. 1–4 mm long) and parts of specimens must be specially prepared and often mounted on microscope slides for higher power magnification. We call this process whole-mount preparation (Page 40 - Whole Mount Preservation).

SAMPLE HOLDING EQUIPMENT AND SUPPLIES

Freezer and Refrigerator

In addition to the usage of alcohol and glass vials as a means of preserving insects in the diagnostic laboratory, samples may also be held using low temperatures. A small refrigerator can be a valuable tool in a diagnostic laboratory by providing short-term storage of samples that may otherwise deteriorate rapidly at room temperature. Samples that contain fresh plant parts are especially sensitive to decomposition. Because diagnosis is linked so tightly with a good quality sample, a refrigerator is valuable to sample preservation.

In addition to increasing the holding time for plant samples, refrigeration also does the same thing for insects. Live insects slow down their metabolisms significantly under cold temperatures and thus may be held longer without damage or spoiling. Diagnosticians often take advantage of the refrigeration of live specimens as a way of decreasing their activity so that they can be readily handled, observed or photographed without them flying or crawling away. Remember that upon reintroduction to room temperatures, however, activity is quickly regained.

Freezing specimens or samples is a quick way of lengthening the holding or turn-around time of samples. During busy times it may not be possible for diagnosticians to preserve specimens for long-term storage. In such cases, a freezer is a very convenient location for short to medium-term preservation. To a point, the lower the temperature, the longer the samples can be stored. It must be remembered, however, that over time at extremely low temperatures specimens gradually desiccate (become freeze-dried) and eventually will be ruined. Even so, a freezer can be used to keep specimens fresh for a month or two and is a great convenience.

Freezers also have the advantage of quickly killing insects before they have a chance of damaging themselves by breaking wings or legs inside a collecting container. Using a freezer to dispatch, as well as to temporarily store insect samples, can be a great boon to the diagnostician.

A small refrigerator/freezer combination can be very valuable in a diagnostic laboratory by using lower temperatures to slow down the natural decomposition of organic samples, as well as providing storage of specimens.

REFERENCES AND RESOURCES

Text Resources

Some might say that diagnosticians are only as good as their reference materials and networking links. It is true that no single person can be expected to know everything there is to know, especially about an area as general as insect diagnostics. Likewise, there is no single resource text or internet site that covers everything that an insect diagnostician might need. Diagnostic laboratories that have been functioning for any time at all will have accumulated a series of reference texts, keys, guides and manuals, as well as a collection of internet sites bookmarked for routine use. Choices of resources are many and varied. Whichever resource provides a diagnostician with the correct answer becomes the correct choice. Differences in resource selection are largely due to personal preferences of the diagnostician, often based on whatever he or she is accustomed to using.

Obviously, not all reference texts, keys and other print resources can be listed here. However, the following list may provide a starting point for new laboratories:

Starter References for New Laboratories

Arnett, R.H., Jacques Jr., R.L., 1981. Guide to Insects. 511 pp. Simon and Schuster, New York.

Bennett, G., Owens, J.M., Corrigan, R.M. Truman's Scientific Guide to Pest Control Operations, 520 pp. Santa Monica, CA: Advanstar Communications.

Beatty, G.H., Beatty, A.F., 1963. Efficiency in caring for large Odonata collections. Proc. North Cent. Branch Entomol. Soc. Am. 18, 149–153.

Borror, D.J., White, R.E., 1998. A Field Guide to the Insects of America North of Mexico. 404 pp. Houghton Mifflin, Boston.

Cranshaw, W., 2004. Garden Insects of North America. 656 pp. Princeton University Press, Princeton and Oxford.

Fichter, G.S., Zim, H.S., 1966. Insect Pests. 160 pp. Golden Press, New York.

Gibb, T.J., Oseto, C.Y., 2006. Arthropod Collection and Identification – Laboratory and Field Techniques. 311 pp. Elsevier/Academic Press, New York.

Jacques, H.E., 1987. How to Know Insects. 205 pp. William C. Brown, Dubuque, IA.

Johnson, W.T., Lyon, H.H., 1976. Insects that Feed on Trees and Shrubs. 556 pp. Cornell University Press, Ithaca, NY.

Kaston, B.J., 1978. How to know the spiders. 272 pp. William C. Brown Co., Dubuque, Iowa.

Levi, H.W., Levi, L.R., 1990. Spiders and their kin. 160 pp. Golden Press, New York.

Mallis, A., 2004. Handbook of Pest Control. 1400 pp. GEI Media, Inc, Toronto.

Metcalf, C.L., Flint, W.P., 1962. Destructive and Useful Insects. 1087 pp. McGraw-Hill Inc, New York.

Mitchell, R.T., Zim, H.S., 1964. Butterflies and Moths. 160 pp. Golden Press, New York.

Peterson, A., 1960. Larvae of Insects. Volume 1, 416 pp. Volume 2, 315pp. Edwards Brothers, Inc, Lillington, NC.

Roe, R.M., Clifford, C.W., 1976. Freeze-drying of spiders and immature insects using commercial equipment. Ann. Entomol. Soc. Am. 69, 497–499.

Shull, E.M., 1972. Butterflies of Indiana. 262 pp. Indiana Academy of Science, Muncie, IN.

Stehr, F.W., 1987. Immature Insects. 754 pp. Kendall/Hunt Publishing Company, Dubuque, IA.

Stein, R., Eisenberg, W.V., Brickey, Jr. P.M., 1968. Extraneous materials. Staining technique to differentiate insect fragments, bird feathers, and rodent hairs from plant tissue. J. Assoc. Off.Anal. Chem. 51, 513–518.

Triplehorn, C.A., Johnson, N.F., 2005. An Introduction to the Study of Insects. 7th ed. Thomson: New York. 864 pp.

Westcott, C., 1973. The Gardener's Bug Book. 689 pp. Doubleday, New York.

Zim, H.S., Cottam, C., 1956. Insects. 160 pp. Golden Press, New York.

Internet Resources

Similarly, many valuable internet sites are available to assist diagnosticians. These are growing in number but because of rapid changes, additions and improvements, a list of web addresses cannot be recommended in this text.

Discussion concerning the use of print and electronic references and resources to identify arthropods is provided in Chapter 5.

PHOTOGRAPHIC EQUIPMENT AND METHODS

Although diagnosticians are usually on the receiving end of digital photography (where digital photos are taken by someone else and sent to the diagnostician), it is important for them to have at least a working understanding of photography equipment and techniques. Additionally, diagnostic laboratories often are equipped with a portable camera of some kind. Diagnosticians must be able to assist clients in understanding how best to obtain diagnostic photographs and to be able to create their own when necessary.

Due to rapid technological changes, recommendations as to the absolute best photographic equipment for a diagnostic laboratory are more a matter of personal preference than in the past. In times past, a 35 mm, handheld, single lens reflex (SLR) camera was nearly the only option for high-quality photography. More recently, however, compact cameras and even telephone cameras have advanced to the point where they can provide remarkably clear photographs.

The advent of digital photography has led to its status as the current method of choice. It offers distinct advantages over film photography to a functioning insect identification laboratory. Because of the advancements and wide range of options currently available, it is safe only to recommend obtaining the best quality digital photographic equipment that a budget will allow.

A huge advantage of modern digital cameras is that they often come complete with auto-focus lenses. These allow the photographer to use the camera in a truly 'handheld' manner, following the insect subject as it moves, rather than relying on a tripod and a still and quiet insect as in macro or close-up photography.

Anyone who has attempted to capture images of insects understands that insect photography is a complex subject and recommendations of photographic equipment can quickly become a matter of personal preference.

Camera Options

A compact digital camera can be used for insect photography if the photographer understands its limitations. In their favor, compact cameras have become very affordable and highly portable (very light and easy to pack and carry). Many photographs submitted to diagnosticians are taken with compact or even cell phone cameras, both because of their portability and because they can very easily be used to send the resulting photos via the internet. That said, digital compacts tend to be limited by their small sensors that are inherently noisier than the larger sensors fitted to digital single reflex (DSLR) cameras. They are also affected by compromises inherent in the design of their lenses and focusing systems. Consequently, such cameras are best suited to the casual insect photographer. Another option is a so-called 'bridge camera,' which has a SLR-style body, but with a zoom lens that is permanently attached to the camera. These function better for general insect photography and can produce excellent results, but they are unlikely to be as good for close work as a DSLR with a dedicated macro lens. Currently, a DSLR in the hands of a knowledgeable user is the best option for insect photography.

Microscopic Photography

Microscopes that have digital photographic capabilities are a great luxury in a diagnostic laboratory. They can greatly assist in education seminars, workshops, and other events that diagnosticians use to help educate their clients. Photography using microscopes can enhance a reference collection of stored images, especially of small insects or key character differences between species. In addition, this technology is beginning to be used to share electronic (even real time) images with both clients and peers to increase the speed and efficiency of insect and mite identification (see DDIS section).

Dedicated digital cameras that are mounted on a microscope can provide excellent images of very tiny subjects. Such a camera/microscope combination is of great value in a diagnostic laboratory for identification, education, and referencing materials (Fig. 2.7).

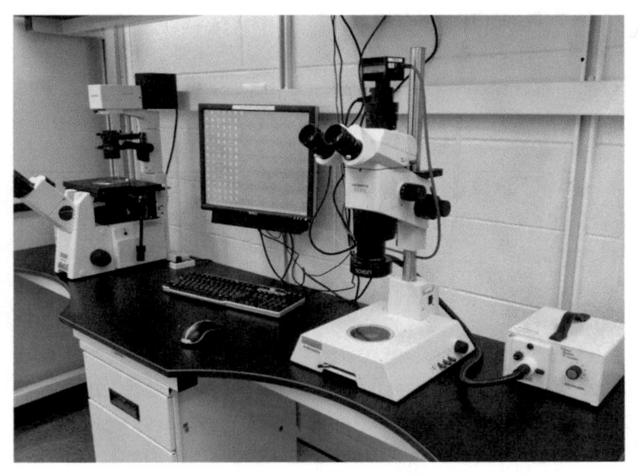

FIGURE 2.7 Integrated computerized microscopic photography equipment.

Undoubtedly, both microscopic photography techniques and technology will continue to advance and will significantly impact how an insect diagnostician functions well into the future.

Photomacrography (Macrophotography)

Close-up photography using an SLR camera can be enhanced by using special equipment. This includes employing true macro lenses, extension tubes, and close-up filters.

A true macro lens is a dedicated lens that does not require any special attachment to achieve 1:1 or life-size magnification. These lenses are 'prime' lenses, meaning that they are of a fixed focal length (usually 50, 100 or 180 mm).

Extension tubes are hollow black tubes that are placed between a lens and the camera body. They simply move the lens elements farther away from the film plane, thus increasing magnification. The resulting magnification using an extension tube is relative to the focal length of the lens used. For example, a 50 mm extension tube on a 50 mm macro lens will effectively double the magnification from 1× to 2×.

Close-up filters are simply diopters that magnify whatever is in front of them and screw into the front of a lens (either macro or non-macro). They are sometimes described as reading glasses for a camera; like reading glasses, they come in various 'powers.'

The trade-off that one achieves with greater magnification comes at the expense of 'depth of field' or the amount of the image that is in sharp focus. This is an inverse relationship.

Specialized photographic techniques include computerized multi-layering focused images of a specimen in a way that makes the final shot appear both magnified and totally in focus (Fig. 2.8).

FIGURE 2.8 Multi-layered photograph Clerdidae: (*Tilloidea*, unknown sp.). (Courtesy of Bobby C. Brown Ph.D. USDA-APHIS Identifier. Purdue University.)

Photographing Live Insects

There is a distinct difference in both the equipment and the skills necessary to photograph 'still' insects in the laboratory compared to photographing 'live' insects in nature. Producing high-quality live shots relies on three elements:

1. Approaching the subject without frightening it.
2. Getting it in focus quickly.
3. Shooting a photograph of high image quality.

Lenses, crop sensors, and depth of field are all adjustments that can influence these three elements. Keep in mind that shooting insects in their natural environment is largely about working distance, because the further away the photographer is from the insect, the less likely it will be disturbed or frightened. The distance between the insect and photographer increases with smaller sensors and stronger telephoto lenses.

The third element, image quality, has much to do with ISO setting, number of megapixels, light quality, and how stationary the camera remains. Producing quality images depends upon a working knowledge of the camera's capabilities in these four areas. Many higher-end cameras automatically adjust for these variables but the auto can be overridden when necessary. Image quality is as much under the control of the photographer as it is of the camera. Like other arts, producing quality photographs is a matter of practice and learning.

Digital images can be cropped, enlarged, lightened and recolored to a degree after they have been taken. This means that the final photo is not identical to that taken in the field. As technology improves, the ability to produce high-quality macro images of insects and mites will also improve.

OBTAINING SAMPLES AND MAKING A REFERENCE COLLECTION

Active Sampling Methods and Equipment

Sampling and collecting arthropods are two concepts that overlap in many ways and are often used interchangeably. However, there is a subtle difference between the two terms. For the purposes of this text we will define sampling techniques as methods for obtaining a specific arthropod for the purpose of identification. Sampling techniques are used or recommended by diagnosticians when they want to determine what or how many arthropods are at work in a certain situation. Sampling ties in with a diagnostician's responsibility for providing management recommendations more closely than does collecting. Collecting insects, i.e., for the purpose of building a reference collection, is a more general activity. Collecting usually involves obtaining a greater number of specimens than does sampling.

Catching specimens by hand may be the most simple and most common method of sampling, but this method is not always productive because of the evasive behavior of many insects, nor advised due to the potential for stings or bites, repulsive chemicals, or urticating setae from contacting certain arthropods. Additionally, some insects are not active at the times and places that are convenient.

In these cases, particular kinds of equipment and special sampling methods are needed. Equipment and methods described here have general application. Advanced studies of specific insect or mite groups may have developed unique procedures for collecting and surveying.

The equipment used for insect or arthropod capture is not necessarily elaborate or expensive. In many instances, a container with a lid is all that is needed. Motivated homeowners generally find a way to capture a specimen in a container without the need of specific recommendations. Use of a pair of forceps, a tissue or other simple tool to allow capture without actually touching the specimen will increase the comfort of most submitters. Very tiny insects may be captured by using a moistened Q-Tip or artist's paint brush, then transferring them to an appropriate container containing alcohol.

Sticky tape is often used to adhere insects and arthropods for submission. While not uncommon, this method is discouraged because it is very difficult or impossible for a diagnostician to remove the specimen from the tape intact, making identification difficult. In addition, preserving arthropods on tape is seldom successful because the glue tends to break down over time.

Specimens that have been knocked down with a flyswatter, shoe or newspaper are often submitted. In many cases, identification is still possible even though the specimen has been damaged. In a similar fashion, an aerosol insecticide can be used to 'knock down' or incapacitate flying insects long enough for capture and submission.

Several sampling tools and devices have been designed to make collection simple and effective. These include aspirators, brushes, and insect nets.

Aspirators are handy for collecting many kinds of small-bodied or agile insects and mites. They are of simple construction and may be obtained via a biological supply house or made by hand (Fig. 2.9).

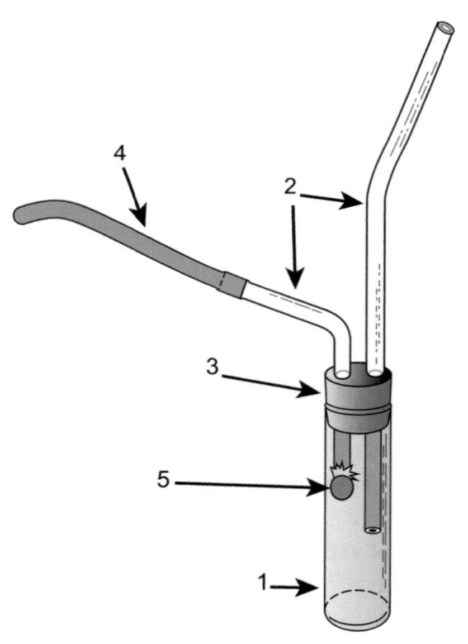

FIGURE 2.9 Components of a vial aspirator. 1. Vial. 2. Glass copper or tubing. 3. Rubber stopper. 4. Rubber or plastic tubing. 5. Mesh cloth and rubber band.

A small, fine brush (Fig. 2.10) (camel's hair is best) aids in collecting minute specimens.

Moistening the tip allows tiny specimens to adhere to the brush as they are quickly transferred to a proper vial.

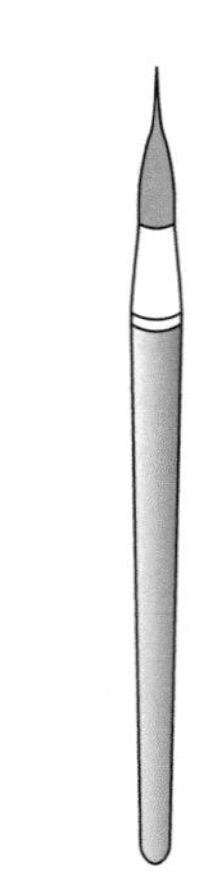

FIGURE 2.10 A fine-tip camel's hair brush.

Insect collecting nets are central in most entomological activities, including insect diagnostics. Nets come in three basic forms: aerial, sweeping, and aquatic (Fig. 2.11).

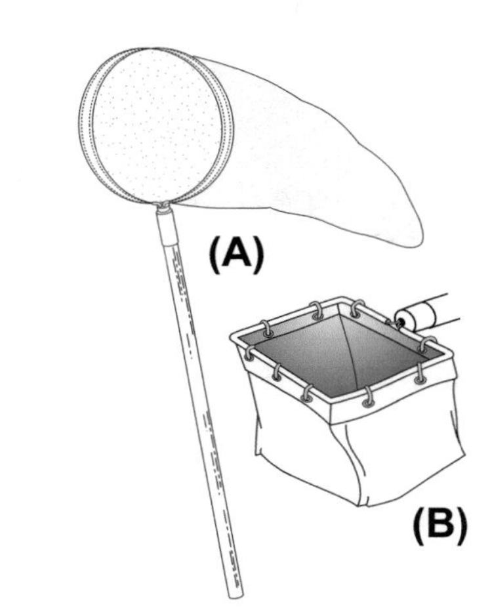

FIGURE 2.11 Insect collecting nets. (A) Aerial or sweep net. (B) Aquatic net.

The aerial net is designed especially for collecting large-bodied flying insects. Both the bag and handle are relatively lightweight. The sweeping net is similar to the aerial net, but the handle is stronger and the bag is more

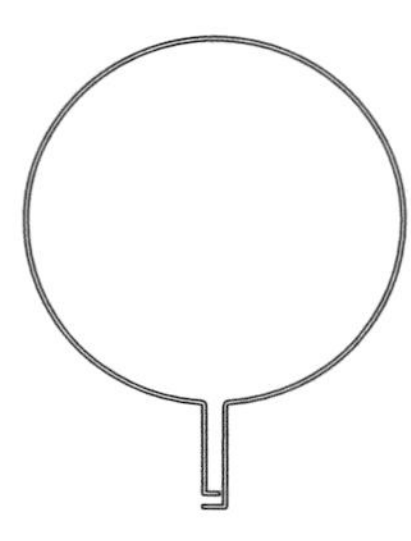

FIGURE 2.12 A common design for the rim of an insect collecting net made from steel wire.

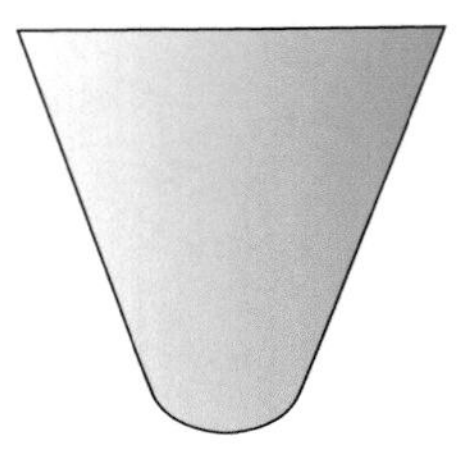

FIGURE 2.13 General shape of the bag on an aerial or sweep net.

durable to withstand being dragged through dense vegetation. The aquatic net is used for gathering insects from water and is usually made of metal screening or heavy scrim with a canvas band affixed to a metal rim. A metal handle is advisable because wooden ones may develop slivers after repeated wetting.

The net chosen depends on the kind of insects or arthropods intended for collection. Several kinds of nets, including collapsible models with interchangeable bags, are available from biological supply houses, but anyone with a little mechanical ability can make a useful net. The advantage of a homemade sweep net is that its size and shape can be adapted to the needs of the user, to the kind of collecting intended, and to the materials available.

Net-constructing materials and directions include the following:

1. A length of heavy (8-gauge) steel wire for the rim, bent to form a ring 30–38 cm in diameter (Fig. 2.12). Small nets 15 cm or smaller in diameter sometimes are useful, but nets larger than 38 cm are too cumbersome for most collecting.
2. A strong, light fabric, such as synthetic polyester, through which air can flow freely. Brussels netting is best but may be difficult to obtain; otherwise, nylon netting, marquisette, or good-quality cheesecloth can be used. However, cheesecloth snags easily and is not durable. The material should be folded double, and should be 1.5–1.75 times the rim diameter in length (Fig. 2.13). The edges should be double-stitched (French seams).

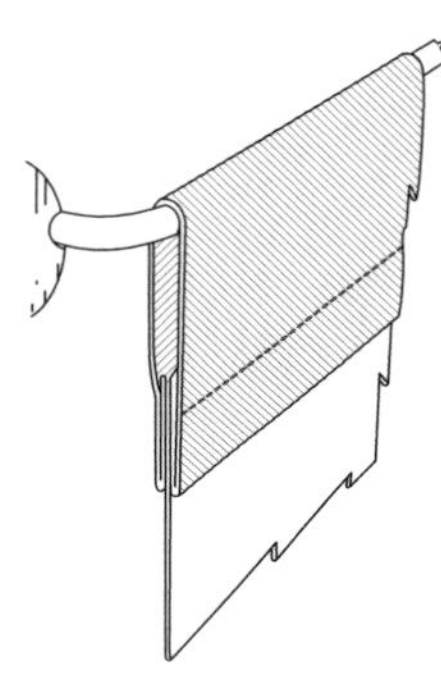

FIGURE 2.14 Attachment of the net to the rim of the collecting net.

3. A strip of muslin, light canvas, or other tightly-woven cloth long enough to encircle the rim. The open top of the net bag is sewn between the folded edges of this band to form a tube through which the wire rim is inserted (Fig. 2.14).
4. A straight, hardwood dowel about 19 mm in diameter and 105–140 cm long (to suit the collector). To attach the rim to the handle, a pair of holes of the same diameter as the wire are drilled opposite each other to receive the bent tips of the wire, while a pair of grooves as deep and as wide as the wire are cut from each hole to the end of the dowel to receive the straight part of the wire (Fig. 2.15).

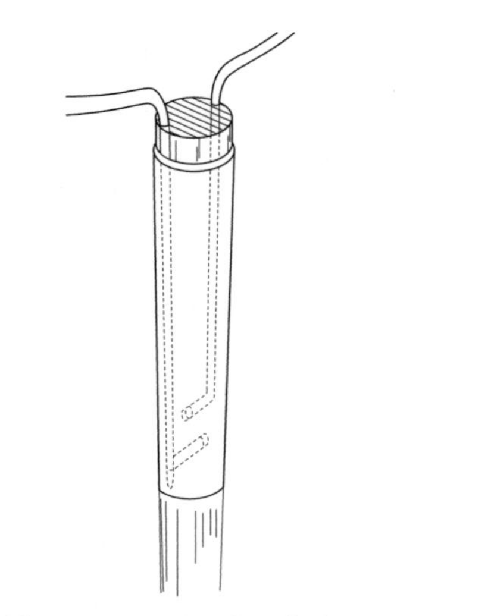

FIGURE 2.15 How to make a handle for an insect net.

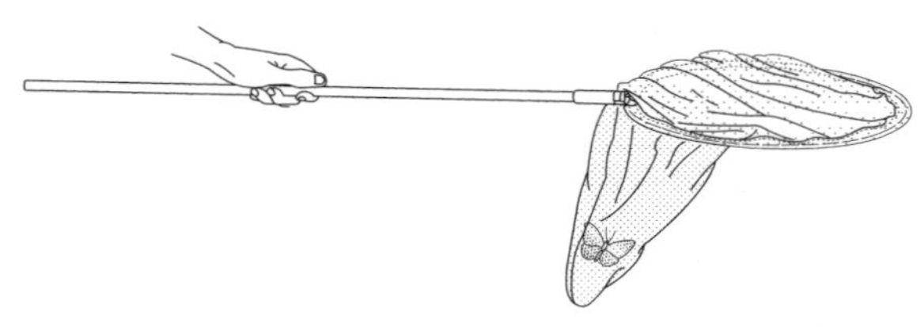

FIGURE 2.16 Proper procedure showing a bag flipped over the rim of a collecting net to prevent escape of the insect.

5. A tape or wire to lash the ends of the rims tightly into the grooves in the end of the handle. This may be electrician's plastic tape or fiber strapping tape commonly used for packaging. If wire is chosen, the ends should be bound with tape to secure them and to keep them from snagging. A close-fitting metal sleeve (ferrule) may be slipped over the rim ends and held in place with a small round-headed screw instead of tape or wire lashing (Fig. 2.15).

After the net has been placed on the rim, the ends of the band should be sewn together and the rim ends fastened to the handle. The other end of the handle should be filed to remove sharp edges. The net is then ready for use.

Efficient use of a net is gained only with experience. Collection of specimens in flight calls for the basic stroke: swing the net rapidly to capture the specimen, then follow through to force the insect into the very bottom of the bag. Twist the wrist as you follow through so that the bottom of the bag hangs over the rim (Fig. 2.16); this will entrap the specimen. If the insect is on the ground or other surface, it may be easier to use a downward stroke, quickly swinging down on top of the specimen. With the rim of the net in contact with the ground to prevent the specimen from escaping, hold the tip of the bag up with one hand.

Most insects will fly or walk upward into the tip of the bag, which can then be flipped over the rim to entrap the specimen. Sweeping the net through vegetation, along the sand and seaweed on beaches, or up and down tree trunks will catch many kinds of insects and mites.

An aerial net may be used in this way, but the more durable sweeping net is recommended for such rough usage. After sweeping with the net, a strong swing through the air will concentrate anything into the tip of the bag, and then, by immediately grasping the middle of the net with the free hand, the catch will be confined to a small part of the bag. Only the most rugged sweeping net may be used through thistles or brambles. Even some kinds of grasses, such as sawgrass, can quickly ruin a fragile net. Burrs and sticky seeds are also a serious problem.

The catch may be conveyed from the bag to a killing jar in a number of ways. Single specimens are transferred most easily by lightly holding them in a fold of the net with one hand while inserting the open killing jar into the net with the other. While the jar is still in the net, cover the opening until the specimen is overcome; otherwise, it may escape before the jar can be removed from the net and closed.

To prevent a butterfly from damaging its wings by fluttering in the net, squeeze the thorax gently through the netting when the butterfly's wings are closed. This will temporarily paralyze the insect while it is being transferred to the killing jar (Fig. 2.17).

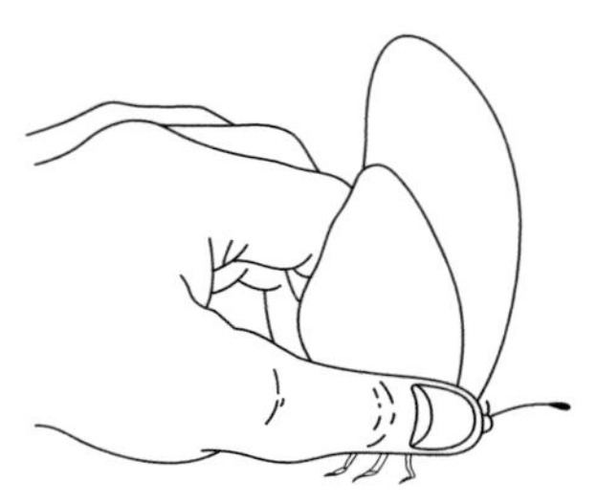

FIGURE 2.17 Technique for paralyzing a butterfly by squeezing its thorax between thumb and index finger.

Experience will teach how much pressure to exert. Obviously, pinching small specimens of any kind is not recommended. When numerous specimens are in the net after prolonged sweeping, it may be desirable to put the entire tip of the bag into a large killing jar for a few minutes to stun the insects. They may then be sorted and desired specimens placed separately into a killing jar, or the entire mass may be placed into a killing jar for later sorting. These methods of mass collection are especially adapted to obtaining small insects not readily recognizable until the catch is sorted under a microscope.

Removal of stinging insects from a net can be a problem. Wasps and bees often walk toward the rim of the bag and may be made to enter a killing jar held at the point where they walk over the rim. However, many insects will fly as soon as they reach the rim, and a desired specimen may be lost. A useful technique involves trapping the insect in a fold of the net, carefully keeping a sufficient amount of netting between fingers and insect to avoid being stung. The fold of the net can then be inserted into the killing jar to stun the insect (Fig. 2.18).

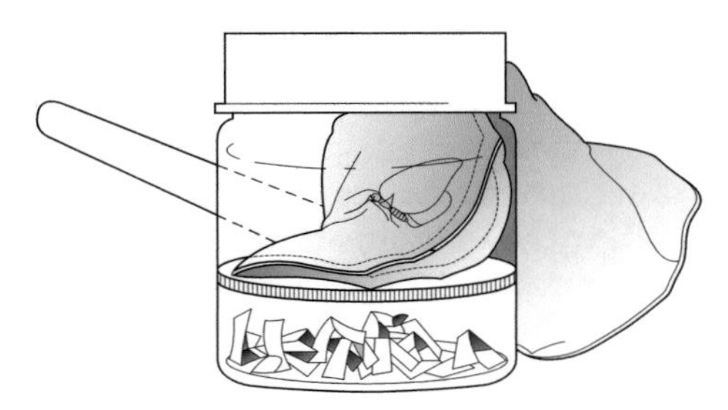

FIGURE 2.18 Technique of stunning insects while in collecting nets by using the killing jar.

After a few moments, the stunned insect may be safely removed from the net and transferred to a killing jar. If the stunned insect clings to the net and does not fall readily into the jar, pry the insect loose with the jar lid or forceps. Do not attempt this maneuver with fingers because stunned wasps and bees can sting reflexively.

Several special modifications are necessary to adapt a net for aquatic collecting. Aerial nets made of polyester or nylon may be used to sweep insects from water if an aquatic net is not available. The bag will dry quickly if swept strongly through the air a few times. A wet net should not be employed for general collecting until thoroughly dry because certain specimens, especially butterflies, may be damaged.

Passive Sampling

Passive insect sampling involves the use of traps. Most insect traps used for obtaining specimens for diagnosis employ some form of sticky substance and are placed where insect activity is likely. These are then left in place for a period of time (passive) while insects move about. Insects that fly, crawl or jump onto the sticky surface cannot extricate themselves and thus remain with the trap. Some sticky traps also contain a food bait or other attraction (sound, temperature, pheromone, odor, CO_2) to artificially lure insects to the trap.

While such traps are very effective in capturing insects that are otherwise difficult to find, they are not suited for obtaining specimens for diagnosis. Specimens stuck in the glue usually cannot be extracted by a diagnostician either for microscopic examination or for preservation. On the other hand, sticky traps can confirm an infestation and sometimes offer a rough indication of its size or location.

Rearing Insects

Diagnosticians occasionally have the need of rearing insects. This is most common when insects in their immature stages are submitted for identification. Because eggs and pupae are notoriously difficult to identify, it is recommended that they be held until the larval or adult stage can emerge, making identification much easier. In addition, a reference collection is much more valuable if a representative of each life stage is identified and linked together.

Rearing insects and observing them as they pass through their various stages of development is beneficial because it enables the diagnostician to more easily recognize and remember the insects. Rearing the insect from the immature stage to the adult stage is not only educational, but results in a helpful addition to the reference collection.

Many kinds of rearing cages are used, most of which are constructed of screen to prevent escape of the adults as they emerge. Most insects are easy to rear if they are provided an environment similar to that in which they were found. For example, when rearing a caterpillar found on oak leaves, it should be caged with a continuous supply of fresh oak leaves to eat until it pupates. Unless one is familiar with the specific habits of the insect being reared, it is best to always provide a few inches of soil in the bottom of the rearing cage because many insects need a moist substrate such as soil in which to pupate and enclose properly.

Rear aquatic insects by placing them in an aquarium with the proper environment. As with fish, the water must be properly aerated. Aquaria and air pumps are purchased from pet supply stores. Aquatic insects also need access to the kind of food on which they naturally feed, such as tiny aquatic plants or other organisms.

Rearing Conditions and Problems

Moisture

The moisture requirements of insects and mites are varied. Examination of the habitat from which specimens were collected should provide clues about their moisture requirements in captivity. Many insects in the pupal stage resist desiccation. All life stages of species that normally infest stored foods also require very little moisture. In fact, many produce their own metabolic water. Most species found outdoors require higher levels of humidity than generally exist indoors. Additional moisture can be added to indoor rearing cages in several ways. To increase the humidity in a cage, keep a moist pad of cotton on top of the screen cover of the cage or place a moist sponge or a small glass vial filled with water in the cage. The mouth of the vial is plugged with cotton and the vial laid on its side so that the cotton remains moist.

Pupae may be held for long periods in moist sawdust, vermiculite, sphagnum, or peat moss. In a flowerpot cage, the water used to keep the plant alive and moisture released through transpiration should provide sufficient moisture for the plant-feeding insects and mites. Spraying the leaves daily also may supplement moisture requirements in rearing cages. Too much moisture may result in water condensation on the sides of the cage, which may trap the specimens and damage or kill them. Excess moisture also enhances the growth of mold and fungus, which is detrimental to the development of most insects and mites. A 2–3% solution of table salt sprayed regularly in the cage will help prevent mold and fungus growth.

Temperature

Of all the environmental factors affecting the development and behavior of arthropods, temperature may be the most critical. Because arthropods are cold-blooded, their body temperature is usually close to the temperature of the surrounding environment, and their metabolism and development are directly affected by increases and decreases in temperature. Each stage of an insect or mite species has a low and a high point at which development ceases. These are called threshold temperature levels.

Most species that are collected and brought indoors for rearing can be held at normal room temperature; the optimum temperature for rearing varies from species to species and with different stages of the same species. As with all rearing techniques, every attempt should be made to duplicate optimum natural conditions. Specimens that normally would overwinter outdoors should be kept during the winter in rearing cages placed in an unheated room, porch, or garage.

Never place an enclosed rearing cage in direct sunlight; the heat becomes too intense and may kill the specimens.

Dormancy and Diapause

Insects and other arthropods cannot control the temperature of their environment. Instead, they make physiological adjustments that allow them to survive temperature extremes. In regions with freezing winters, insects and mites have at least one stage that is resistant to low temperatures. The resistant stage may be egg, larva, nymph, pupa, or adult. When winter arrives, only the resistant stage survives.

Dormancy is the physiological state of an insect or mite during a period of arrested development. Diapause is the prolonged period of arrested development brought about by such adverse conditions as heat, drought, and cold. This condition can be used to advantage in rearing. For example, if rearing cages must be unattended for several days, then many specimens can be refrigerated temporarily to slow their activity and perhaps force diapause. This measure should be used with caution because the degree and duration of cold tolerated by different species vary.

The reverse situation, that of causing diapause to end, is equally useful.

Overwintering pupae that normally would not develop into adults until spring can be forced to terminate diapause early by chilling them for several weeks or months, and then bringing them to room temperature so that normal activity will resume. Often, mantid egg cases are brought indoors accidentally with Christmas greenery. The eggs, already chilled for several months, hatch when kept at room temperature, often to the complete surprise and consternation of the unsuspecting homeowner.

Light

Most species of insects and mites can be reared under ordinary lighting conditions. However, artificial manipulation of the light period will control diapause in many species. If the light requirements of the species being reared are known, then the period of light can be adjusted so that the specimens will continue to develop and remain active instead of entering diapause. Light and dark periods can be regulated with a 24-hour timing switch or clock timer. The timer is set to regulate light and dark periods to correspond with the desired lengths of light and darkness. For example, providing 8–12 hours of continuous light during a 24-hour cycle creates short-day conditions that resemble winter; providing 16 hours of continuous light during a 24-hour cycle creates long-day conditions that resemble summer. Remember that many insects and mites are very sensitive to light; sometimes even a slight disturbance of the photoperiod can disrupt their development.

Food

The choice of food depends upon the species being reared. Some species are detritovores and will accept a wide assortment of dead or decaying organic matter. Examples include most ants, crickets, and cockroaches. Other insects display food preferences so restricted that only a single species of plant or animal is acceptable. At the time of collection, carefully note the food being consumed by the specimen and provide the same food in the rearing cages.

Carnivorous insects should be given prey similar to that which they normally would consume. This diet can be supplemented when necessary with such insects as mosquito larvae, wax moth larvae, mealworms, maggots, and vinegar flies or other insects that are easily reared in large numbers in captivity. If no live food is available, a carnivorous insect sometimes may be tempted to accept a piece of raw meat dangled from a thread. Once the insect has grasped the meat, the thread can be gently withdrawn.

The size of the food offered depends on the size of the insect being fed. If the offering is too large, the feeder may be frightened away. Bloodsucking species can be kept in captivity by allowing them to take blood from a rat, mouse, rabbit, or guinea pig. A human should be used as a blood source only if it is definitely known that the insect or mite being fed is free of diseases that may be transmitted to the person.

Stored-product insects and mites are easily maintained and bred in containers with flour, grains, tobacco, oatmeal or other cereal foods, and similar products. Unless leaf-feeding insects are kept in flowerpot cages where the host plant is growing, fresh leaves from the host plant usually should be placed in the rearing cage daily and old leaves removed.

Artificial Diets

Some species can be maintained on an artificial diet. The development of suitable artificial diets is complex, involving several factors besides the mere nutritional value of the dietary ingredients. Because most species of insects and mites have very specific dietary requirements, information regarding artificial diets is found mainly in reports of studies on specific insects or mites.

Special Problems and Precautions in Rearing

Problems may arise in any rearing program. Cannibalism, for instance, is a serious problem in rearing predaceous insects and necessitates rearing specimens in individual containers. Some species resort to cannibalism only if their cages become badly overcrowded. Disease is also a problem and can be caused by introducing an unhealthy specimen into a colony, poor sanitary conditions, lack of food, or overcrowding.

Cages should be cleaned frequently and all dead or clearly unhealthy specimens removed. Exercise care not to injure specimens when transferring them to fresh food or when cleaning the cages. Mites and small insects can be transferred with a camel's-hair brush.

Specimens reared by a diagnostician should be labeled with that information. Descriptions of methods, time required, host plants or other foods provided and general conditions of the rearing process are valuable should other scientists find a need to repeat the process.

PRESERVING ARTHROPODS

Killing an insect or other arthropod quickly is paramount to the preservation of a specimen that can be identified and can be added to a reference collection. If an insect is allowed to fly or crawl about inside a container for an extended time, it will inevitably harm its wings, break legs and antennae, or lose color.

A specimen is preserved much better if it is dispatched quickly and in a manner that preserves its appearance and maintains its integrity.

Alcohol

The general rule of thumb for preserving and submitting arthropods to an insect diagnostic laboratory is to contain them in alcohol. For purposes of general diagnostic laboratories, ethyl alcohol mixed with water (70–80% alcohol) is generally considered the best general killing and preserving agent. Dropping most arthropods into alcohol will dispatch them quickly and preserve them in the same process. Even so, preservation of insects in alcohol is a subject about which there is some controversy. Many insects collected in alcohol are later pinned for placement in a permanent collection. Hard-bodied insects such as beetles, flies and wasps can be pinned directly after removal from alcohol. On the other hand, adult bees should not be collected in alcohol because their usually abundant body setae become badly matted. Adult moths, butterflies, mosquitoes, moth flies, and other groups with scales or long, fine setae on the wings or body may become severely damaged if collected in alcohol, regardless of the concentration.

Some insects may bleach out or fade in alcohol; however, color is usually not relied upon heavily for diagnosis. Certain arthropods, especially larvae, shrink and turn black when placed directly into alcohol. This can be prevented by submerging them in hot water (near boiling temperature) for 1–5 minutes, depending upon their size, prior to placing them in alcohol. This method 'fixes their proteins' and prevents discoloration to some degree.

Many histological or molecular preservation techniques call for preservation in nearly 100% (absolute) alcohol. Absolute ethyl alcohol is often difficult to obtain, thus some collectors use isopropanol (isopropyl alcohol) with generally satisfactory results. Two advantages of isopropanol are that it does not seem to harden specimens as much as ethyl alcohol and it is more readily obtained, especially in an emergency.

Although controversy exists concerning the relative merits of ethyl alcohol and isopropanol, the choice of which to use is not so important as what concentration to use. Recommended concentration depends on the kind of insect or mite to be preserved. Parasitic Hymenoptera are killed and preserved in 95% alcohol by some professional entomologists. This high concentration prevents the membranous wings from becoming twisted and folded, setae from matting, and soft body parts from shriveling. This concentration also may be desirable if large numbers of insects are to be killed in a single container, such as in the killing jar of a Malaise trap, because the insect body fluids will dilute the alcohol. On the other hand, soft-bodied insects (such as aphids and thrips, small flies, and mites) become stiff and distorted when preserved in 95% alcohol. These specimens should be preserved in alcohol of a lower concentration.

Other Liquid Preservatives

Remember that water is not a preservative. Even alcohol that is diluted with water to less than 70% may allow specimens to begin to decompose. Likewise, large specimens or small specimens that have been crowded into one vial should be transferred to fresh alcohol within a day or two to reduce the danger of diluting and discoloring the alcohol with body fluids.

For some groups of arthropods, preservation is better if certain substances are added to the alcohol solution. Thrips and most mites, for example, are best collected in an alcohol–glycerin–acetic acid (AGA) solution. Larval insects may be collected into a kerosene-acetic acid (KAA) solution.

Formalin (formaldehyde) solutions are not recommended because it causes specimens to become excessively hardened and then difficult to handle. However, formalin is a component of some preservatives.

Glycerin is an excellent preservative for many kinds of insects, and it has characteristics that should not be overlooked by diagnosticians. Glycerin is stable at room temperature and does not evaporate as rapidly as alcohol. The periodic 'topping off' or replacement of alcohol in collections is expensive and time-consuming. Glycerin-preserved material does not require the attention that alcohol-preserved material does. Additionally, when specimens are preserved in glycerin, some colors are preserved better and longer, and the specimens remain supple.

Even more importantly for diagnosticians, when specimens are removed from glycerin, they may be inspected for relatively long periods under a microscope without shriveling or decomposing. This makes for easier handling and more consistent analysis of specimens.

Larvae and most soft-bodied adult insects and mites can be kept almost indefinitely in liquid preservatives. However, for a permanent collection, mites, aphids, thrips, whiteflies, fleas, and lice usually are mounted on microscope slides.

Freezing Insects

One of the preferred methods for killing as well as short to medium-term preservation of insects is freezing. Collected insects can be killed effectively and with minimum damage by placing them immediately into a freezer. This method has three distinct advantages:

1. No messy and potentially dangerous chemicals are needed.
2. Insects subjected to cold decrease activity immediately and thus do not thrash around in a container and break wings, legs and antennae.
3. Insects may be left in the freezer for longer periods of time and need only be thawed before final preservation.

Convenience alone makes the freezer method attractive to many diagnosticians.

Freezing is not without its disadvantages, however. If specimens are to be left in a freezer for an extended period of time, they must have a certain amount of moisture so that they do not become 'freeze-dried.' Conversely, too much moisture with a sample allows for condensation inside the bottle and subsequent freezing of the insect to the vial. These issues can be minimized by wrapping the specimen in dry absorbent paper prior to freezing.

Freeze-Drying

Modern techniques for artificially drying specimens preserved in fluid involve freeze-drying and critical point drying. Many soft-bodied insects and other arthropods may be preserved by freeze-drying in a very lifelike manner and with no loss of color. Freeze-drying is a great improvement over traditional methods of preservation in fluids, especially with lepidopteran larvae, in which important characters of color and pattern are largely lost in all the traditional methods. Unfortunately, the cost of freeze-drying equipment is high, amounting to several hundred dollars (Blum & Woodring, 1963; Roe & Clifford, 1976).

Well-funded laboratories may also have options of preserving insects and other arthropods using a designated freeze-drying device. Preserving biological specimens this way most commonly employs carbon dioxide (CO_2), using acetone as an intermediate fluid to transition fluid from liquid to gas without surface tension damage.

Briefly, the procedure consists of killing the insect by first freezing it in a natural position and then dehydrating it under vacuum in a desiccator jar kept inside a freezer at –4 to –7°C. With a vacuum of 0.1 μm at –7°C, a medium-sized caterpillar will lose about 90% of its moisture and about 75% of its weight in 48 hours. Freezing it in mounting position initially helps to prevent distortion while drying. The time required to complete the freeze-drying process is variable, at least a few days with small specimens and more than a week with larger specimens. When dry, specimens can be brought to room temperature and pressure and then permanently stored in a collection.

Like all well-dried insect specimens, they are rather brittle and must be handled carefully. Freeze-drying has also yielded excellent specimens of plant galls formed by insects and sometimes other insect-related materials.

Critical Point Drying

Critical point drying (CPD) (Gordh & Hall, 1979) is a procedure that was developed to prepare specimens for study under the scanning electron microscope. Like freeze-drying, CPD is relatively expensive and sophisticated because special equipment and chemicals are required. Nevertheless, the results are ideally suited for specimens collected into or preserved in fluid. The technique involves passing specimens through an intermediate fluid (acetone, ethyl alcohol, Freon) and then into a transitional fluid (CO_2, Freon, nitrous oxide), and then subjecting them to a critical temperature and pressure. The CPD procedures are simple and relatively rapid.

Specimens in alcohol are placed in a small mesh-screen basket with a lid. Specimens then are dehydrated through a series of increasing concentrations of ethyl alcohol (ETOH), ending in a solution of 100% ETOH. The specimens within the basket are passed through two washes of 100% ETOH. Specimens that have been preserved in 70% ETOH can be taken through the alcohol series without rehydration. Leave the basket of specimens in concentrations of alcohol for 20–30 minutes. Recently collected specimens preserved in alcohol or small Lepidoptera larvae may require 1–2 hours in each concentration. Specimens can be left for longer periods in concentrations of alcohol above 50%. After removal from the last wash of 100% ETOH, the specimens in their basket are placed within the chamber of the critical point drier and are processed according to drier instructions.

Several critical point driers are manufactured. Liquefied CO_2 (research grade) is the ideal transitional fluid because it is easy to use, comparatively inexpensive, and less noxious than others. Critical point drying has distinct advantages over conventional methods of specimen preparation, including the following:

1. Many specimens can be handled simultaneously, which saves considerable time.
2. Pigmentary colors remain lifelike.
3. Specimens do not collapse or shrivel.
4. Appendages can be manipulated and are more supple than in air-dried specimens.

Critical point drying is most effective for very small, soft-bodied insects, such as parasitic Hymenoptera and Diptera. As freeze-drying and critical point drying become more popular and available, we may see dry specimens of immature insects in entomological collections alongside pinned adults.

Dry Preservation

Occasionally, a diagnostician will come across dry specimens that warrant preservation. In some cases it has become standard practice to hold specimens for an indefinite period, allowing them to become dry with the express intent of rehydrating them at a later date in preparation for mounting and pinning. While this practice is not advised because storing soft-bodied insects by such methods allows the specimens to become badly shriveled and subject to breakage, it is not impossible to save such specimens.

Almost any kind of paper or cardboard container may serve for dry storage. Small boxes, paper tubes, triangles, or envelopes are preferred over tightly closed, impervious containers made of metal, glass, or plastic. These latter should be avoided because mold may develop on specimens if even a small amount of moisture is present. Little can be done to restore a moldy specimen.

Even properly dried stored specimens will need to be relaxed in a relaxing chamber prior to final pinning and mounting (see relaxing methods later in this chapter).

Papering

Like any dry storage, papering is an option that is advised only when pinning specimens when they are fresh is not an option. Papering has long been used successfully for larger specimens of Lepidoptera, Trichoptera, Neuroptera, Odonata, and some other groups. Papering is a traditional way of storing unmounted butterflies and is satisfactory for some moths. Unfortunately, moths often will have their relatively soft bodies flattened, legs or palpi broken, and the vestiture of the body partly rubbed off.

To save space in most large collections, Odonata are sometimes permanently stored in clear cellophane envelopes instead of pinning them. Cellophane is preferred over plastic because cellophane 'breathes.'

Papering consists of placing specimens with the wings folded together dorsally (upper sides together) in folded triangles or in small rectangular envelopes of glassine paper, which are the translucent envelopes familiar to stamp collectors. Glassine envelopes have become almost universally used in recent years because of the obvious advantages of transparency and ready availability. In many collections, glassine has become partly superseded by plastic. However, many collectors still prefer folded triangles of a softer, more absorbent paper, such as ordinary newsprint (Fig. 2.19).

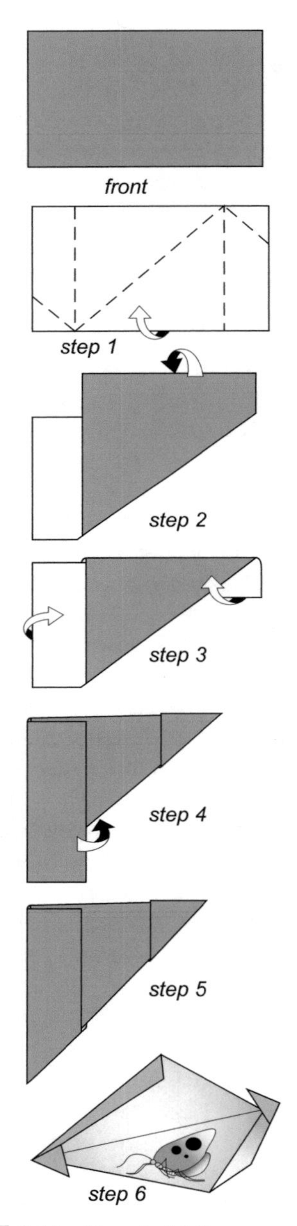

FIGURE 2.19 Steps in folding paper triangles.

Specimens can become greasy after a time, and the oil is absorbed by paper such as newsprint but not by glassine. Moreover, glassine and plastic are very smooth, and specimens may slide about inside the envelopes during shipping, losing antennae and other brittle parts. Although softer kinds of paper do not retain creases well when folded, this shortcoming may be circumvented by preparing the triangles of such material well before they are needed and pressing them with a weight for a week or so. Triangles are easy to prepare if the paper is folded as shown below.

Some Lepidoptera are most easily papered if first placed in a relaxing box for a day or two. The wings, often reversed in field-collected butterflies, may then be folded dorsally without difficulty. Avoid crushing the specimens by not packing them too tightly before they are dried, and by writing collection data on the outside of the envelopes before inserting the insects. Again, to prevent molding, do not store fresh specimens immediately in airtight containers or plastic envelopes.

Mounting Insects

Specimens are mounted so that they may be handled and examined with the greatest convenience and with the least possible damage. Well-mounted specimens enhance the value of a collection; their value for research may depend to a great extent on how well they are prepared. Standardized methods have evolved in response to the aesthetic sense of collectors and to the need for uniformly prepared research material of high quality.

The utility of a mounted specimen, especially in a reference collection, is determined by how well it is preserved and how easily it can be accessed for examination. Utility supercedes beauty in reference collections.

Ideally, specimens should be freshly obtained. The body tissues of fresh specimens have not hardened or dried, and thus the specimen is easier to manipulate. Specimens that are dried already must be specially treated before mounting. Dry specimens usually must be relaxed (Weaver & White, 1980), and those even temporarily preserved in liquid must be processed so that they will dry with minimal distortion or damage. Some specimens may be kept permanently in a liquid preservative or in papers or envelopes, as discussed earlier.

Preparing Dry Specimens

RELAXING

As a rule, specimens should be spread and pinned when fresh; however, in cases where this is not an option, dry insects must be re-hydrated before pinning. The procedure of re-moistening a dry specimen is called relaxing. Relaxing softens the body so that the body will not shatter when the mounting pin is inserted into the thorax. In addition, relaxed body parts such as head, legs and wings may be rearranged or repositioned without breaking.

Insects, particularly Lepidoptera, whose wings are to be spread, should sometimes be relaxed even if they have been killed recently because the strong flight muscles tend to stiffen, even in a matter of minutes, and make adjustment of the wings difficult. Treatment in a relaxing chamber is a method of softening them.

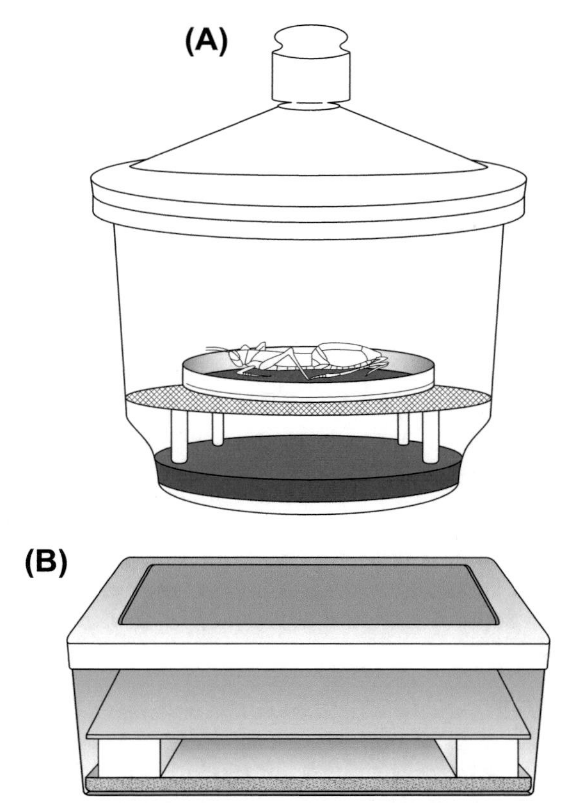

FIGURE 2.20 Insect relaxing chambers. (A) Glass relaxing chamber. (B) Homemade box relaxing chamber.

CONSTRUCTION OF A RELAXING CHAMBER

A glass relaxing chamber (Fig. 2.20) may be purchased from a biological supply house that specializes in insects, but may also be readily constructed in the laboratory.

Glass dishes and jars with covers (low, wide-mouthed jars and casserole dishes are excellent) and earthenware crocks make excellent relaxing chambers. Glass and earthenware containers are less immediately affected by fluctuations in temperature and thus may relax insects more evenly, but they are less available and more costly than plastic storage containers. In either case, containers 5–15 cm deep are most convenient because they will accommodate large specimens and even specimens already on pins that may require remoistening and adjusting. A layer of damp sand, peat, or crumpled paper toweling is placed in the bottom of the container and covered with a layer of cotton, cellulose wadding, or similar material that will not absorb water readily and serves to prevent direct contact between the insects and the moisture beneath. Some diagnosticians object to the use of cotton because of the tendency for insect legs to become entangled in it and break off. However, this problem can be prevented by covering the cotton with a piece of soft tissue paper. This is also advantageous for very small specimens, especially if the material has a smooth surface. Heavy paper, such as blotting paper or cardboard, may be used in place of cotton if supported at least 1 cm above the moist bottom layer to avoid direct contact. Wooden or plastic strips or fine-mesh plastic screen may be used for this purpose.

Most specimens held in the airtight 100% humidity environment of a relaxing chamber usually soften in 8 hours but larger specimens may require 24–36 hours. Care must be exercised to ensure that the specimens do not become wet during humidification. The maximum length of time that insects may be left safely in a relaxing chamber partially depends on the temperature. At 18–24°C, the specimens may be left for about three days; beyond that time they will begin to decompose. If the relaxing chamber is placed in a refrigerator at 3–4°C, the specimens may be kept for two weeks, although they may be slightly damaged from excessive condensation over time. If relaxing chambers are placed in a deep freeze at −18°C or lower, the specimens will remain in a comparatively fresh condition for months.

MOLD

Growth of mold must be avoided because it will destroy specimens left too long in a relaxing chamber. In temperate and arid regions, mold probably will not be a problem if insects are relaxed for no longer than two days at normal room temperature. However, relaxing chambers in regular use must be kept clean, with frequent renewal of the contents. If mold is likely to develop, as may happen with large specimens held more than two days, a few crystals of naphthalene, paradichlorobenzene, phenol, or chlorocresol may be sprinkled in the bottom of the relaxing chamber. Such chemical mold inhibitors may damage plastic boxes and must be used with care.

Occasionally, it is necessary to relax and reposition only a part of an insect, such as a leg that may conceal characters needed for identification. This is accomplished by putting a drop or two or Barber's Fluid or ordinary household ammonia directly on the leg. Most household ammonia contains detergent, which helps wet and penetrate insect tissue. After a few moments, and perhaps after adding a little more fluid, the body part may be moved slightly if pried carefully with a pin. When the part moves easily, it may be placed in the desired position and held there with a pin fixed in the same substrate that holds the pin on which the specimen is mounted. Leave the positioning pin in place until the insect re-dries thoroughly.

A few methods for relaxing insects involve heat. These and other methods are summarized in a reference describing a steam-bath method (Weaver & White, 1980).

Preparing Liquid-Preserved Specimens for Mounting

Specimens with hard exoskeletons, such as beetles and some bugs (Pentatomidae, Cydnidae), may be removed from the preserving fluid and mounted directly on pins or points. However, other specimens preserved in fluid must be removed from the fluid in a way that minimizes distortion or matting of setae as the specimens dry. Cellosolve and Xylene baths are designed to leave surface pile, setae, and bristles in a loose, unmatted, natural condition.

This is accomplished by removing the specimens from their liquid preservative and placing them on blotting paper to dry for a few minutes and then transferring them into a Cellosolve bath for about three hours or slightly longer for large insects. This extracts water and other substances from the specimens. However, Cellosolve does not evaporate readily, so it must also be removed. This is done by again placing specimens briefly on blotting paper and then placing them in a Xylene bath for one to four hours, depending on size of the specimen.

Specimens subjected to this treatment become somewhat pliable and appendages may be repositioned slightly during the pinning process. However, since such specimens do not cling as firmly to the pin as those pinned while fresh, a small amount of adhesive should be placed around the site where the pin protrudes from the body of the specimen.

Direct Pinning

Direct pinning refers to the insertion of an insect pin directly through the body of an insect in such a way that the insect can be mounted, preserved and handled. Only specially designed (standard) insect pins should be used; ordinary straight pins are too short and thick and also have other disadvantages. A well-made insect pin is considered essential in diagnostic laboratories.

Insect pins are made of either spring steel (which is called 'black') or stainless steel with a blued or a lacquered (japanned) finish.

The rustproof nature of stainless steel pins makes them most desirable for use in permanent collections.

Standard insect pins are 38 mm long and range in thickness generally from size #000 to #7. Sizes refer to the diameter of the pin. Normally the very slender pins (#000–#1) are reserved for rare occasions. Size #2 (0.46 mm diameter)–#4 in diameter are used for most general insect mounting. Sizes #4–#7 are for large specimens with heavy bodies.

Insect pins are unique because they have heads made by either mechanically squeezing out the end of the pin, or by pressing a small piece of metal or nylon onto the pin. These allow a person to grip and handle the end of the pin in such a way that does not bump or damage the specimen.

For handling a large number of pinned specimens, pinning or dental forceps may be helpful (Fig. 2.21). Their curved tips permit a diagnostician to grasp the pin below the data labels and to set the pin firmly into the pinning-bottom material of the box or tray without bending the pin. Unfortunately, pinning forceps are impractical in handling large-winged specimens, such as many Lepidoptera.

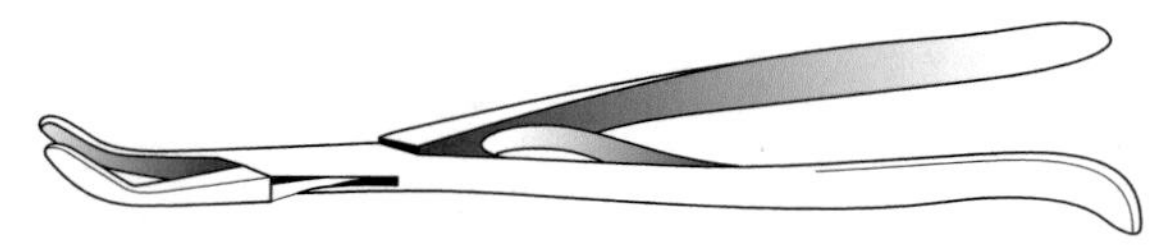

FIGURE 2.21 Pinning forceps.

Insects should be pinned vertically through the body with a pin of appropriate thickness (size) and in a position that is slightly unique for each insect order.

PINNING TIPS

1. Choose the appropriate pin based on the size of the specimen to be mounted.
2. Exercise care so that the pin does not damage the legs as it passes through the body.
3. Do not attempt to pin specimens unless they are relaxed or recently killed, because inserting a pin into a dry specimen may cause it to shatter.

Standard pin placements for some of the more common groups of insects are illustrated below in Figure 2.22.

Many insects are pinned to the right of the midline so that all the characters of at least one side of the thorax will always be visible.

1. **Orthoptera.** Pin the specimen through the back of the thorax to the right of the midline. For display purposes, one pair of wings may be spread. Some orthopterists prefer to leave the wings folded because of limited space in most large collections (see Beatty & Beatty, 1963).
2. **Large Heteroptera.** Pin the specimen through the triangular scutellum to the right of the midline. Do not spread the wings.
3. **Large Hymenoptera and Diptera.** Pin the specimen through the back of the thorax or slightly behind the base of the forewing

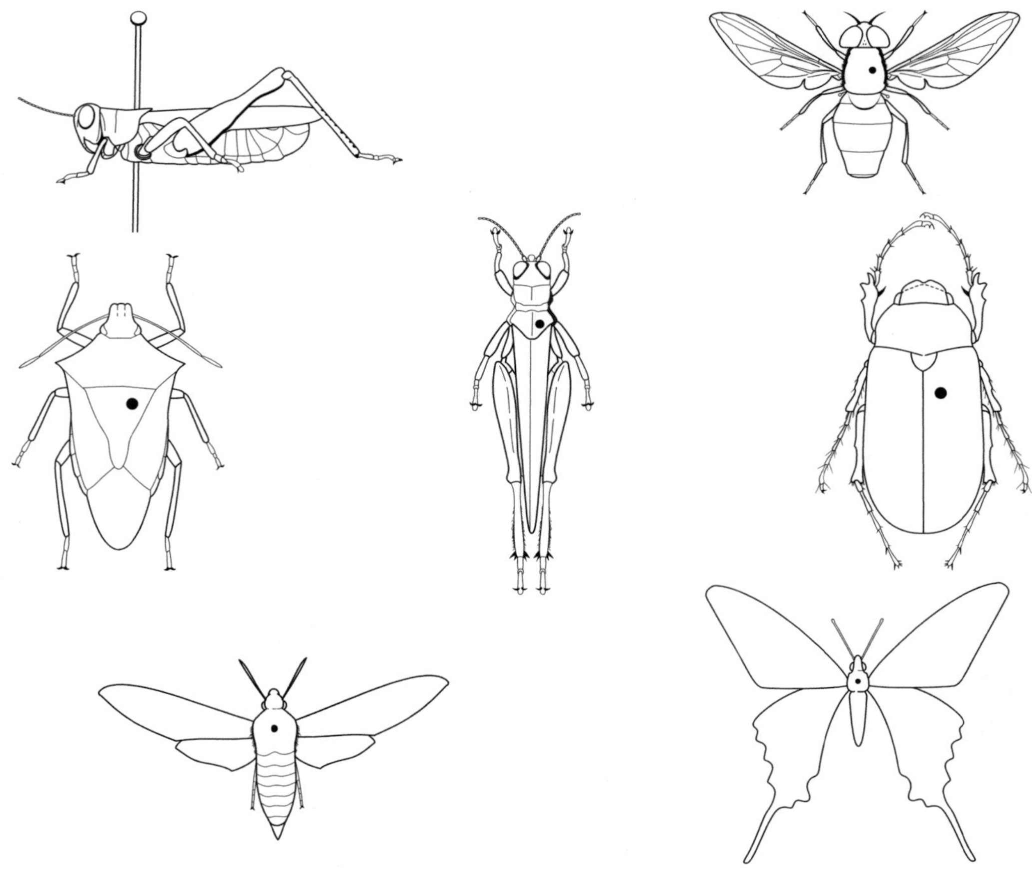

FIGURE 2.22 Proper pin positions in common insect orders; black dots indicate pin placement.

and to the right of the midline. Characters on the body must not be obscured. Legs should be pushed down and away from the thorax; wings should be turned upward or sideways from the body. The wings of most Diptera and many parasitic Hymenoptera will flip upward if the specimen is laid on its back before pinning and pressure is applied simultaneously to the base of each wing with a pair of blunt forceps. Wings should be straightened if possible so that venation is clearly visible. Folded or crumpled wings sometimes can be straightened by gentle brushing with a camel's hair brush dipped in 75% alcohol. Peterson's XA mixture (equal parts of xylene and ethyl alcohol) is recommended for treating Hymenoptera wings.

4. **Large Coleoptera.** Pin the specimen through the right forewing (wing cover or elytron) near the base. Do not spread the wings of beetles.
5. **Large Lepidoptera and Odonata.** Pin the specimen through the middle of the thorax at its thickest point or just behind the base of the forewings.

The proper height of an insect on a pin will depend somewhat on the size of the specimen. As a guideline, enough of the pin should

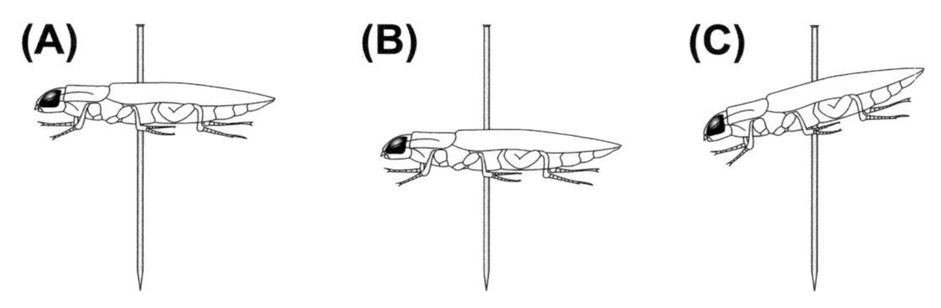

FIGURE 2.23 Pin heights. (A) Correctly pinned specimen. (B) Specimen mounted too low. (C) Specimen not level.

be exposed above the thorax to grasp the pin without the fingers touching the specimen. A specimen mounted too high on a pin probably will be damaged in handling. If a specimen is positioned too low on a pin, the specimen's legs may be broken when the pin is inserted into a tray or box. Also, insufficient space may be left for labels if a specimen is mounted too low on a pin. Figure 2.23 illustrates correct and incorrect examples of pinning.

After the pin is inserted and before the specimen is dry, the legs, wings, and antennae should be arranged so that all parts are visible. With some insects, the legs and antennae simply can be arranged in a life-like position and allowed to dry. In most cases, the appendages must be held in place with additional insect pins until the specimen dries. Long-legged species or specimens with drooping abdomens can present problems. Temporarily mounting insects on a piece of Styrofoam board or similar material is recommended. When the primary pin is pushed deep into the Styrofoam, legs and abdomens may be supported until dry. A series of strategically placed support pins or a piece of stiff paper pushed up on the pin from beneath help to keep the specimen in position. When the specimen is dry, supports can be removed.

Double Mounts

Entomologists double mount insects that are too small to be pinned directly on standard pins yet should be preserved dry. The term double mounting refers to a method by which very tiny insects are first attached to a tiny pin or glued to a paper point which is then attached to the standard insect pin. The insects are thus 'double' mounted, first on the tiny pin or on the paper point and then onto the standard insect pin.

Minuten pins of different sizes are available from supply houses (Fig. 2.24). They are generally constructed of stainless steel and are finely pointed at one end and headless on the other end. The headless end of the minuten is first inserted into a small cube of soft, pithy material, such as fine cork, balsa wood, fine textured plastic or silicone rubber. (Sometimes it is preferable to mount an insect on a minuten before inserting the minuten into the mounting cube.) The pointed side of the minute is

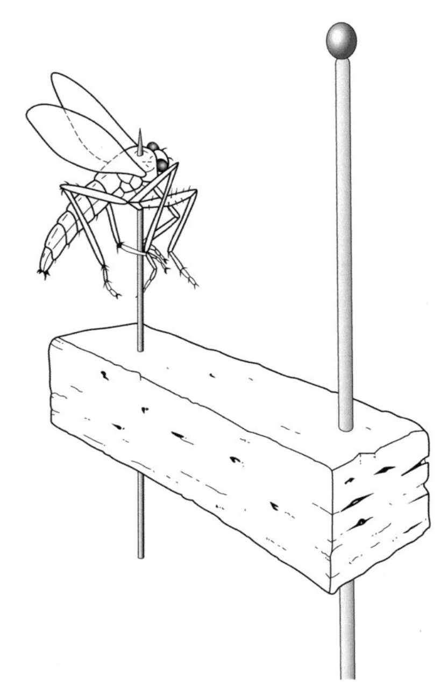

FIGURE 2.24 A properly double-mounted insect on a minuten pin.

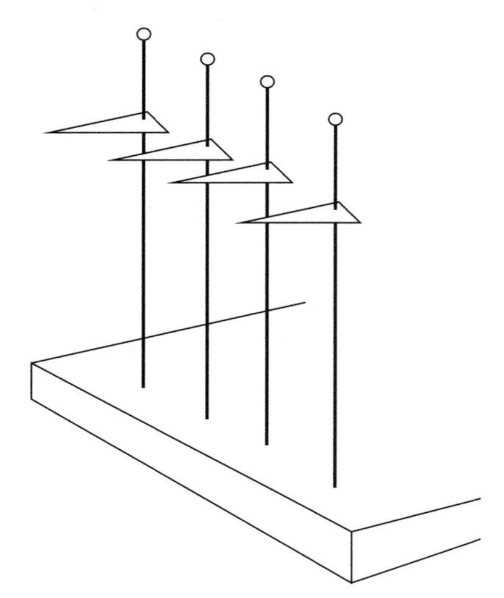

FIGURE 2.25 Proper mounting of card points.

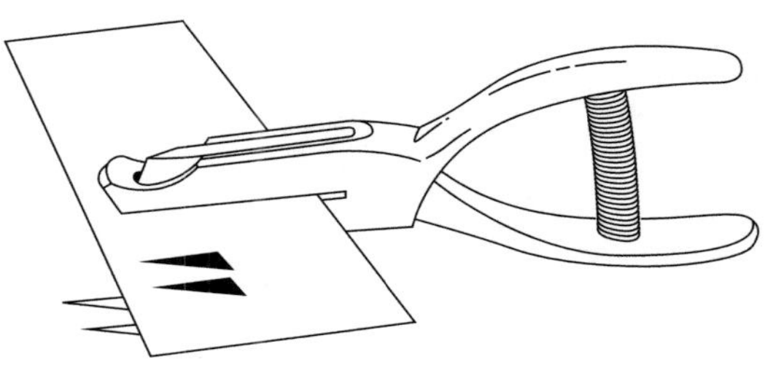

FIGURE 2.26 Point punch.

then run through the body of the small specimen, and the mounting cube is pinned by the standard pin.

Many entomologists prefer to mount insects on a minuten in a vertical position, with the minuten parallel to the main pin. The insect lies in an excellent position for examination under a microscope and is least likely to be damaged in handling.

Points are slender, small triangles of stiff paper or card stock of high-quality, acid-free paper. The paper of choice is used in herbarium mounts. The triangles are pierced through the broad end (Fig. 2.25) with the standard insect pin, and the insect is then glued to the narrow tip of the point. Card points may be cut with scissors from a strip of paper or, more commonly, via a special punch for card points (Fig. 2.26), available from entomological supply houses.

For most insects, the card point is attached to the right side of the specimen, with the left side and midventral area clear. For better adhesion with some insects, the tip of the card point may be bent downward at a slight angle to fit against the side of the specimen (Fig. 2.27).

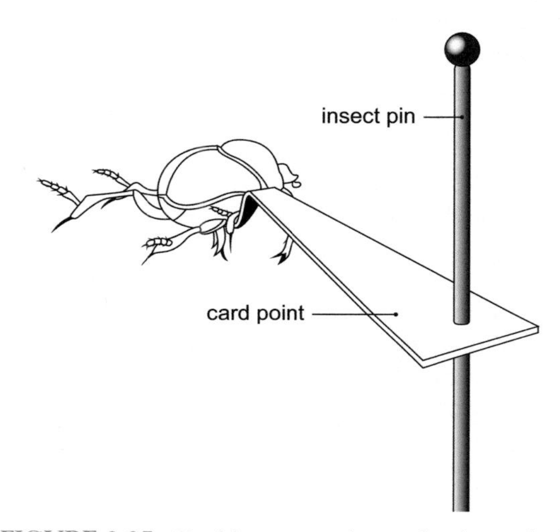

FIGURE 2.27 Double mount using card point method.

Only a very small part of the point should be bent, and this is accomplished with fine-pointed forceps. The bend increases the surface area of the point that makes contact with the specimen. The card point should be attached to the side of the thorax, not to the wing, abdomen, or head. Some insects, such as small flies and wasps, are best mounted on unbent points. The specimen is mounted directly onto the card, with the right side of the thorax contacting the adhesive.

The selection of adhesive for mounting insects on card points should be based on bonding characteristics, solubility, and availability as follows:

1. The adhesive should bond to the specimen so that the specimen will not fall from the card with time.

2. The adhesive should be soluble in a solvent that is inexpensive, non-toxic, and available in case the specimen should need to be remounted at a later date.

Ordinary white glue or carpenter's glue is readily available in the United States and is most often used. Clear fingernail polish is also commonly utilized for pointing insects. Only a small amount of adhesive should be used to glue the specimen to the card point. Excessive glue may obscure certain sutures or sclerites necessary for identification, just as the card point itself may conceal certain ventral structures if allowed to extend beyond the midline of the insect.

If specimens are in good condition and are well prepared, they may be kept in museum collections for a long time, perhaps for centuries.

WHEN TO DOUBLE MOUNT

Opinions differ on when to use direct pinning and when to use a double mount. Perhaps the decision is best determined through experience. As a general rule of thumb, do not use a double mount if you can mount a small insect on a #0 or #1 pin without damaging the specimen.

WHEN TO USE MINUTE PINS AND WHEN TO USE POINTS

Some specimens, such as moths, should never be glued to points; other specimens should never be pinned with minutens. The following suggestions will serve as a guide.

1. Small moths, caddisflies, and neuropteroids. Mount the specimen on a minuten inserted through the center of the thorax with the abdomen positioned toward the insect pin. The mount must be sufficiently low so that the head of the pin can be grasped easily with fingers or pinning forceps. Do not glue moths to points. Ideally, such specimens should be spread in the conventional manner despite their small size.
2. Mosquitoes and small flies (recently killed). Pin the specimen with a minuten through the thorax with the left side of the specimen positioned toward the main pin. Note that the minuten is vertical. This is more advantageous than if it were horizontal because the specimen is less liable to come into contact with fingers or pinning forceps. Placing a small amount of glue on the tip of the minuten before piercing the specimen will help hold soft-bodied insects.
3. Small wasps and flies (not recently killed). Mount the specimen on an unbent card point, with the point inserted between the coxae on the right side of the insect (keeping clear of the midline), or glue the tip of the point to the mesopleuron.
4. Small beetles, bugs, leafhoppers, and most other small insects should be glued to the card point with its tip bent down on the right side of the specimen.

Mounting Aids

Pinning Blocks

Pinning blocks (Fig. 2.28) allow for insects and labels to be easily mounted at uniform heights on the pin. Specimens should be mounted on the pin such that the pin can be grasped between the thumb and index finger without touching the specimen and yet the specimen is kept above the bottom of the box or tray.

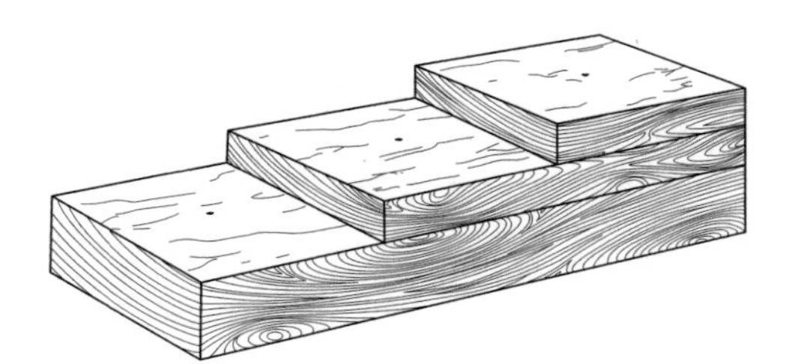

FIGURE 2.28 Three-step wooden pinning block.

After piercing the insect through the appropriate spot on the specimen, insert the pin into the deepest hole in the pinning block until the pin goes no further. This pushes each specimen to the same height on each pin.

Repeat this process for making label heights on pins uniform (Fig. 2.29). Use of pinning blocks improve the general appearance of a collection and help preserve specimens from breakage.

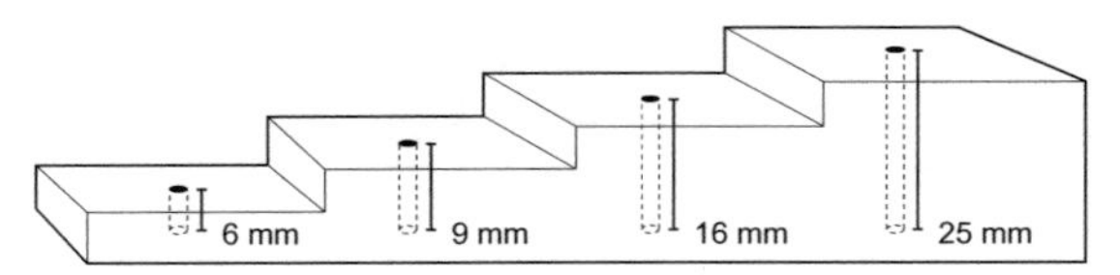

FIGURE 2.29 Plastic pinning block illustrating hole depths.

Double mounts should conform to the same rule as direct pinning. Do not place a double mount too high or too low on the pin. Double mount cubes or points may be adjusted at any time, whereas a directly pinned insect is virtually impossible to move without damage after it has dried. If points become loose on the main pin, place a little adhesive at the connection.

Spreading Boards and Blocks

The appearance of an insect collection can be enhanced if specimens are pinned in a life-like position. Most entomologists find that a mounting board made of Styrofoam, cork, or cardboard is convenient (Fig. 2.30). Pins on which fresh specimens have been mounted should be inserted into the mounting board deeply enough so that the specimen rests slightly above the flat surface. The legs, wings, abdomen, and antennae can then be conformed into the desired position by strategically placing holding pins. Leave the specimen in place until dry (sometimes 1–2 weeks). Once dry, the specimen will hold this shape indefinitely.

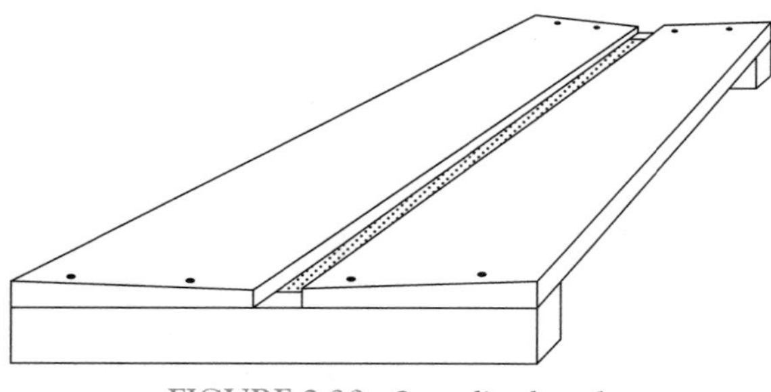

FIGURE 2.30 Spreading board.

In reference collections, mounting boards allow an insect diagnostician to prominently display certain specific characters used in identification.

Insects to be preserved with their wings spread uniformly are generally set and dried in this position on spreading boards or blocks. These pinning aids vary greatly in design, but the same basic principle is inherent in all of them: a smooth surface on which the wings are spread and positioned horizontally, a central longitudinal groove for the body of the insect, and a layer of soft material into which the pin bearing the insect is inserted to hold the specimen at the proper height. An active collector will need several spreading boards because the insects must dry for a considerable time (about 2 weeks for most specimens, even very small ones) before being removed from the boards. Spreading boards may be purchased from biological supply houses or may easily be made as described here if the proper materials can be obtained. When purchasing spreading boards, avoid the following:

1. Too hard or too soft a material for the pinning medium under the central groove.
2. Too hard an upper pinning surface.
3. Top pieces without the same thickness at the center (an especially common fault in beveled boards). This defect may be corrected by sanding down the higher side; evenness is especially critical when working with small specimens.

A spreading board or a block may be necessary for moths, butterflies, and other insects that should be mounted with their wings spread. Spreading boards may be obtained from biological supply houses or may be constructed in the laboratory.

Spreading Wings

Before spreading specimens, the spreading boards and the following materials should be at hand.

1. Pins (called setting pins) of size 00 or 000 for bringing wings into position.

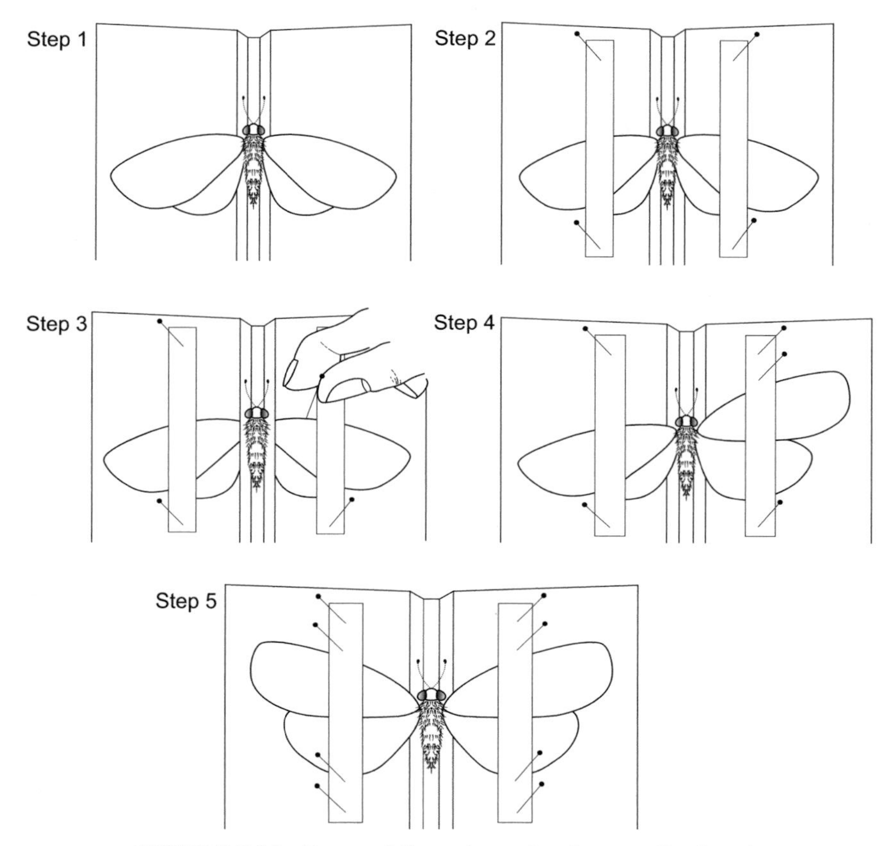

FIGURE 2.31 Steps to follow when using the spreading board.

2. Strips of glassine or tracing paper. The strips of tracing paper should be long enough to extend from the base of the wings to a little beyond the end of the wings of the specimens being spread. Strips about 25 mm long are convenient for spreading most Lepidoptera.
3. Standard #2 or #3 insect pins with nylon heads are most often used.

With this equipment ready, a diagnostician is prepared to mount and spread the specimens. The specimens must be properly relaxed, even the recently collected ones, before any attempt is made to spread the wings.

Insert an insect pin of appropriate size through the middle of the thorax, leaving at least 7 mm of pin above the specimen. The pin should pass through the body as nearly vertically as possible to avoid having the wings higher on one side than on the other. Pin the specimen into the central groove of the spreading board so that the wings are exactly level with the surface of the board. Carefully draw each wing forward, with the point of a setting pin inserted near the base and immediately behind the strong veins that lie near the front of the wings (Fig. 2.31). If care is taken not to tear the wing, the fine setting pins should leave holes so small that they are barely visible. The hind margin of the forewings should be at right angles to the groove in the board. Bring the hindwing into a natural position, with its base slightly under the forewing. The setting pins will hold the wings in position until they can be secured with the paper strips.

The strip is placed near the body of the insect, and a glass-headed pin is inserted in the middle of the strip just outside the margin of the forewing. The pin may be tilted slightly away from the wing to keep the strip down against the wing. The tip is then carefully stretched backward and another pin placed just behind the hindwing. A third pin in the notch where the forewings and hindwings meet is usually enough. None of the three glass-headed pins on each side of the specimen should pass through the wings. After the paper strips are in place, the setting pins may be removed.

To prevent the abdomen from drooping as the specimen dries, support it with a pin on each side, crossing beneath. Pins may also be used to arrange and hold the antennae and legs in position until they dry. The appearance of many insects may be improved by gently blowing on them before spreading, to remove extraneous loose scales and to straighten the setae or, with small moths, the fringes of the wings.

When working with small insects, a large magnifying lens mounted on an adjustable stand may be very helpful. Relaxed specimens should be spread without delay because they dry so quickly that antennae may break or the wings curl if the spreading is not completed promptly. If the wings do not move readily under gentle pressure, do not force and possibly break them. Return the specimen to the relaxing chamber for additional time.

Drying Time

Specimens on spreading boards should be placed in a warm, well ventilated place to dry for at least two weeks. Even very small moths require that long to dry completely and become stiff. If they are placed in a low-temperature oven, such as is used for drying plant specimens, two days may suffice.

Specimens relaxed from a dry condition, as already noted, dry quickest, but even they should be left for several days. Fresh specimens, even large ones, may be dried in two days or less with heat.

The spreading boards must be kept where they are safe from mice, bats, dermestid beetles, lizards, psocids (booklice and barklice), and ants, especially in tropical climates. One preventive measure that is sometimes advisable involves placing the boards or blocks on bricks set in pans of water. If they are hung from the ceiling, then a mosquito net around them may be necessary.

Spreading is a highly individualistic skill, subject to wide variation. Nearly everyone, with practice, evolves his or her own technique, so that two workers may appear to follow different procedures and yet produce equally good results. There is no single standardized technique with respect to the fine points of spreading.

Riker Mounts

Sometimes it is desirable to prepare special specimens for reference in such a way that they may be relatively roughly handled and closely examined without risk of damage. Riker mounts have long been used for this purpose. Many different Riker mounts are available for purchase from biological supply houses but each is typically a flat cardboard box about 3 cm deep, filled with cotton and with a pane of glass or plastic set into the cover (Fig. 2.32).

Height and width dimensions of the box vary widely and are chosen based on the size of the sample to be enclosed.

No hard and fast rules are in place for preparing Riker mount specimens. Typically, however, unpinned insect specimens are held in their desired position and allowed to dry (about two weeks) whereupon they are placed inside the Riker mount. When the appearance of the sample is acceptable, the box is closed.

FIGURE 2.32 Riker mounts.

Each Riker can be sealed completely to prevent access by potentially damaging pests such as dermestid beetles, or a large set of Riker mounts may be occasionally fumigated together to avoid damage due to pests.

A distinct advantage of Riker mounts in a reference collection is that insect-damaged plant material or associated insect frass may also be dried in place with the specimens to make valuable educational references.

Riker mounts are practical only for relatively large specimens. Tiny specimens or specimens that must be presented for examination from all angles under magnification do not lend themselves to Riker mounting.

FIGURE 2.33 Riker cabinet.

HOUSING RIKER MOUNTS

Riker mounts should be stored out of direct sunlight, which may cause fading of colors and general deterioration, and should be kept in a dry location. A functional Riker reference collection is one that is clearly marked with identifying labels, organized, and stored in such a manner that retrieval is simple and easy (Fig. 2.33). Various systems of labeling the mounts and using electronic databases can be used. The most important feature is ease of use.

Riker mounts available for purchase include those with a white bottom where identifications and detailed descriptions of the specimen, its habitat, damage, biology or other information can be summarized. These can be written directly on the bottom of the Riker mount and can be very valuable to the diagnostician.

Embedding

Preservation of various kinds of biological specimens in blocks of polymerized transparent plastics was popularized several years ago and is still of some interest (Fig. 2.34).

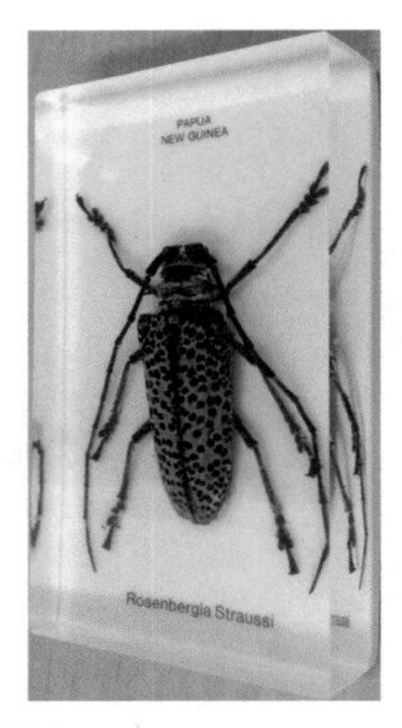

FIGURE 2.34 Embedded sample.

The process is rather complicated and laborious, but if carefully done it will yield useful preparations, especially for exhibits and teaching. Directions for embedding insect specimens may be found in the following references and in directions furnished by suppliers of the materials: Foote (1948); Fessenden (1949); Hocking (1953).

Whole Mount Preparation

Whole mount preparation is similar to slide preparation of tissues for histological purposes, although there are some distinctions.

The techniques and materials used to prepare specimens for high-power microscopic examination may vary slightly with the kind of insect, mite or body part to be preserved, as well as the diagnostician's preferences and objectives; however, the basic concepts and principles involved in preparing whole mounts are provided below.

The process of whole mount preparation can best be described step by step as follows.

STEP 1: MACERATION

Maceration is the chemical removal (dissolving) of muscles and other soft tissues in a specimen while leaving the sclerotized or chitinized parts that are needed for identification. This process is normally accomplished by immersing the specimen in a 5–10% solution of sodium hydroxide (NaOH) or potassium hydroxide (KOH). (These chemicals are strongly caustic and must be handled carefully to avoid damage to the skin and eyes. If spattered on the skin, immediately wash off the chemical with water.)

KOH may be used for maceration of most larger insects, but delicate specimens should be cleared in NaOH to avoid damage.

The length of time the specimen is subjected to maceration is dependent upon the size and physical characteristics of the specimen, the concentration of the chemical agent, and the temperature at which the treatment occurs. Overnight treatment at room temperature is generally adequate for most specimens; more refractory specimens may require longer periods at higher temperatures.

STEP TWO: WASHING

When the specimen is adequately macerated, remove the caustic macerating agent from the specimen by washing it in distilled water, when available, or tap water. This is a quick process that effectively stops the maceration.

Dehydration

Specimen dehydration is usually necessary before the specimen is permanently mounted on a microscope slide. Dehydration is accomplished by immersing the specimen in a graded series (50%, 70%, 80%, 90%, 95% and two washes at 100%) of ethyl alcohol (ETOH). Each wash extracts water from the tissues such that the final treatment with absolute ethyl alcohol leaves the specimen dehydrated. The amount of time in each wash varies with the size and nature of the specimen.

Staining

Specimen staining is sometimes necessary with insects and mites because their immersion in the mounting medium may make colorless and transparent tissues virtually invisible if the medium has a refractive index close to that of the tissues of the specimen (Stein et al., 1968). However, if a phase-contrast microscope is available, then staining (even with colorless specimens) is not necessary. Several kinds of stains and techniques for using them are available from biological supply houses.

Bleaching

If specimens are too dark to reveal sufficient detail after maceration, they may be bleached in a mixture of one part strong ammonia solution to six parts hydrogen peroxide solution. The length of time the specimen is left in the ammonia-peroxide solution depends on the amount of bleaching needed.

Slide Mounting

An assortment of slides, cover slips and mounting media are available from biological supply houses. Selection of a mounting medium is very important. Mountants are substances in which a specimen is placed for observation beneath a coverslip for microscopical observation. Mountants may be classified as temporary or permanent.

A temporary mount made with lactic acid, glycerin, or other medium on a 2.5 by 7.5 cm cavity slide may be kept for a year or more without damage to the specimen. The specimen is placed near the edge of the cavity and wedged into position by using a fine needle and manipulating a coverslip over the cavity and the specimen. After the specimen is positioned and the coverslip is centered, a commercial ringing compound, such as transparent fingernail polish or quick-drying cement, should be applied around the edge of the coverslip and the adjacent area of the slide to seal temporary mounts.

Specimens may be mounted permanently on slides using either Canada balsam or Euparal (both obtainable from entomological supply houses). Specimens must be dehydrated through a graded series of ethyl alcohol concentrations before mounting them in Canada balsam. Euparal is a satisfactory mounting medium for most insects (other than scale insects and thrips) and has the advantage of not requiring dehydration of specimens before mounting. They may be taken directly from 80% ETOH.

To place specimens in the medium, put one or more drops of the medium in the center of a clean glass slide. Place the cleared and washed (also stained or bleached, if necessary) specimen in the medium on the slide. Make certain that the specimen is completely immersed and that air bubbles are absent. Arrange the specimen in the desired position with a fine needle, then gently lower a coverslip onto the specimen with forceps, holding the coverslip at a slight angle so that it touches the medium first at one side to prevent air entrapment. Apply gentle pressure with the forceps to fix the position of the specimen. Prepare multiple specimens in more than one position (e.g., dorsal side up and dorsal side down). Canada balsam, Euparal, and other permanent mounts do not require ringing.

Curing

The final stage in preparing permanent mounts is thorough drying or hardening of the mounting medium. Allow slides to dry completely while they remain in a horizontal position. Dry for a minimum of 48 hours and if possible for 3–4 weeks in this position. Handling slides before the mountant is dry can cause the specimen to move.

LABELING

To have maximum scientific value, specimens (including slide mounts) must be accompanied by a label or labels giving, at the very minimum, information about where and when the specimen was collected, who collected it, and the host or food plant.

Samples taken for forensic evidence must be sealed and fully labeled. Chain of custody must also be documented, including the name of the person collecting the specimen, dates of collection, and similar information, each time the sample is transferred from one person to the next, verified by signatures.

Paper

The selection of an appropriate paper for labels is very important. Ordinary paper contains acids that weaken the cellulose fibers and subsequently break the paper down over time. Paper manufactured from rags is superior because cotton fibers in the paper have superior strength and hold up over time and in solution (alcohol). The best paper has 100% rag content. The paper used for making labels should be heavy so that the labels remain flat and do not rotate loosely on the pin. Lined ledger paper (100% rag and 36-pound weight) or high-rag-content paper that is used for professional-grade herbarium sheets (obtained from biological supply houses) are best. Labels made from poor-quality paper become yellow and brittle with age, tend to curl, disintegrate in liquid preservatives, and are generally unsatisfactory.

Pens

Handwritten labels are made with 'rapidograph,' or technical drawing pens. Several brands (Staedtler, Koh-i-nor, Rotring) are available from art stores, craft shops, and bookstores. Technical drawing pens come in several sizes that produce lines of different widths. For most label information, 0.25 mm (no. 000) to 0.30 mm (no. 00) are suitable sizes for achieving small, legible writing. Proper care and maintenance of technical writing pens is necessary for optimal performance.

Ink

The ink should be a superior grade of India ink that is permanent and will not 'run' if the labels are placed in jars or vials of liquid preservative. Be sure the ink is completely dry before placing the label in the liquid. Labels created using a ballpoint pen or hard-lead pencil soon fade and become illegible, especially when placed in liquids. Labels may be lettered carefully by hand with a fine-pointed pen. Personal computers and printers also can provide a rapid, inexpensive, and professional label for museum-quality specimens (Inouye, 1991). Use four-point type or similar. Sheets of labels also may be printed in advance, with blank spaces left for the date, or other unknowns, to be added in at a later time.

Print on only one side of a label.

Size of Labels

The size of the label is important. Labels must be neither too large nor too small. In determining the size of labels, a relationship must be established between the size of the insect on a pin and the amount of data a label will hold. It is best to make all labels no larger than about 7 by 18 mm. Use several labels if need be for each specimen to accommodate all of the information that must be presented.

One advantage of moderately large labels for small insects is that if a pin with such a label is accidentally dropped, the label will often keep the insect from being damaged. On the other hand, large labels may damage nearby insects in a box when the pin holding the label is removed from the box. If capital and lowercase letters are used, then spaces between words are not necessary. If there is any chance of ambiguity, then use full spellings. With only one line of data, the label should be wide enough so that when the pin is inserted, all data are legible.

Label Data

For museum specimens, the information provided on the data label is as important as the specimen on the pin. The information must be concise, accurate, and unambiguous. Indispensable data must answer the questions of where and when the specimen was collected and include the name of the collector. This kind of data, usually known as locality, date, and collector data, should be given as follows.

Locality

The collection locality should be given in such a manner that it can be found on any good map. If the place is not an officially named locality, then it should be given in terms of approximate direction and distance from such a locality.

Alternatively, the coordinates of latitude and longitude may be given. The Smithsonian Institution (U.S. National Museum) recommends that for localities in the United States and Canada, the name of the state or province be spelled in capital letters, such as ALBERTA, UTAH, KANSAS, INDIANA, and NO. DAKOTA. This method should also be used for foreign countries, such as NETHERLANDS, CHINA, EQUADOR, and HONDURAS. The next subordinate region should be cited in capitals and lowercase letters, such as counties and parishes in the United States and Canada and provinces elsewhere.

Date

Avoid ambiguity in providing the date of collection. Cite the day, month, and year in that order. Use the international convention of writing day and year in Arabic numerals and the month in Roman numerals without a line over and under the numerals. Place a period or hyphen between each number: 4.VIII.14 (= August 4, 2014), or 16.V.14 (= May 16, 1914).

For reared specimens, the dates of collection of the immature stages and of adult emergence should be cited. 'Pupa 10-XI-2013, emer. 14-VIII-2014' indicates that the pupa was collected on 10 November, 2013 and that the adult emerged on 14 August, 2014.

Collector

The name of the collector is regarded as invaluable information on data labels. Spell out the last name of the collector or collectors, using initials for given names. If the last name is a common one, such as Keene, Sharp, or Asay, always include middle-name initial. If a collecting party consists of more than three collectors, then use the leader's name followed by et al.: T.J. Gibb et al.

Other Data

Many kinds of information may be important but not relevant or available for all insects. For instance, cite the hosts of parasitic insects and the host plant for phytophagous insects when this information is known. Details of the habitat elevation, ecological type, and conditions of collection) are important and are usually put on a label in addition to the primary data. Such data may include 'swept from *Salsola kali*,' 'pheromone trap in orange grove,' 'at light,' '3200 m,' 'sandy beach,' and 'under bark dead *Populus deltoides*.'

The presumed nature of the association between insect and plant should be clearly indicated, for example, 'Resting on flowers of *Vaccinium* sp.' The word 'ex' (Latin for 'out of') should mean that the insect was observed feeding on or in or was bred from the mentioned plant, for example, 'Ex seed *Abutilon theophrasti*.'

Forensic evidence also should include additional data, such as agency of collector, case number, victim or subject's name, a brief description of the item, and name and signature of the person sealing the evidence.

As mentioned earlier, it is a good idea for collectors to keep a notebook in which details of locality, habitat, and other important data are recorded. Each individual locality may be assigned a notebook or code number with which the collecting jars and vials are marked until the specimens can be prepared. However, citation of such a number on permanent labels is not recommended.

Placing the Labels

For double-mounted insects, insert the pin through the center of the right side of the label (Fig. 2.35), with the long axis of the label oriented in the same direction as the card point. Take care that the pin is not inserted through the writing on the label. For specimens mounted by direct pinning, the label is centered under the specimen, with the long axis of the label coinciding with the long axis of the specimen. The left margin of the label is toward the head of the insect. An exception to this is when specimens have the wings spread, such as Lepidoptera.

In these cases, the label always should be aligned transversely, at right angles to the axis of the body, with the upper margin toward the head. Labels may be moved up the pin to the desired height by using a staging or pinning block.

The middle step of the block will give about the right height if only one label is used. With more than one label, space the labels on the pin beneath the specimen so that the information on the labels can be read without moving any of them.

Labeling Vials

Specimens preserved in fluid should be accompanied by one rather large label that includes all collection data. Do not fold the label because small specimens may be damaged or lost when the label is removed. Multiple labels or small labels that float in the vial may also damage specimens. Further, when two labels lie face to face, they cannot be read.

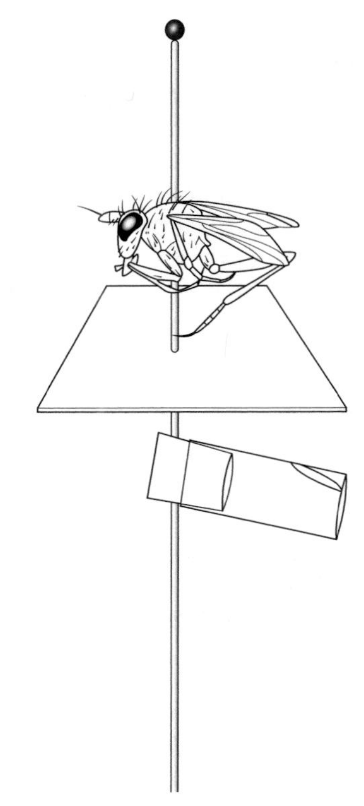

FIGURE 2.35 Specimen mounted with microvial containing genitalia.

Always place labels inside the vial. Labels attached to vials may become defaced, destroyed, or detached, regardless of the method or substance used to affix them to the vial.

Labeling Microscope Slides

Preserving specimens on microscope slides is time-consuming and tedious. The labels for slide-mounted specimens should be given equal care and consideration. Labels made expressly for this purpose can be obtained from biological supply houses. Labels with pressure-sensitive

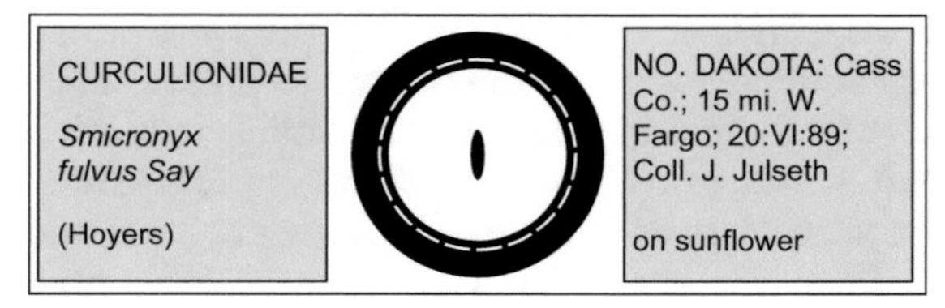

FIGURE 2.36 A properly labeled, microscope slide mounted insect.

adhesive are now available that seem durable with time. Modern adhesives are superior to the older glues, which often failed with time or were consumed by cockroaches or psocids. Labels should not be affixed to slides with transparent tape.

All microscope slide labels should be square (Fig. 2.36). Specimens should be centered on the microscope slide and a label placed on either side of the specimen. Never put labels on the underside of a microscope slide. The label on the left side of the specimen should contain taxonomic information. The label on the right side of the specimen should include collection information. The sequence of information should also be standardized. Put as much data on the label as possible, including the kind of mounting medium in which the specimen is preserved. This kind of information is important when specimen remounting is required (Waltz & McCafferty, 1984).

Identification Labels

When specimens are sent to an expert for identification, they should be accompanied by permanent collection labels giving all essential data. If associated field notes are available, copies of these should accompany the specimens. When the identification has been made, the scientific name of the specimen and the name of the identifier should be printed on a label associated with the specimen. On pinned specimens, this information is always printed on a separate label placed below the collection label or labels on the same pin. When a series of specimens consists of the same species, the identification label is often placed only on the first specimen in the series, with the understanding that all other specimens to the right in that row and in following rows belong to the same species.

Identifications for specimens preserved in alcohol or on slides may be written on the same label as the collection data or on a separate label, depending on the preference of the collector or person making the identification.

MAKING A REFERENCE COLLECTION

One of the most valuable tools that a diagnostician can acquire and use is a reference collection. When self-made, its value increases significantly. Nearly every experienced diagnostician has a reference collection of some sort or another, as it is a natural result of work in diagnostics. Diagnosticians want to retain and preserve samples that may aid in future identifications as well as in training.

A reference collection can include dry insect specimens that are spread and mounted in the same way that museum specimens are mounted, but may also include wet specimens, meaning that they are contained in vials of alcohol or other preserving solutions. Reference collections are unique because they also include samples that typically do not belong in a museum collection. For example, insect-damaged wood or leaves or other materials that have been damaged and can be preserved are of value for reference. Reference collections also include samples of frass, plant host material, accompanying photos of bites, stings, and other impacts, and any series of insects as found by submitters and accompanying testimonials that offer insight into how, where, or why a sample was obtained (Figs 2.37 and 2.38).

Such samples are often, but not always contained in Riker mounts to make it easy to see and handle them. Sometimes insect parts, including

FIGURE 2.37 Riker mounted reference samples.

FIGURE 2.38 Insect damage reference sample.

wings, legs, or genitalia, are mounted and stored separately. Slides are often used to preserve reference materials that are very small, including mites, ticks, tiny insects and insect parts. These are also of great value in a reference collection.

With the use of microscopes, digital photography and computer-assisted storage and retrieval capabilities, digital representations of these physical samples can be contained in an electronic reference collection as well.

The information that accompanies samples in a reference collection also differs somewhat from a museum specimen. While museum specimens have a fairly discrete set of required information (date, location, collector) that must accompany the specimen, reference collections are not strictly bound by these rules. An acceptable and valuable specimen in a reference collection may include as little as a common name, or as much as a paragraph describing the background of the sample, its scientific name and a description of where, how and why the sample was submitted.

Storing Specimens

Like collection tools and supplies, the storage of a reference collection can be as simple or as complex as resources allow. Often the collection is started in a simple fashion, using only a few insect drawers and a box for holding vials. Over time this can be upgraded and increased to state of the art storage, similar to an insect collection in a museum.

A searchable database can add enormously to the utility of the reference collection. Computerized storage of digital images requires the same type of organization needed for storage of physical specimens. The value of the database lies in its ability to search for samples very quickly and from several fields. For example, a search may be requested for all specimens that fit designated criteria such as all beetle larvae, or all buprestid beetle larvae, or all buprestid beetle larvae that harm fruit trees, or all buprestid beetle larvae that harm the trunks of peach trees.

If reference materials are unorganized or difficult to access, they will not be used regularly and their value depreciates measurably.

Storage of wet specimens is best done using vial racks and appropriate cabinets to hold the racks. Drawers that open to expose a series of racks seem to be most functional. Specimens are best arranged by scientific identity, much like museum collections; however, they may also be arranged by commodity group or life stage. Well-constructed databases can make searching for specific specimens easy regardless of how they are arranged.

Larger reference samples and Riker mounts can be stored in drawers or totes that make them easy to access. When each is identified by association with a specific arthropod these can be very valuable to a diagnostician.

Most specimens in a diagnostic reference collection will be dry, pinned insect specimens. Pinned specimens should be kept in any standard, commercially available insect drawer (Fig. 2.39). These standards are named U.S. National Museum, California Academy of Sciences, Cornell, or Schmitt sizes (Fig. 2.40). The exact fit of the glass top to the drawer frame and its tight-fitting wooden construction discourages infestation by damaging pests. Polyethylene or Plastazote foam are most commonly used for drawer bottoms. These materials are particularly adaptive to receiving insect pins and holding them in place. Boxes may or may not use the unit-tray system, with various sizes of unit trays made to fit into a drawer.

Usage of tightly-closing steel insect cabinets, designed specifically to house standard insect drawers, is recommended to house specimens in a reference collection (Fig. 2.41). These cabinets house standard insect drawers that are built

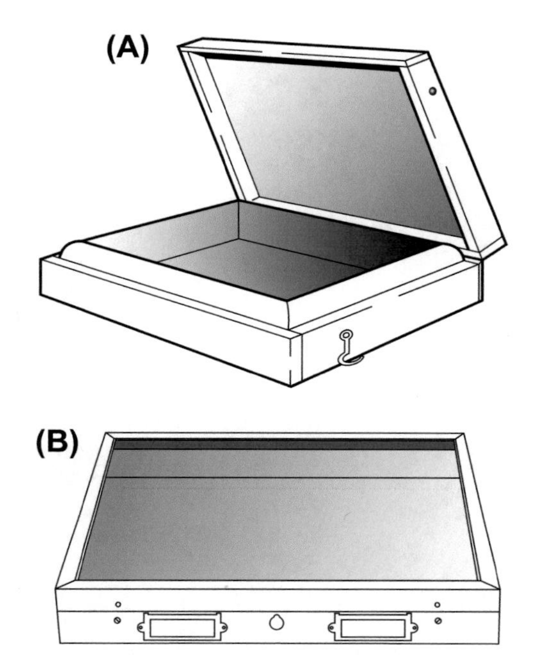

FIGURE 2.39 Storage containers for pinned insects. (A) Schmitt box. (B) Cornell drawer.

especially for insect collections, provide optimum protection for valued specimens, are easy to access and make individual specimens easy to observe without actually removing them each time.

Care of the Reference Collection

If care is taken and a few basic precautions are followed, a collection of insects or mites can be maintained indefinitely. Material preserved in liquid usually needs little attention beyond the occasional replacement of preservative and sometimes lids.

Microscope slides are usually stored in wooden or plastic boxes obtainable from biological supply houses (Fig. 2.42).

The inner sides of the boxes are slotted to hold the slides vertically and to separate them from one another. Preferred slide boxes are designed so that the slides rest horizontally. Small plastic

FIGURE 2.40 Cornell drawer.

FIGURE 2.41 Insect cabinet.

slide boxes, usually made to hold five slides, are convenient for keeping slides in a unit-tray system along with pinned specimens. This is particularly advantageous when genitalia or other insect parts are mounted on slides, because they are readily accessible when examining the pinned specimens.

Any box used to store insect specimens must be nearly airtight to keep out museum pests. The most commonly encountered pests include dermestid beetles, psocids (booklice), and silverfish. In a period of a few months, these insects (and certain other pests) can devour specimens, chew labels, or otherwise ruin a collection made over many years. These pests are uncanny in their ability to find their way into the best boxes or insect drawers. Constant vigilance is necessary to prevent them from destroying valuable specimens and priceless scientific information.

The periodic fumigation of all insect storage boxes is necessary. The best-made insect drawers provide for chemical fumigants. In the past, paradichlorobenzene (PDB) was used in museums as a fumigant. However, this material has demonstrated chronic health effects and is under EPA investigation for possible carcinogenic effects on humans. Naphthalene is widely used as a fumigant, but in a technical sense it is regarded as a repellent. These should not be placed loose in a box of pinned specimens. If crystals or flakes of napthalene are used, a small quantity should be placed in a cloth bag or in a pillbox whose top is perforated with tiny holes. This container is pinned firmly into one corner of the

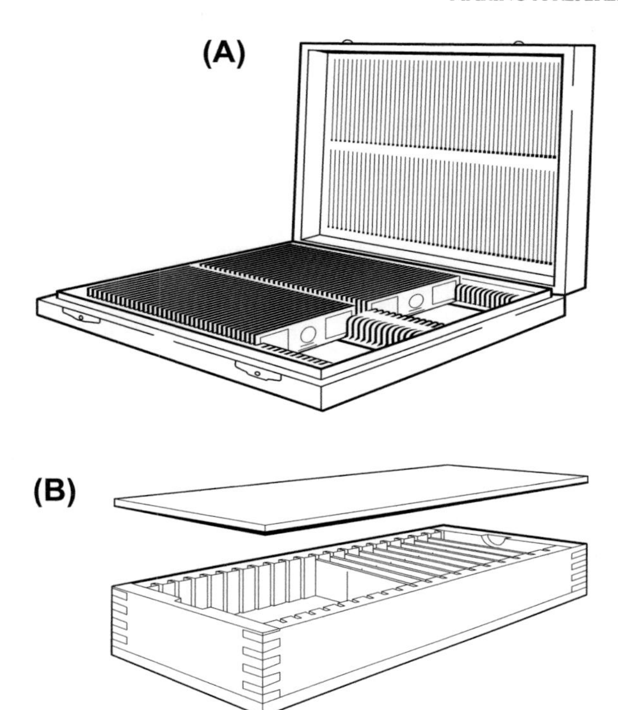

FIGURE 2.42 Storage containers for microscope slides. (A) Plastic slide storage box. (B) Wooden slide storage box.

box of specimens. Napthalene mothballs may be pinned in a box by attaching the mothball to the head of an ordinary pin. This is done by heating the pin and forcing its head into the mothball. When moving boxes, be careful that the mothballs and fumigant containers do not come loose and damage the specimens. To keep pests out of Riker mounts and other display cases, sprinkle naphthalene flakes on the cotton when the mount is prepared.

Another useful method to protect a collection against pests is to cut strips of dichlorvos (738 Vapona strips, No-pest Strips, Vaponite, Nuvan) into small pieces, and secure them in the insect drawers. Dichlorvos is registered as non-restricted use, but only pest strips should be used. This method gives a fairly rapid kill while avoiding the hazards of using flammable liquids. Under conditions of high humidity, Vapona strips corrode metals and dissolve polystyrene plastic.

Mold is another serious problem with insect collections, especially in moist, arm climates. Mold is a fungus that readily attacks and grows on insect specimens. Once a specimen has become moldy, it is extremely difficult to restore it. If only a few filaments or hyphae of mold are present on a specimen, they may be removed carefully with forceps or with a fine brush and the specimen dried well in a warm oven and then returned to the collection. Only keeping the collection in a dry place will prevent mold. In humid climates it is sometimes necessary to keep insect collections in rooms with artificial dehydration.

CHAPTER

3

Submitting Samples to a Diagnostic Laboratory

SUBMITTING SAMPLES

Preparing, Preserving, Packaging and Submitting

A wide range of insect diagnostic facilities are in use today. Public universities often employ a team of diagnosticians to answer questions relating to insects or other pests, such as weeds or diseases. This team of specialists normally serves the general public, as well as private companies and enterprises. In a university setting, diagnosticians are considered General Insect Specialists, as compared to commodity specialists such as field crop, urban, or landscape specialists, and therefore deal directly with a wide range of clientele groups. By extension, they also work with a wide variety of different samples submitted for diagnosis.

Some large corporations employ their own diagnosticians as quality assurance supervisors. These diagnosticians are often responsible for identifying, managing and preventing any foreign contamination (including insect parts) in processing or packaging. Samples handled by these diagnosticians are quite different in nature and in quality than those provided to other diagnosticians.

Pesticide manufacturers or agriculture enterprises and large pest management companies also employ their own diagnosticians. These are expected to serve the company's interests, but rarely diagnose samples for others. As a result, the range of insects that they must deal with is comparatively narrow.

Regardless of size, specific function, or location of an insect laboratory, one element that remains constant is that in order to provide an accurate diagnosis, an adequate sample must be received and accompanying background information must be provided. These two provisions are the key upon which a valuable diagnosis pivots.

> Importance of Sample Submission
>
> A proper sample allows for an accurate diagnosis which allows for confident recommendations which allow for effective pest management.

A traditional insect sample submitted to a laboratory is sent through the postal system. Ideally it consists of a properly collected and preserved specimen, packaged and shipped such that it

Contemporary Insect Diagnostics
http://dx.doi.org/10.1016/B978-0-12-404623-8.00003-X

arrives intact and without delay, and is accompanied by all pertinent background information such as where and when it was collected (what the habitat was like), what it was doing when collected (biting, boring, flying, crawling, burrowing etc.), how many other insects were present, and some history of the encounter (was it a new problem or had it been persisting for some time?). However, times have changed and modern technology allows for samples to be submitted in a variety of non-traditional ways.

What remains constant is the absolute need for properly collected samples together with an adequate description and background of the problem. These are critical, regardless of the method in which they are submitted. In this chapter we will discuss preparing, preserving, packaging and submitting samples to a diagnostic laboratory.

Sample Submission Forms

Whatever method of submission is chosen, essential information must accompany the sample. At minimum, this includes:

- submitter name and contact information,
- a clear request for services (desired level of identification, control recommendations, life history and biological information),
- essential background information. [A diagnosis is based on a description of the problem such as where it was found, what level of damage (if any) is apparent, and how long the problem has been present.]

A sample submission form is designed to elicit background information about a sample that will aid a diagnostician in identifying the specimen, as well as recommending appropriate management practices. Many different submission forms are used. These may differ based on the diagnostician's specific responsibilities, location and personal preferences. In a large public diagnostic laboratory a common submission form is often used for insect as well as weed, disease and horticultural identifications. Where electronic databases are employed, a standard submission form is normally the first step in the data entry. If fees are charged, the submission form doubles as a vehicle for tracking and accounting.

A submission form should:

- elicit important background information about a sample;
- provide a directive of what is expected of a diagnostician;
- give the contact information for the submitter;
- contain essential information for a historical database;
- allow for sample tracking and accounting if fees are charged.

A carefully crafted submission form will often extract the needed information from a client. Most laboratories have a standard submission form that is required for each sample submission. Some even have specific submission forms for digital samples.

Below are three examples of submission forms currently used by public diagnostic laboratories. Note that while each is slightly different, all include the basic information needed by diagnosticians (Figs 3.1–3.3).

What information to include:

Each sample is unique. Some require more or different bits of informational background than others. At minimum, sufficient information must be provided to allow for identification, sample tracking and proper recommendations. A thoroughly completed submission form meets these needs.

A single form can be used for a series of insects collected on the same plant or at the same specific site. However, a separate form should be used for different insects or when they are collected on separate plants or sites.

Utah Plant Pest Diagnostic Lab

Submission form for INSECTS, ARACHNIDS, PLANT DISEASES and associated damage

For diagnosis, mail this completed form with sample and *$7.00* cash or check to:

Utah Plant Pest Diagnostic Lab, Department of Biology, 5305 Old Main Hill, Logan, UT

UtahState UNIVERSITY extension

Fee Information: All samples submitted to the UPPDL must include $7.00 cash or check. Plant disease samples that cannot be positively diagnosed in-house can be outsourced to an out-of-state lab for additional testing at the consent and expense of the client ($55.00-$90.00 minus $7.00 already paid). Refunds will not be given for samples with an inconclusive diagnosis.

SUBMITTER INFORMATION

Directions for Submission: Insect/Spider or Disease

Name:

Address:

City: State: Zip: County:

Phone: ( ) Cell: ( )

Email:

Lab use only

Submission Date: ______

Sample Number: ______

Contact County Agent: ☐ Y ☐ N **Agent Name:** **Agent Email:**

CONTACT ME BY: ☐ Regular Mail ☐ Phone ☐ Email

INFORMATION REQUESTED ☐ Identification ☐ Control Recommendations

COLLECTION INFORMATION: Where was the sample collected?

Host Plant Common Name:

Host Plant Scientific Name/Variety:

If not collected from a plant, then please describe:

PLANT INFO:

Plant age: Size:

Number of plants affected:

Percent of plants affected:

Problem worsening: ☐ Y ☐ N

PLANT SYMPTOM(S) DESCRIPTION: Describe in detail what is wrong with the plant and what parts are affected

PEST/SYMPTOM IMAGES: If possible, please provide photos of the affected plant(s). A range of photos from landscape-level shots to close-ups are ideal. Including photos can aid diagnosis. Digital images can be emailed to ryan.davis@usu.edu.

I included pictures as: ☐ Printed photos submitted with the sample ☐ Digital images ☐ No images submitted

PEST INFORMATION

Number of Specimens in Sample: Number of Individuals Observed:

Describe the Appearence of the Pest:

EXTRA INFORMATION: Use the back of this sheet to include additional information: e.g. irrigation type and rate, fertilization type and rate, pesticides, fungicides or herbicides applied, map/schematic of property and location of affected plants including directional arrows (N, E, S, W), general care/maintenance or situation of plants, planting depth or method, progression of symptoms, etc.

DIAGNOSIS (Lab use only)

Common Name: ______ Family: ______

Genus: ______ Species: ______

RESPONSE:

[ ] Regular Mail [ ] Email

[ ] Phone [ ] Walkin

Date: ______

Time: ______

FIGURE 3.1 Submission form, Utah University, http://utahpests.usu.edu/uppdl/files/uploads/Submission_Form.pdf.

Plant & Pest Diagnostic Laboratory
LSPS – Room 101, Purdue University
915 W State St, West Lafayette, IN 47907-2054
765-494-7071 FAX: 765-494-3958
http://www.ppdl.purdue.edu

PURDUE UNIVERSITY

(PPDL-1-W) 1/14

Office Use Only: Date received: ________
Sample #: ________
Account #: ________

Date: ________

Submitter's Name ________	**Client's Name** ________
Business ________	Business ________
Address ________	Address ________
City/State/Zip ________	City/State/Zip ________
County ________ Phone ________	County ________ Phone ________
Fax ________ Email ________	Fax ________ Email ________

Please include a check or money order (payable to Purdue University) for $11 per sample ($22 out-of-state clients). DO NOT SEND CASH.
Send invoice to ☐ Submitter ☐ Client

☐ Perform only routine diagnosis ($11 in-state/$22 out-of-state)
☐ Please notify submitter if additional fees for advanced testing are needed
☐ Perform additional advanced testing if necessary (up to $50)

Mail reply to: ☐ Submitter ☐ Client
Fax reply to: ☐ Submitter ☐ Client
Email reply to: ☐ Submitter ☐ Client
☐ Copy Extension Educator

Information about Submitter/Client (please check one each for submitter and client)

Submitter	Client		Submitter	Client	(continued)
____	____	Extension Educator	____	____	Pest Control Operator
____	____	Homeowner	____	____	Nursery
____	____	Farmer	____	____	Lawn or Tree Care Co.
____	____	Dealer/Industry Rep.	____	____	Garden Center
____	____	Golf Course	____	____	Consultant
____	____	Landscaper	____	____	Purdue Specialist
____	____	Greenhouse	____	____	Other ________

Check information desired:
____ Problem identification
____ Specimen identification
____ Control recommendations
____ Other ________

Plant and Pest Information

Plant or Host: Cultivar/Variety:

Location (choose one):
____ In dwelling
____ Tree/Shrub
____ Turf/Lawn
____ Golf Course
____ Flower bed
____ Vegetable garden
____ Field/Farm
____ Greenhouse
____ Nursery
____ Orchard
____ Animal/Human
____ Aquatic
____ Stored grain/Food products
____ Other ________

Degree of Damage (choose one):
____ Heavy
____ Medium
____ Light

Insect Problem? (choose one):
____ Damaging plant
____ Biting/Stinging
____ Infesting food
____ Nuisance

for Plant/Weed Identification Only

Plant type:
____ Tree ____ Deciduous
____ Shrub ____ Evergreen
____ Vine
____ Groundcover
____ Herbaceous

Plant size:
____ Height
____ Width

Flowers:
____ Color
____ Month(s)
____ Size

Fruits:
____ Color
____ Month(s)
____ Size

Plant age:
____ Annual
____ Perennial (# years ____)

Unique features (bark, leaves, odor, thorns, etc.): ________

Additional Plant and Site Information

Approximate age: ________ Height: ________ Number of years at present site: ________
Exposure: ____ Full sun ____ Partial shade ____ Full shade ____ Windy ____ Protected Irrigation frequency: ________
Root disturbance from: ____ sidewalks/driveway construction activities (describe): ________
Size of planting: ________ % of plants affected: ________ Date first noticed problem: ________
Date planted: ________ Tillage practices: ________ Previous crop: ________
Chemicals/fertilizers applied (past 2 years)(include rates): ________

Soil type: ____ sandy ____ clay ____ silt ____ loam ____ organic Soil pH: ________

DESCRIBE THE PROBLEM (Include symptoms, plant parts affected, pattern of occurrence, etc. Attach separate sheet if necessary):

____

Your tentative diagnosis/ID: ________

Print Form

FIGURE 3.2 Submission form, Purdue University (Purdue university www.ppdl.purdue.edu/ppdl/pubs/PPDL-1-W.pdf).

UNIVERSITY OF MINNESOTA
PLANT DISEASE CLINIC

Sample Submission Form

Clinic Contact Information

Mailing Address: Plant Disease Clinic
495 Borlaug Hall
1991 Upper Buford Circle
St. Paul, MN 55108

Delivery Address: 1519 Gortner Ave
105, Stakman Hall
St. Paul, MN 55108

Phone: (612) 625-1275
Fax: (612) 625-9728
Email: pdc@umn.edu
Website: http://pdc.umn.edu

Client Information:

Submitter	Grower/Homeowner
Please: (☐ mail ☐ email ☐ fax) results and bill.	Please: (☐ mail ☐ email ☐ fax) results and bill.
Name: ______	**Name:** ______
Company Name: ______	**Company Name:** ______
Address: ______	**Address:** ______
City: ______ **State:** ______	**City:** ______ **State:** ______
Zip: ______ **Phone #:** ______	**Zip:** ______ **Phone #:** ______
Email: ______	**Email:** ______
Fax: ______	**Fax:** ______

Submitted Plant Information:

Date Submitted: ______ Species/Variety: ______
Plant/Crop: ______
Specific Test Request ______

Problem (i.e. symptoms, plant part affected, pattern of occurrence, other plants affected, when symptoms were first discovered*): ______

Additional Information (i.e. site information, pesticide/fertilizer applications, soil description, watering routine, other plant species affected*):

*Please use reverse side or attach additional sheets if needed

FIGURE 3.3 Submission form, University of Minnesota (Minnesota http://pdc.umn.edu/prod/groups/cfans/@pub/@cfans/@pdc/documents/asset/cfans_asset_337282.pdf).

PHYSICAL SAMPLES

Most insects are submitted to a diagnostic laboratory in the traditional manner by collecting the insect, preparing it for transport and sending it to a diagnostic laboratory. Samples that can be physically seen and handled by a diagnostician are preferred because they allow the greatest opportunity to make an accurate identification.

Capturing an insect without damaging it is the first step. Insects killed by using a vacuum, flyswatter or shoe are seldom in a condition that allows for proper identification. It is preferable to submit a series of insects rather than just a single specimen. Physical samples almost always consist of dead insects, either preserved in liquid (wet) or submitted dry.

Wet Samples

Most diagnosticians prefer that samples be submitted wet. Submitting insects that have been preserved in liquid is also the easiest method of ensuring that they arrive in a manner that lends itself to identification.

Small glass vials of various sizes and construction are a staple in diagnostic laboratories. These are valuable both for submission of wet samples and also for permanent storage of specimens for reference (see Chapter 2, Preserving arthropods). Usually vials are made of clear glass to allow both the specimen and the label to be seen without removing them from the container. Sample vials come in various sizes and styles and can be obtained from several biological supply houses across the country.

Alcohol or other preservatives are sealed in the vial by rubber stoppers or, more commonly, screw caps with special plastic inserts to prevent leakage.

Diagnosticians may contact those who submit samples regularly and either provide vials or recommend a source where they can be purchased.

Recommendations of specific preserving liquids are currently under debate. Use of 70% rubbing alcohol has been a standard for many years. It preserves insects very well, is clear (allowing quick identification) and is likely the preservative that diagnosticians will use to permanently house the specimen in their museum or reference collection. However, some have expressed concerns regarding the safety of shipping potentially flammable materials, including alcohol, formalin and formaldehyde, and therefore recommend the use of household vinegar, which is inexpensive, readily available and non-flammable. Others argue that the amount of alcohol (isopropyl or ethanol) in a vial is so small as to practically mitigate any concern about flammability.

Whatever medium is chosen should be disclosed to the diagnostician.

Insects should never be submitted in water. Water has no preservative properties. Insects, spiders or mites submitted in water decompose and discolor very quickly and seldom arrive in condition that allows for accurate identification.

Immature insect specimens decompose very quickly when they die. For that reason, caterpillars, maggots, grubs and nymphs, as well as spiders and mites, should almost always be submitted wet. This effectively stops the decomposition process and also serves to protect the specimens from being broken or lost. Some immature insects (caterpillars, maggots and grubs) and spiders may still discolor, or turn black, if placed directly into alcohol. However, color by itself is not as important to a diagnosis as most laypeople may assume. Very few diagnoses are based on color alone. Even so, color loss can be largely prevented by first placing the specimen into boiling hot water before transferring into alcohol.

When preparing vials for shipping, clients should ensure that each vial is completely filled with alcohol to avoid undue sloshing, which could potentially ruin the specimen inside the vial. Precautions should also be taken to ensure that vials will not leak while specimens are in transit. Screw-top vials should be firmly closed; if further sealed with a turn and a half of plastic adhesive tape around the lower edge of the cap and neck of the vial, they are nearly certain to remain intact.

When submitting more than one vial, package them such that no piece of glass can come into contact with another piece of glass during transit. This may be accomplished by wrapping each vial with cotton, tissue, foam padding, paper toweling, or similar material. Several individually wrapped vials may be bound or held together with tape or rubber bands as a unit, or they may be tightly placed in a small cardboard box with enough packing material to ensure that they are not shaken around. Placing all of the packed vials in a self-sealing plastic bag will further contain fluids in the event that a vial should break.

Dry Samples

In some situations dry insects must be submitted. Insects submitted dry are at risk for two reasons. One potential problem in submitting dry insects is that, if they are not completely dry, they will decompose. There are few things more unpleasant for a diagnostician than opening an airtight bag or box containing rotting insects. The smell is powerful and worse, the chances of properly identifying a decomposed insect are slim. Small insects dry out very quickly but large insects, even in the adult stage, may not dry quickly and therefore can be a problem if contained in an airtight package too soon.

If insects are to be submitted without a liquid preservative they must be in the adult stage. Adult insects have a fully developed exoskeleton that remains in position even though the internal part of the insects dry and shrivel after death. Fortunately, it is the external morphology that is used in most insect identifications.

A second risk in submitting dry insects is that they are very fragile. Entomology instructors often teach insect morphology by dissecting an insect into its many parts: head, thorax, abdomen, six legs each consisting of femur, tibia, and tarsi with several parts, two multi-segmented antennae and two or four wings. Diagnosticians joke that such dissection services are rendered freely by the postal service if the insects are submitted dry and not packed correctly (Fig. 3.4 A-B).

To avoid breaking, insects must never be submitted in tape (Fig. 3.4 C). Specimens must be protected. Small crush-proof containers are neces-sary for the protection of dry insects during submission.

Containers may be constructed of cardboard, plastic, or metal. In many cases expired pill or medicine bottles serve this function nicely and are readily available. The specimen should be supported by partly filling the container with soft tissue paper or other padding, such as cotton, to keep specimens from becoming damaged. Absorbent tissue is readily available and is a suitable medium for packing dry insects inside a container. Some diagnosticians do not recommend the use of cotton in sample containers because dried specimens may become entangled in the fibers and impossible to extricate without damage. However, others (such as collectors of minute or fragile insects) find that specimens stored in a few wisps of cotton are best protected from damage.

Pinned (mounted) specimens are not commonly submitted by the public to a diagnostician for identification. However, diagnosticians may find it necessary to submit insects that they have pinned to specialists, taxonomists or professional identifiers for verification. Therefore, the following instructions are useful.

Pinned specimens should always be securely placed in a small box with a pinning bottom made of polyethylene or polystyrene foam, or a similar material. Insert the pin firmly into the pinning bottom, leaving sufficient space between pins so that the specimens can be removed easily. Pinned specimens placed too

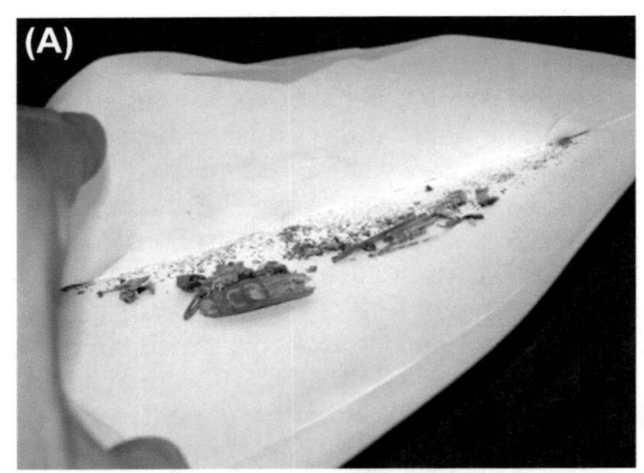

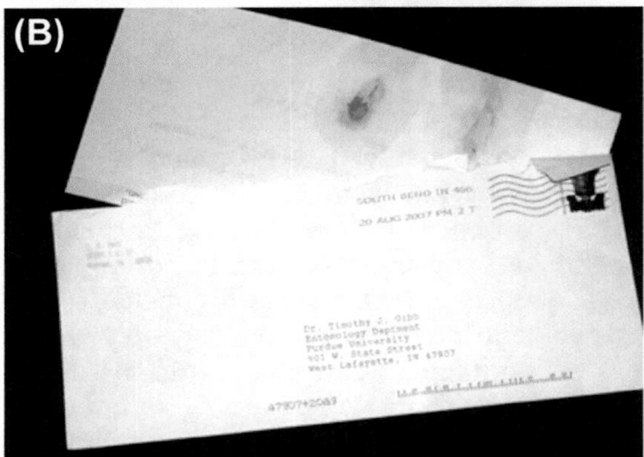

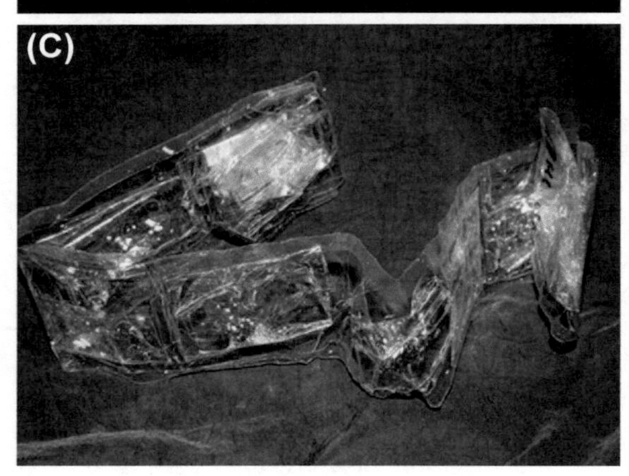

FIGURE 3.4 A-C Poorly submitted insect samples (A) Crushed insects in envelope. (B) Live insects submitted in a letter. Insects stuck to tape

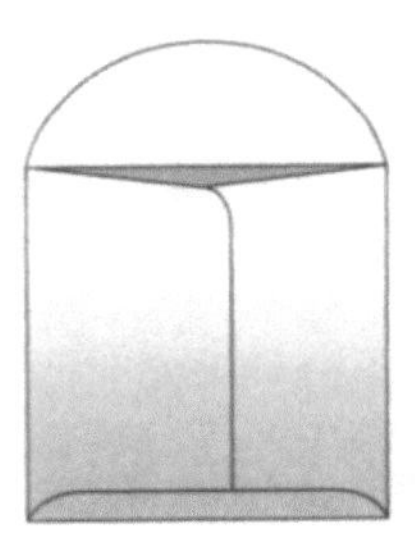

FIGURE 3.5 Glassine envelope designed for field storage of moths and butterflies.

closely together are sometimes damaged by adjacent specimens during removal.

Place bracing pins on both sides of heavy or long-bodied specimens to prevent them from rotating (cartwheeling) on their pins and damaging adjacent specimens. Bracing pins positioned along the sides of data labels or card-mounted specimens also prevent cartwheeling.

Butterflies and moths have wing scales that rub off easily if touched. Specially designed glassine envelopes will prevent dislodging of scales, and become an acceptable method of submission. These envelopes are available from most entomological and biological supply houses (Fig. 3.5).

The box or container holding the pinned specimens should be well wrapped and placed within a larger carton with at least 5 cm of lightly compressed packing material between it and all sides of the carton. The box should have a tight-fitting or a flap-type lid held in place with sufficient packing tape to ensure security.

Submitting Plant Parts, Soils, or Frass

Sometimes samples of plant parts, such as stems, roots, leaves or fruit, must be submitted for diagnosis. Samples of soil, wood or insect frass are also submitted on occasion. These samples do not lend themselves to traditional methods of submission. While each case must be considered individually, the following general recommendations are useful.

When submitting plant parts, place the sample flat between layers of dry newspaper or toweling and avoid excessive folding of the leaves and flowers so that they arrive in the most natural state. A piece of cardboard placed with the sample can help keep the sample flat.

Be aware that moisture is an enemy to organic samples. Always submit plant samples wrapped in absorbent material – never directly in plastic bags. Over time, any moist sample in a container will mold and rot.

Fruit samples are very perishable because of their high water content, so extreme caution must be exercised in submitting them. Wrap whole, uncut fruit specimens in paper, place in a sturdy box, and pack with additional paper to prevent crushing.

Soil or samples of loose frass that may spill should be packaged in a dry container that will not break. Plastic containers are less desirable than cardboard because moisture inside the sample tends to condense on the plastic, facilitating mold growth.

If moist organic samples are submitted, the diagnostician should first be alerted that the sample is to be expected and arrangements such as expedited delivery, overnight or 3-day delivery should be discussed. Upon arrival, samples should be unwrapped and held under refrigeration to preserve them until the first possible opportunity for examination.

Submitting Live Specimens

To protect American agriculture, federal law prohibits the importation and movement of live pests, pathogens, vectors, and articles that might harbor such live organisms, unless authorized by the United States Department of Agriculture (USDA). In cases where it becomes necessary to ship live specimens, contact the Animal and Plant Health Inspections Service (APHIS) within the USDA for instructions regarding federal regulations.

In addition to meeting federal laws, the shipment of some species also must be approved by state officials. Be sure to comply with all federal, state, and local regulations. Shipments of live insects and associated materials without valid permits may be seized and destroyed.

Always package live insects such that they will neither be crushed nor expire en-route. Remember that freshly dead arthropods deteriorate and decompose very quickly, making identification difficult.

Pupae and larvae should be placed in tightly closed containers without vent holes. These insects require only a minimum of air and will seldom suffocate during delivery. Pupae should be packed loosely in moist (not wet) moss or similar material. Larvae should be provided with enough food material to last until their arrival. Most beetle larvae and many caterpillars should be isolated and packaged individually because they can be cannibalistic.

To prevent excessive accumulation of fecal material and moisture, do not overload containers. As described above, organic material held without ventilation tends to become moldy, especially when kept in plastic bags. For this reason, pieces of the host plant bearing such insects as scale insects should be partially or completely dried before being placed in a container, or they should be packed in a container such as a paper bag, which will permit drying to continue after closure. Other small, active insects are killed easily by condensate inside a container. In this case, it is advisable to make several tiny holes in the container to allow for air movement, or place a fine mesh screen over one end of the container when shipping.

If live specimens are to be mailed, they should be packaged such that they cannot escape under any circumstance. Multiple layers of secure packaging are recommended. Some containers designed to hold living insects are strong enough to be mailed without additional packing, but generally containers should be enclosed in a second carton with enough packing material to prevent damage to the inner carton. In all cases, affix a permit for shipping live insects in a conspicuous place on the outside of the shipping

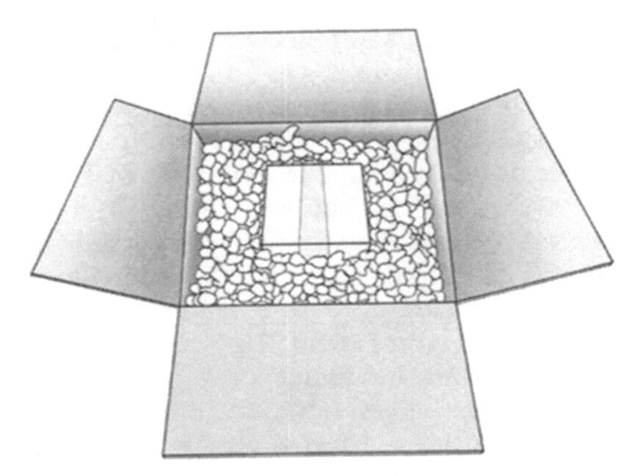

FIGURE 3.6 Packaging specimens for shipment.

container. It is also wise to attach a 'Fragile – Handle With Care' sticker, or print it in bold lettering on the outside of the container.

Insects shipped internationally fall into a different category.

In response to concerns about trade in rare or endangered species, 'wildlife' regulations concerning the shipment of insect specimens have changed markedly. Previously, most insects were excluded from the category of 'wildlife,' but recent rules have been expanded to include insects in this definition. Even shipping dead insects to foreign countries (or the importation of specimens from foreign sources) may now require the filing of U.S. Fish and Wildlife Service Forms. It is advisable to check with Fish and Wildlife Service officials at the nearest expected port of entry (usually either a major airport or seaport) to find out what local requirements exist, as there can be variation from port to port.

Packing and Shipping

Samples are at high risk of damage when they are in the mail. Both the length of time in transit and the amount of jostling and bouncing that the sample is subjected to are largely out of the control of either the submitter or the diagnostician. Thus, packing specimens is an important step that requires considerable thought and attention.

Mailing cartons should be constructed of strong corrugated cardboard or other stiff material. Screw-top mailing tubes are ideal for small samples and wet specimens. All containers must be sufficiently large to allow for ample packing material to minimize the effects of jarring. A rule of thumb suggests a minimum of 5 cm on all sides between the sample and the inner surface of the mailing carton. Remember that the packing material is intended to absorb shock (Fig. 3.6).

Containers holding specimens should not be packed so tightly that this objective is negated.

Suitable packing materials include excelsior, shaved wood, crumpled newspapers, Styrofoam chips, or plastic bits. Clear plastic, bubble wrap or blister packing is even better. This material is very lightweight and has excellent shock-deadening properties.

Samples can be sent to a diagnostic laboratory in several ways. Personal or hand delivery (walk-in) is an option in situations where driving distance is not a concern. Most walk-in samples can be delivered very quickly and with minimal damage. Clientele normally hope for an instantaneous diagnosis or for a chance to speak personally with the diagnostician. While his works in some few instances, other circumstances dictate that the sample be left with the diagnostician who will then contact the client via telephone or e-mail when a diagnosis is made.

More commonly, a sample is submitted to the diagnostician through the postal system. To maintain the integrity of the sample, it should be packed as described above in a sturdy box to prevent shifting. Be certain that proper postage and addresses are on the box before mailing. If the sample is time-sensitive or perishable, the post office or package delivery service should be consulted about estimated time of delivery and reception instructions. Sometimes a physical street address is mandatory and a person must be available to receive and sign for the package. Absent this, undue delays may result.

Including a sample submission form with the sample is preferred but in cases where the form is not available, be certain to enclose background information and a clear written request with each submission. Each sample must be labeled with submitter contact information. Return telephone, fax or e-mail contact address is required at minimum.

Traditionally, a client may bring or send a physical sample into the diagnostic laboratory where the diagnostician appraises it, develops a response and then returns a written response via postal mail. This process has been tried and tested over many years and, until recently, has been the preferred method of sample submission. The biggest drawback of physical sample submissions is lengthy turn-around time (time between when a sample is submitted and the response is received). It may take up to two weeks to completely process a physical sample. While pests are at work, structures continue to be damaged, crops are in peril, plants are injured, and people suffer. Answers are needed quickly and waiting two weeks may be unacceptable. Physical samples also have a risk of deteriorating or becoming lost in the mail. Fortunately, there are various other ways of requesting help from an insect diagnostician.

TELEPHONE INQUIRIES

Telephone queries have an advantage over submitting a physical sample because they are given in real time. When a client only needs very basic information, telephone inquiries can be best. The advantage of a telephone conversation is that a diagnostician can aid a client in providing pertinent information through the questions that are asked. If an insect can be diagnosed based only on a verbal description, considerable time can be saved.

Of course, diagnoses made over the telephone are somewhat unreliable because no physical specimen can be examined. Diagnosticians must rely on descriptions of an insect or other arthropod, given by a layperson who is often recounting the description from memory. Identifications are tenuous.

Telephone clients often have unrealistic expectations. Many do not appreciate that there are more than 900,000 insect species of insects known to mankind and that simply describing a specimen as 'small and black' does little to narrow this list down. In many, but not all cases, such a telephone call ends with a request to collect and submit a physical sample or digital photo.

On the other hand, when a common insect is described, some confidence in the diagnosis can be held. Diagnosticians become amazingly skilled at interpreting verbal clues and, because of their knowledge of arthropod biology coupled with a keen awareness of time and place, they are often able to provide the identification and other information that a client needs. When this occurs by telephone, the advantage is an immediate sample turn around and a satisfied customer.

On the other hand, telephone inquiries are an interruption for the diagnostician and depending on the caller, may be very time-consuming. Compared to physical samples, telephone inquiries may require as much or more time to diagnose, especially when they result in a request for a physical sample. A productive insect diagnostician soon learns the value of telephone management, voice mail messaging and e-mail.

An informative voice mail message can be a great way to save time and still get needed information to the client very quickly. E-mail communication has nearly become a standard among diagnosticians. These are not only common but they allow for the following:

- a conversation that is near real time;
- dialogue (questions to be asked and answered);
- photos and fact sheets that can be shared through attachments;
- forwarding and copying responses or questions to other individuals;
- creation of a written record and conversation trail.

These points make e-mail a very powerful and widely used method of communication between clients and diagnosticians.

Similarly, social media (Facebook, chat groups, Twitter accounts, group messaging) can be an effective means of transmitting information about insect pests. The value of social media is the speed at which information travels within the group. Questions can be asked and answered in near real time. Discussions about predicted occurrences of pests or pest management recommendations can be had and disseminated to other interested parties instantaneously. When monitored and used effectively, these forms of communication with a diagnostician may be very worthwhile.

ELECTRONIC SAMPLE SUBMISSION (DISTANCE DIAGNOSTICS)

Circumstances vary and times change. Diagnosticians must likewise adapt to the possibilities that submissions may be made in still other non-traditional ways. Digital photography and the Internet make for a perfectly acceptable way to submit a sample for diagnosis. While viewing a photo does not allow for as much precision as identifying a physical insect sample, the speed with which an image can be obtained, submitted, and responded to make the use of electronic sample submission very attractive.

Submitting Electronic Images

Like physical samples, digital image samples also must be accompanied by a submission form. This can be included electronically as part of the submission or follow as a separate e-mail. Without a submission form linked to an electronic image, samples can be lost or misidentified.

As with the submission of physical samples, the submission form affords the diagnostician important information about the background of the specimen, while also supplying vital information needed to properly respond to the client.

Digital photographs submitted electronically may not only simplify the submission process, but also may enhance the speed at which a response is returned to the client. Reduced turnaround time can result in:

1. more direct and interactive communication with the client such that follow-up questions and answers in both directions is much more likely;
2. peace of mind, - waiting days for a written response is difficult;
3. earlier implementation of management practices;
4. reduced chance of handling error or loss of specimen due to in-transit deterioration or mail mix-up.

Concerns regarding the submission of electronic samples include the risks that:

1. specimens cannot be identified by photos alone;
2. poor photographs limit diagnostic capabilities;
3. resolution of images is often lacking;
4. images may be submitted out of context (without accompanying information about the sample).

Diagnosticians who accept digital samples must recognize that limited confidence in the identification is almost always the result. Keying out photographs is dubious unless the photographer already has an understanding of insect taxonomy, as specific key morphological characters must be observable and these differ from one diagnosis to another.

Fortunately, in most electronic sample cases only a general identification is required. Often family-level identification is possible with electronic images and this level of identification is sufficient to determine pest status and pest management recommendations.

For management information to be given, at least some information should accompany each digitally submitted sample, including:

- where the insect was found;
- approximately how many were present;
- what the insect was doing (behavior);
- what information is requested.

These details should be submitted with the digital sample. If they are absent or incomplete, the diagnostician must request them.

When all goes according to plan, a diagnostician may receive electronically transmitted photographs, together with background information and a detailed request, and can have a response back within hours or minutes. Follow-up interactions also can be immediate.

EQUIPMENT NECESSARY FOR DISTANCE DIAGNOSTICS

Digital Photographs

Types of cameras and other photographic equipment, including cell phones, were discussed in Chapter 3. Smart phones with surprisingly high-quality cameras have the built-in capability of immediately sending images and requests by text or e-mail to a diagnostician.

Submitting printed images or images via email can expedite a proper diagnosis. The accuracy of insect identification from digital images is dependent on the image: the sharper the image, the greater the likelihood of a correct diagnosis.

Certain features of an insect are used for species identification. Often an image of the entire insect, together with close-ups of certain morphological features, is required. Some insects are identified by wing venation, while others are best identified by leg morphology or other features on the head, thorax or the genitalia. Communication with the diagnostician can help determine what a digital image should focus on.

General Digital Photo Tips

- Always use the best photography equipment and techniques possible.
- Take several photos from a range of perspectives (close-ups through landscape-level photos).
- Carefully review the images before sending them. What you see is exactly what a diagnostician sees. There is no magic button in a diagnostic laboratory that resolves blurry or dark photos.
- Select only a few (max of five) photos to submit per sample or question. Choose photos that show different aspects of the sample. There is no advantage to submitting five identical photos.
- When possible, include a size reference such as a ruler or a coin.

Include a whole insect image, as well as close-ups of insect parts if you know that they are important for diagnosis.

- Submit high resolution photos with as much magnification as possible.
- Include photos of insect signs and symptoms if they are also present.

Computer and Internet Access

A computer which is set up to receive a request (i.e., e-mail), as well as the electronic image, is the first requirement of handling digitally submitted samples. It goes without saying that a quality computer is essential in receiving, responding to and storing digital image samples. Some laboratories have special submission forms specifically for submitting digital samples.

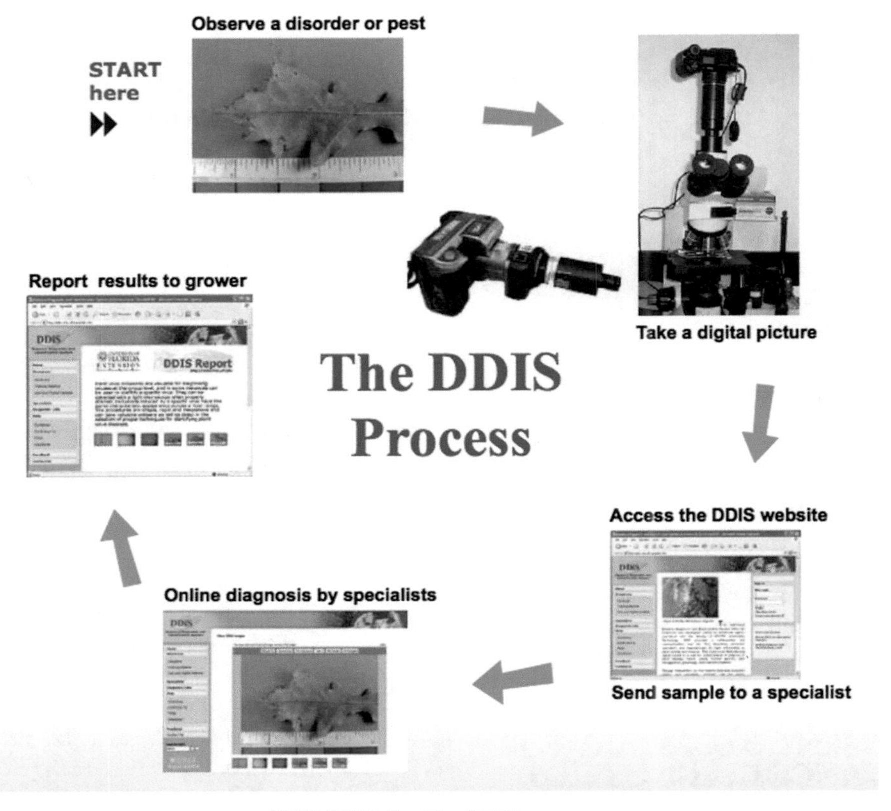

FIGURE 3.7 The DDIS process.

Digital Distance Diagnostics Programs (DDDS)

Progressive diagnostic laboratories have recently developed a web-based system that takes advantage of current and emerging technologies to handle submissions of digital images to their diagnostic laboratory.

For example, the Institute of Food and Agriculture Sciences (IFAS) at the University of Florida has developed a Distance Diagnostic and Identification System (DDIS) that it describes as follows:

> DDIS is a system designed specifically for university agricultural specialists and diagnostic laboratories or clinics. DDIS provides a collaboration and communication platform for first detectors, extension specialists and diagnosticians to share information on plant insects and diseases. The system uses field data and digital media as tools for enhancement of diagnosis of plant diseases, insects, weeds, invasive species, plant management, physiology, and nutrient problems.
>
> Through interactions on the Internet between extension agents and specialists, problems can be communicated immediately and assessed. Specialists around the state can perform diagnosis and identification and provide the best management practice recommendations to the users. The archived DDIS database becomes a resource for research, educational programs, and classroom teaching.
>
> The flow chart [in Figure 3.7] shows the process of DDIS sample submission and diagnosis.
>
> First, a user observes a pest or symptom and captures an image through a digital camera with or without the use of a microscope. After capturing the 'digital samples,' the user then signs on to the DDIS

website to submit these with the pertinent field data to a database. The on-line submission forms mimic the hard copy paper forms also used by the diagnostic laboratories. After the user submits a sample to the DDIS server, it automatically notifies the specialist(s) that an identification and diagnosis is needed. Specialists can then retrieve the submitted sample via the website, and determine an identification or diagnosis along with appropriate control recommendations. After identification and diagnosis are complete, specialists submit their findings and recommendations to the database. The DDIS system automatically notifies the sending user that the diagnosis and recommendations are available to view in the DDIS website. Users then retrieve a report of the diagnosis and recommendations.

One of the major advantages of a distance-based digital diagnostic system is that a user may send a sample simultaneously to several specialists for diagnosis. For example, a sample image can be sent to an entomologist and a plant pathologist so that they may collaborate on an answer. Multiple or combined responses by the different specialists are returned to the user.

The entire process of diagnosing pest problems through DDDS can often be completed in a matter of a few hours, rather than many days as would be required using conventional or overnight mail.

From a national security standpoint, DDDS programs support the needs and responsibilities of first detectors and first responders. If serious pests are intentionally or unintentionally introduced that threaten public safety, food supply, water or health, these can be identified and counter measures taken very quickly.

CHAPTER

4

Insect Identification Techniques

NAMING ARTHROPODS

History

Man has always had an urge to describe his environment. This has largely been accomplished by assigning names. Names have been given to plants and animals since the beginning of time. It is human nature.

Naming arthropods is no exception. Arthropods have probably always been of interest to people. No one seems to be neutral about them. Most are slightly wary of them, a few have a debilitating horror of them, and a smaller, usually strange, few of us have developed an inordinate fondness for them. It is no wonder that they have been assigned many names.

The history and science of naming arthropods is very interesting. In some cultures, even today, they are included in a larger group, which also includes mice and rats, and is simply called 'vermin.' Vermin, as a group, historically included all pests that were dirty, filthy, loathsome and harmed people, regardless of whether they had four legs or six, laid eggs or gave birth to living young. When arthropods were first recognized as a group of their own – small, often annoying creatures – they were given the name 'bug' which meant 'tiny animal that is ugly or dirty.' Interestingly, those who looked even more closely found that not all bugs were the same. Some were more of a pest than others. Variations of 'bugginess' were noted. For example, a ladybug was considered slightly different than an ordinary bug because, though possessing some of a 'bug's' characteristics, it was somewhat less loathsome. The name 'lady' was then added to modify 'bug.' The name stuck and we still use that colloquial name today. Such names arose from all parts of the world. Colloquial names were given based on the appearance, origin or behavior of the bug in question. The dragonfly was named for its strongly reticulated wings, large head, enormous eyes, and long body resembling a dragon (which term was derived from the Greek 'drakon'). The name 'gadfly' was used to refer to any biting fly, as 'gad' was the Anglo-Saxon term for 'to goad or to sting' and the original insect was known for its propensity for stinging cattle. 'Beetle' comes from the Anglo-Saxton spelling of 'bitel,' derived from 'bitan,' which means 'bite.' 'Cricket' comes from the Welsh 'criciad' based on 'criculla,' or to chirp. The Anglo-Saxon form of the term 'bee' was 'beo.' 'Humble bee' is a corruption of the German 'hummel bee' or 'huzzy,' i.e., 'buzzing bee' colloquially rendered as 'busy bee.' The word 'wasp' or 'waspa' is an Anglo-Saxon derivative of the Latin 'vespa' which includes stinging insects. The Hessian fly was so named because it was first introduced to England by Hessian troops during the Revolution of 1688. The list goes on.

Contemporary Insect Diagnostics
http://dx.doi.org/10.1016/B978-0-12-404623-8.00004-1

Colloquial names became the accepted way to reference various individual arthropods. If based on physical or biological characteristics, these names were valuable in that they described the arthropod in some way. Not all were simply bugs. When colloquial names endured the test of time and became generally accepted, they became common names. Common names were used by the population at large to identify different insects. However, simply recognizing that distinct differences exist between individual 'bugs' and assigning separate names to them did not mean that there was not still confusion. Various countries and languages still had reason to discuss arthropods and sometimes multiple, but vastly different, common names were in use.

This confusion and the need to specify more exactly which organism was under discussion ultimately led to a more formal classification system.

Linnaeus is recognized as the first biologist to begin organizing or classifying into distinct groups the great number of plants and animals that he observed, based on their apparent relationship with each other. He recognized that if organisms are to be discussed scientifically, a separate name must be given to each group so that all would recognize the organism in question. Still, he knew that using a name that described the animal and its unique differences was important. Using a single and universal language was key; consequently, Latin was chosen.

Lamarck, who was also a proponent of naming organisms based upon their relatedness with one another, may have been the first biologist to see the power in using these relationships as a way of differentiating and ultimately of assigning names. He institutionalized the use of a key for plant identification and formalized the rules for constructing dichotomous keys (Voss, 1952; Walter & Winterton, 2007).

In dichotomous keys, two sets of contrasting statements, called leads, present the user with a mutually exclusive form an either/or option (couplet). Such statements refer to unique characters or distinct differences between organisms. The choice of one lead takes a user to either (1) a succeeding couplet that again ends in another either/or statement; or (2) ultimately to an end point with the identification of the specimen in a scientific manner, called the scientific name.

Thus, the scientific name 'arthropod' in Latin refers to animals with external skeletons, segmented bodies and jointed limbs (arthro = joint, pod = leg or foot). Technically, all mites, spiders and insects are included in this group. A smaller subset of arthropods is the insect group. The word 'insect' is an abbreviation of the Latin 'insectum,' derived from 'insectus,' the pluperfect of 'insectare' or to 'cut off.' An insect therefore is an articulate animal (having jointed legs) whose body in its mature state appears to be cut or divided into three distinct parts (the head thorax and abdomen). As relationships become stronger, successively smaller groups can be identified as having similar characteristics, until a single insect can be described as unique unto itself.

The identification and naming of insects and all arthropods has since become a science that is very complex and rigid: scientific nomenclature. Diagnosticians rely heavily on scientific nomenclature to identify arthropods. A large part of what they do depends upon finding the exact identity or scientific name of a specimen. However, that is not to say that colloquial names have vanished. Scientists may use scientific nomenclature but the lay public does not. Colloquial names still exist in common usage.

Colloquial Names

The term 'worms' is a colloquial name used to refer to both animals in the annelid group, including earthworms, as well as to any caterpillar-like (Order Lepidoptera) larvae, the two groups being only distantly related.

'Daddy long legs' is another example of a name that may be used to describe a cellar spider (Pholcidae) but also a group of arthropods called harvestmen (Order Opiliones) that are not spiders at all. Sometimes crane flies are also called Daddy long legs simply due to the length of their legs.

One of the challenges that a diagnostician must deal with is the use of colloquial names by laypeople and the uncertainty that this brings to a diagnosis. For example, if a client calls to say that they have a problem with 'stink bugs' a diagnostician in the western United States would almost certainly think of the black bombardier beetles that stick their abdomen up in the air when disturbed and release a pungent odor in self-defense. In the eastern United States, however, 'stink bug' does not refer to a beetle at all but rather a green or brown-colored insect of the pentatomid family that releases a foul smell when handled. Neither is technically wrong, because there are no established rules regarding colloquial names.

Diagnosticians will hear the word 'waterbug' used to refer to large cockroaches found in the home, as well as to completely unrelated hemipterans that live in lakes and ponds. Termites are sometimes called white ants or even maggots when they are found in a home.

'Gnats' may refer to any of a large number of tiny flying insects that occur in large numbers. Unfortunately, there are many of these and simply calling them gnats does little to help a diagnostician narrow down exactly what insect is at work, why it is where it is and how to deal with it. To many people a bee is any flying insect capable of stinging, regardless of where it is found or how it differs from other hymenopteran or even dipteran insects.

It seems that the longer our history and the closer our intimacy with insects has been, the more colloquial names exist for them. Searching for colloquial names used to refer to the common bed bug, for example, reveals a long list in which the following are included:

Bat bug
Bed louse
Cloth bug
Crimson rambler
House bug
House louse
Cinches
Mahogany bug
Mahogany flat
Night crawler
Red coat
Wall-louse
Wallpaper flounder

Interestingly, each name is valuable in that it in someway describes the insect, where it lives or what it looks like, but ultimately adds to the confusion by including so many different names.

Colloquial names are a problem for insect diagnosticians who will sometimes be asked to provide biological and management information about the wrong pest because the name of the pest was confused. For example, a client may ask for help with a 'bagworm' problem, when the insect they really meant was not a bagworm at all, but rather a tent caterpillar. The confusion originates because the tent caterpillars live in prominent, silky nests that are reminiscent of bags. The biology of bagworms (Family: Psychidae) and tent caterpillars (Family: Lasiocampidae) is not very similar at all, except that they both live in trees. Proper management recommendations differ significantly as well.

Experienced diagnosticians learn to recognize colloquial names after a time and, with a few follow-up questions to the client, can quickly determine if the name is being used correctly or not.

Common Names

As stated above, the average person seldom uses scientific names, but refers to insects by colloquial or common names. This is certainly understandable and a diagnostician, in order to communicate, must also use common names.

Common names are not only used for individual species but also refer to various groups of insects, which further adds to the confusion. For example, consider the names, 'beetle,' 'blister beetle,' and 'black blister beetle.' In this example, the common name 'beetle' may refer to any of thousands of insects belonging to the order Coleoptera. 'Blister beetle,' on the other hand, is a common name for all members of the family Meloidae, a subset of the order Coleoptera and 'black blister beetle' is the common name for *Epicauta pennsylvanica*, a particular species of insect.

Common names are most valuable when they are generally accepted and used to describe a specific insect. They are the preferred term for most laypeople who know nothing of scientific nomenclature. Even so, with so many different insects occurring in so many different places, assigning common names is arbitrary and sometimes confusing (Gurney, 1953).

To reduce this confusion, the Entomological Association of America (ESA) has developed a list of accepted common names to be used for individual species of insects. This list (Common Names of Insects and Related Organisms 1997) includes many, but not all, common names and goes a long way to reduce confusion.

The history, mission and objective of the ESA Standing Committee on Common Names of Insects Common Names is stated as follows:

> In zoology, the formation of a scientific name for an organism follows a strict set of rules adopted by The International Commission on Zoological Nomenclature. This International Code of Zoological Nomenclature serves to promote the stability, accuracy, and universality of an organism's scientific name. Every proposed scientific name must be unique and distinct from all other names. In this way, every named organism is a distinguishable entity. However, scientific names are not always stable. Rules regulating dates of priority, formation of names and the use of Latin for forming names are some of the reasons that scientific names may change. To add to this confusion, taxonomic studies could dictate that an organism be reassigned to a different taxon, resulting in a change in its scientific name.
>
> In contrast, common names of organisms, while not governed by such strict rules, remain more stable than the scientific names. However, the lack of standardization in common names, which often originate from repeated usage by workers in a particular area, may result in one organism being known by several different common names. *Halysidota tessellaris* J.E. Smith, for example, is known as the checkered tussock moth in the UK and as the pale tussock moth in the USA and Canada. Similarly, *Acarus siro* Linnaeus is called the flour mite in Australia and the grain mite in the USA and Canada.
>
> The importance of correctly identifying an organism to a non-scientific community, as well as to fellow researchers, made it apparent that a standardized common name would allow everyone to know what organism was being discussed in a particular locality. In response, several countries adopted lists of authorized common names for those organisms most commonly found in that country, including insects. The need for a common name list for insects became apparent in the United States early in the twentieth century.
>
> In 1903, the American Association of Economic Entomologists (AAEE) formed a Committee on Nomenclature to assure the uniformity of names of common insects; and in 1908, the AAEE published its first list, Common Names of Insects Approved for General Use by American Association of Economic Entomologists. It contained 142 common names. The current revised list of common names of insects and related organisms supersedes all earlier lists published since 1908. The list has grown exponentially throughout its existence. In 1925, 541 names appeared on the list; in 1927, another 42 names were added as a supplement. In 1931, the list had grown to 874 common names, with the addition of another 22 supplemental names. Concomitantly, the Entomological Society of America (ESA) established a Standing Committee on Nomenclature in 1907. At its annual meeting in 1935, the AAEE changed the name of the Committee on Nomenclature to the Committee on Common Names of Insects. In 1940, a list of 1108 common names approved by the joint committees of the AAEE and ESA was published. In 1950, 1294 names appeared on the list. In 1953, the AAEE merged with the ESA, and the increase in the size of the common names list continued. In 1970, another 299 names were added; and in 1978 the list was computerized by ESA Headquarters. At the same time, the name of the list was changed to Common Names

of Insects and Related Organisms to reflect the other arthropods and invertebrates included in the list. The edition published in 1989 contained 2,018 names (Stoetzel, 1989). The final printed edition, published in 1997, added another 28 names to the list (Bosik, 1997). This new online version has another 46 names added since 1997. From now on, new names will be added to the online database as they are recommended by the committee and approved by the Governing Board. For a comprehensive history of the common names of insects, see Chapin, 1989.

Rules and Guidelines for Proposing a Common Name

The Common Name Proposal Form for submitting potential common names is available online. The following set of rules has been established to guide an individual in the selection and formation of a common name:

1. Included species, in most cases, will inhabit the United States, Canada, or their possessions and territories. In special cases, other species may be added.
2. The list of common names is intended to include those insects and other invertebrates commonly of concern or interest to entomologists because of their economic or medical importance, striking appearance, abundant occurrence, or endangered status, or for any other sufficient reason.
3. A common name should consist of three words or fewer, but four are permissible if justifiable.
4. Most names have two parts, one indicating the family or group, and the other a modifier. In the case of names having two parts with one of them being a group name, the group name will be a separate word when used in a sense that is systematically correct, as in 'house fly' and 'bed bug.' If the group name is not systematically correct, it must be combined into a single word with a modifier, as in 'citrus whitefly' and 'citrus mealybug.' The modifying part of the name should be based on some outstanding characteristic of the organism itself, its damage, host, or distribution. Hyphens between modifying words should be used only if the meaning is otherwise obscure.
5. The use of parts of the scientific name in the common name is undesirable unless the words involved are well documented by usage as a common name.
6. Non-geographic, proper names will be in the nominative case.
7. Only in special cases should a species have more than one common name.
8. In petitions regarding taxa warranting special concern in both the larval and adult stage, the preferable name should be one suggested by the appearance or habits of the more important or better-known stage, or the one for which usage has become more established.
9. In all petitions for the adoption of new common names or changes in those previously established, the fullest consideration should be given to past usage and probable future usage. When practical, the opinions of entomologists experienced with the taxa concerned should be obtained before names are proposed. Members who wish to recommend new names or changes in existing names should accept the responsibility for making the necessary investigation. All available documented evidence, both for and against, concerning the need for each proposed name should accompany the petition when submitted.

Scientific Names

Even though common and colloquial names are very important and widely used, not all of them refer to individual insects, nor do all insect species have a common name. Consider that the ESA-established common name list includes approximately 2000 names, roughly only 0.22% of known insect species. The list was designed to include only the most common insects and primarily includes those that have a significant impact on people. Obviously, this is just a small fraction of the total number of existing insect species.

The common names list established by the ESA was designed to bridge, but not replace, the use of scientific naming of insects. Scientific nomenclature will always be the native language of entomologists. It allows for growth as well as change. As taxonomists learn more about the various taxa, and as molecular techniques for confirming relationships become more advanced, it is certain the scheme that we have now will change. Species, genera, families and even orders will continue to change. Professional entomologists know that what they once knew as a certain taxon may be revised, receive a different name, be joined to a different group, or be given a different name altogether. The dynamics inherent in taxonomy can be overwhelming, especially if one is not current with the published literature. By contrast, common names are actually more stable, even though they have their own set of constraints. That is why understanding the process of standardized scientific naming is necessary to professional entomologists such as insect diagnosticians and taxonomists.

Arachnids, insects, crustaceans, millipedes and centipedes all belong in the PHYLUM Arthropoda. Diagnosticians spend most of their time working with the CLASSES Arachnida, which includes spiders, mites and ticks, and Hexapoda (also called Insecta). Within the class hexapoda, many ORDERS of insects exist. Within each order, many FAMILIES exist. Within families, many GENERA occur and, likewise, within genera, many SPECIES exist.

A general understanding of the scientific nomenclature and rank order of associated names is important for diagnosticians. For example, a diagnostician should know that the scientific identification of a black blister beetle (Fig. 4.1) is as follows:

Phylum = Arthropoda
Class = Hexapoda
Order = Coleoptera
Family = Meloidae
Genus = *Epicauta*
Species = *pennsylvanica*

An order is a name applied to a large group of insects having similar characteristics. All butterflies and moths belong to the order Lepidoptera; all beetles, regardless of size, shape, or color, belong to the order Coleoptera; all flies belong to the order Diptera, and so forth.

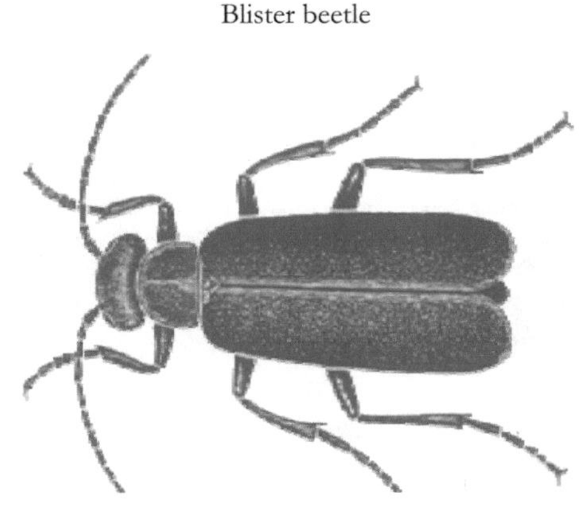

FIGURE 4.1 Blister beetle.

For professional entomologists and those who have sufficient interest, each group of insects can be further broken down. Orders are divided into families, the families into genera, and the genera into species (Mayr, 1969). Professional entomologists generally study insects at the genus and species level.

IDENTIFICATIONS TOOLS

Diagnosticians must rely on tools to assist them with identification of specimens. Depending upon the specific sample and the expertise of the diagnostician, various tools may be employed and often these are used in combination.

The most common tools include; descriptive keys, dichotomous keys, illustrated keys, matrix-based keys, molecular diagnostics and sight identifications.

Practicing diagnosticians collect a series of keys, sometimes found in text books, but also published independently, either in hard copy or electronically. These, and other tools, are essential for diagnosticians to perform their craft.

The following section provides examples of each of these tools and while not exhaustive, may serve to illustrate the breadth of diagnostic tools.

Descriptive Keys

A study of arthropods begins with classification. Classification is central to any biological study. Diagnosticians must rely on tools to help identify the arthropods that are submitted. In this section we will consider several tools that diagnosticians can use to help classify arthropod samples. These include illustrated morphological and behavioral descriptions, dichotomous keys, illustrated dichotomous keys, matrix based keys, molecular diagnostics and sight identification and reference comparisons. Written texts accompanied by illustrations may be the the most common method of separating groups of arthropods and many traditional entomology text books rely upon these tools.

The following llustrated descriptions of arthropod classes, followed by insect orders contains explicit morphological and behavioral descriptions as a means to separate each grouping. Selected line-drawings serve to further assist a diagnostician in distinguishing the various groups. It is arranged according to the preceding synopsis. Early classifications of the hexapods place great emphasis on mouthparts and wings and included those with primitive mouth-parts (Entognatha) as separate orders (Protura, Collembola, Diplura) within the class Hexapoda. This emphasis is reflected in the following discussion of order names.

Recent Changes to Insect Classification

Occasionally changes in classification systems are necessary such as when new insect groups are discovered, such as the recently discovered order of insects (gladiator insects) from West Africa (2002). These have been classified as an entirely new order Mantophasmatodea. (Klass et al., 2002).

Changes in insect taxonomy and classification also occur when previously separate orders are either separated (split) or combined (joined). For example, the previously recognized order Dictyoptera, is no longer in use, raising each of the two groups that were previously part of Dictyoptera (Mantodea and Blattodea) to ordinal levels.

In contrast, three previously recognized orders Neuroptera, Raphidioptera and Megaloptera are now all considered part of the single order Neuroptera.

Similarly, termites (previously recognized as belonging to their own order, Isoptera) are now joined with their close relatives into the single order Blattodea.

Sucking and chewing lice are no longer separate orders (Anoplura and Mallophaga), but rather belong to the common order Phthiraptera.

The long embraced, 2-order classification for all 'true bugs' (Hemiptera and Homoptera) has been recently reduced into a single order called Hemiptera.

Changes at lower levels (Family, Genus and Species) occur more frequently. For instance, the family Lygaeidae recently has been split to form 10 separate families, whereas insects previously known as Cicindellidae are now part of the family Carabidae, Lyctidae is now included in Bostrichidae, Buchidae is considered part of Chrysomelidae, Nymphalidae includes Danaidae as well as Satyridae and Apidae now includes both Anthophoridae and Bombidae.

These are but a few of the recent family changes to contemporary insect classification. Genus and species level taxonomy is even more fluid. The ever changing science of taxonomy and systematics makes insect diagnostics both challenging and rewarding.

ILLUSTRATED DESCRIPTIONS OF ARTHROPOD CLASSES

Diplopoda, Chilopoda, Pauropoda and Symphyla were once combined as one class, the Myriapoda, and are still known as the myriapodan classes. All millipedes (Diplopoda) are phytophagous, but most species do not seriously injure plants. The centipedes (Chilopoda) are predaceous and possess a pair of strong, poison-supplied fangs called *toxicognaths* used to hold and kill their prey. Toxicognaths are modified legs, and the venom they inject can be painful. Centipedes are fast moving, in contrast to the slow-moving millipedes. The minute Pauropoda are of little direct agricultural importance. One member of the Symphyla, *Scutigerella immaculata* (Newport), is sometimes a pest in greenhouses; the species is whitish, about 8 mm long, and may become very abundant.

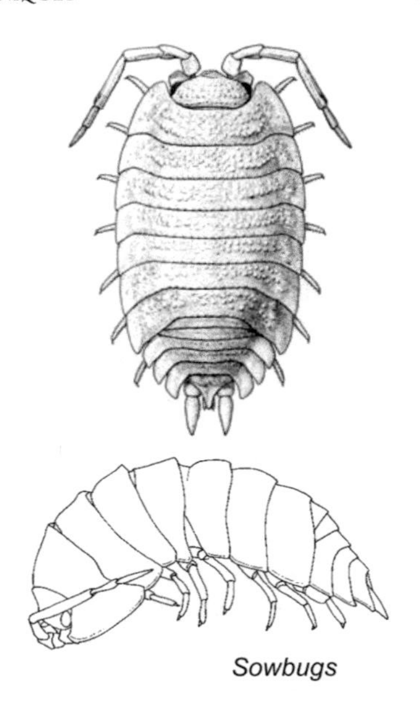

Sowbugs

CLASS CRUSTACEA

The Crustacea are a large class divided into eight subclasses and about 30 orders. Most species of Crustacea are aquatic and are found especially in salt water. Many crustaceans, including crayfish, lobsters, and prawns, are important as human food. Sowbugs (including pillbugs) belong to the subclass Malacostraca and the order Isopoda. Sowbugs are terrestrial isopods that are the only Crustacea of agricultural importance. Sowbugs are found under stones, logs, and debris on the ground. They feed on vegetable matter and may become pests of tender plants. Sowbugs are distinguished by a depressed body, seven pairs of walking legs, and a shield-shaped head: the other body segments extend to the side. The first pair of antennae (antennules) are vestigial; the conspicuous antennae represent the second pair. Eggs are carried by the female in a brood sac on the underside of the body. Breathing involves paired gills on the lower hind part of the body. Because the gills must remain moist, sowbugs cannot withstand drying. Pillbugs are capable of rolling themselves into a tight ball for protective purposes. Crustaceans may be best killed and preserved in 70–80% ETOH.

CLASS HEXAPODA (INSECTA)

All Arthropoda (including insects) are characterized by hardened or sclerotized parts of the integument. After the arthropod emerges from the egg stage, the animal must molt or shed the hardened cuticle to grow and increase in size. These hardened parts (called *sclerites*) restrict growth. The process of molting must take place several times during the insect's period of growth (the nymphal or larval stage). The number of molts varies among species of insects. In some instances the number of molts is species-specific. In other instances the number of molts may be influenced by the sex of the individual, nutrition, or environmental conditions.

The term *instar* refers to the immature insect between molts. The term *stadium* refers to the amount of time between molts. During the molt, each instar of the growth stage increases in size over the preceding instar. During the molt, the animal also may change its form. The amount of change in form can range from insignificant to profound. The transformation in shape or form is called *metamorphosis*. The primitive hexapods include the Entognatha and the division Exopterygota of the Pterygota. These groups undergo gradual metamorphosis in which successive instars differ relatively little in form. Some aquatic exopterygote immatures are called *naiads* and may be entirely different from the adult because the two stages have different life habits. The immature stages of terrestrial insects with gradual (simple or incomplete) metamorphosis are called *nymphs.* The term *larva* is applied to the feeding stages of insects with complete metamorphosis, although some authors (McCafferty, 1981; Merritt et al., 1984) use *larva* to refer to the immature stages of aquatic insects. Stehr (1987) reviews the use of *larva, nymph,* and *naiad.*

The wings of insects with "incomplete" metamorphosis develop externally in the later instars but do not become functional until the adult stage. An exception to this rule involves the Ephemeroptera, which experience a molt after the wings become functional. The technical term for adult insects is *imago;* the first imago (adult) of the Ephemeroptera is called a *subimago.*

The wings of insects with complete metamorphosis typically develop internally. Insects with complete metamorphosis change radically in form from one stage to the next. A "resting stage" called a *pupa* (pl., pupae) is interpolated between the larva and adult stages. The pupal exoskeleton protects the insect during the virtual reorganization of its entire body. The stages in the development of insects with complete metamorphosis include the egg, several larval instars, sometimes a prepupal stage, the pupa, occasionally a subimago, and the imago or adult.

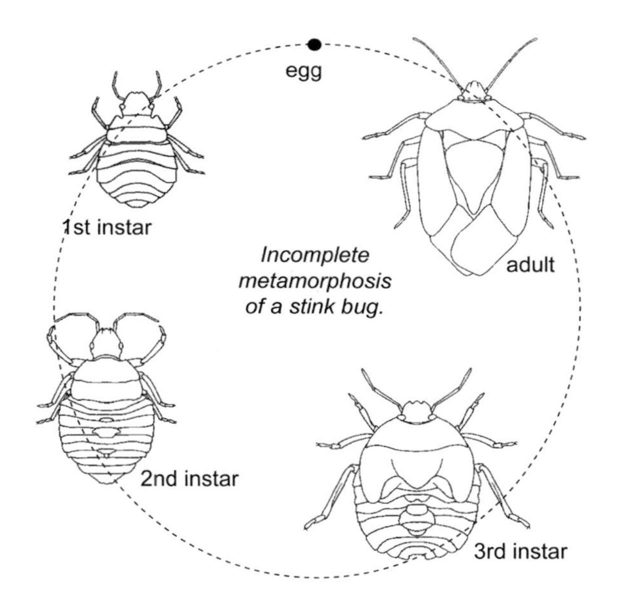

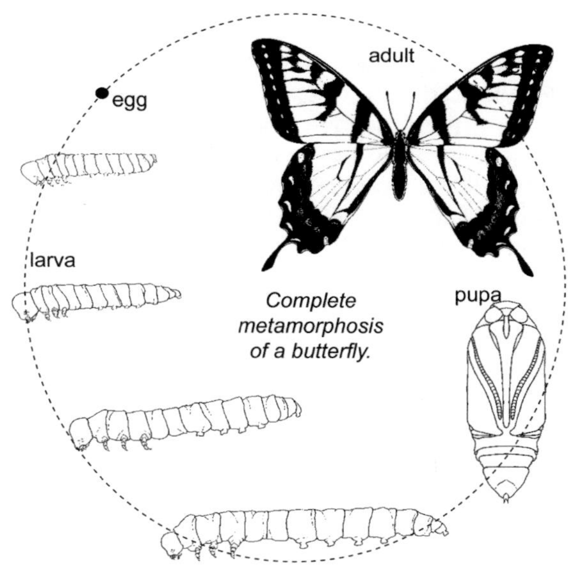

Some insects are wingless (either primarily or secondarily) in the adult stage and thus are difficult to distinguish from immature stages. Both the immature and the adult insects are treated in the key to the orders. Considerable knowledge of the various groups of insects is needed to determine whether a specimen is an immature or an adult. If functional wings are present, the specimen is certainly an adult (or a subimaginal mayfly). Compound eyes are not found in immature endopterygote insects.

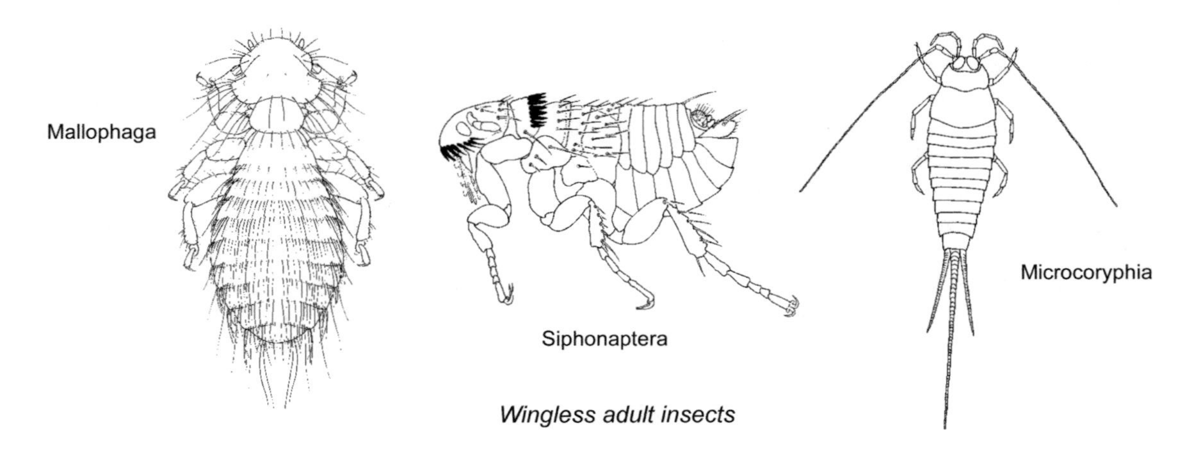

The classification of the insects into orders (ordinal classification) adopted here is a conservative one used in some textbooks. Recent studies on various aspects of insect morphology and physiology have led to the proposal of a considerably different classification, but only time and further study will determine its acceptance or rejection.

A fundamental dichotomy, based on the presence or absence of wings, divides six-legged arthropods (hexapods) into the subclasses Apterygota and Pterygota. The Apterygota of earlier authors included four orders: Protura, Collembola, Diplura, and Thysanura (Richards & Davies, 1977). In recent classifications the term Apterygota may be restricted to the Microcoryphia (= Archaeognatha) and Thysanura (Watson & Smith, 1991). Other classifications have abandoned the traditional dichotomy of Apterygota-Pterygota and adopted a neutral viewpoint. In the opinion of some phylogeneticists, a more fundamental feature of evolutionary importance involves the ability to extend and retract the mouthparts. Hexapods that maintain their mandibles concealed when not in use are called Entognatha; hexapods whose mandibles remain exposed and not extensible are called Ectognatha. Entognathous Hexapoda include the orders Protura, Collembola, and Diplura; ectognathous Hexapoda include apterygotes of the orders Microcoryphia and Thysanura (Boudreaux, 1979; Borror et al., 1989) and the pterygotes.

ILLUSTRATED DESCRIPTION OF MAJOR INSECT ORDERS

PROTURA (FIG. 4.2)

Protura is an order of entognathous Hexapoda (sometimes considered a primitive order of minute Insecta). The name Protura is derived from Greek (protos = first; oura = tail). Protura are cosmopolitan in distribution and include about about 180 named species. Characteristically, proturans are small-bodied (less than 2 mm long) and elongate. The head is prognathous, with entognathous piercing mouthparts and well-developed maxillary and labial palpi.

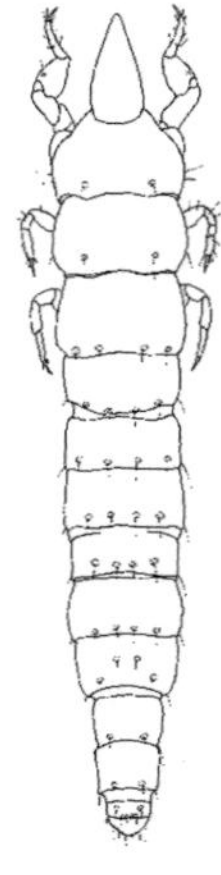

FIGURE 4.2 Proturan.

The antenna and compound eye are absent, and the thorax is weakly developed. The legs display five segments, and the pretarsus shows a median claw and empodium. The adult abdomen includes 12 segments; sterna 1–3 possess small, eversible styli, but cerci are absent, and the gonopore is positioned between segments 11 and 12. Development of Protura is anamorphic. That is, the animal emerges from the egg with nine abdominal segments and adds a segment with each of the first three molts. The lifestyle of Protura is cryptic. They inhabit leaf litter, soil, or moss or may be found beneath rocks or under bark. Protura are most frequently found in Berlese samples.

DIPLURA (FIG. 4.3)

Diplura are a numerically small, cosmopolitan order of entognathous, epimorphic hexapods whose position in relation to the Insecta is questioned. The name is derived from Greek (diploos = two; oura = tail). Diplura consist of about 700 named species included in four families: Campodeidae, Japygidae, Procampodeidae, and Projapygidae. Diplura usually are small in size, but a few species measure up to 50 mm long. The body is narrow, eyes are absent, and the antennae are moniliform, with each segment containing intrinsic musculature. The legs each include five segments, and the pretarsus displays a pair of lateral claws and occasionally a median claw. The abdomen shows 10 segments; some abdominal segments contain styli and eversible vesicles. Cerci are usually present but variable in development; cerci are annulate in campodeids and forcipate in japygids. The gonostyli and gonopore are positioned between segments 8 and 9. Internal fertilization is not known in Diplura. The male manufactures a spermatophore, which is placed on the substrate. Specimens live beneath logs and rocks; a few records associate Diplura with ants and termites. Campodeids are regarded as phytophagous; japygids are reported as predaceous and probably use their cerci to capture prey.

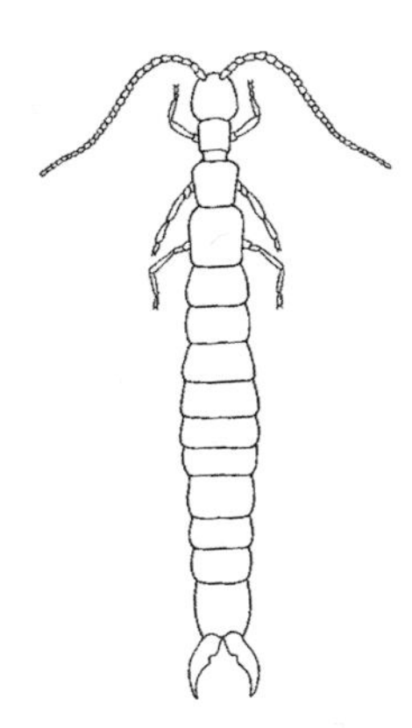

FIGURE 4.3 Dipluran.

COLLEMBOLA [SPRINGTAILS, FIG. 4.4]

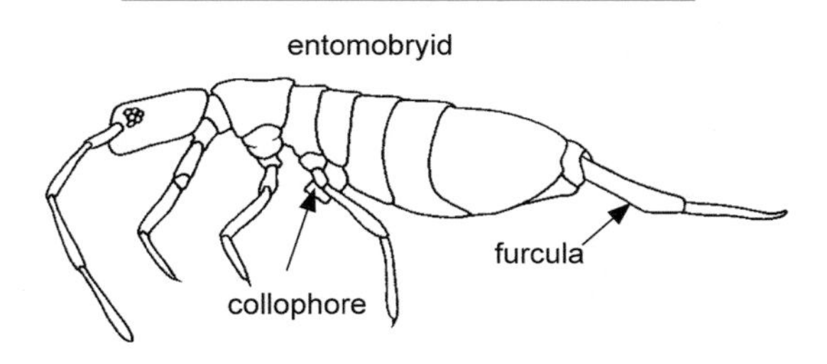

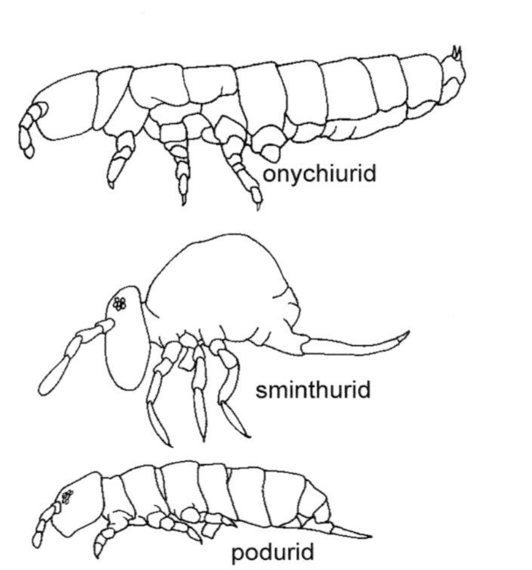

FIGURE 4.4 Collembola.

The scientific name of the order Collembola is derived from Greek (kolla = glue; embolon = peg) and refers to a tubular appendage (collophore) on the first abdominal sternum that secretes a sticky substance. The common name *springtail* refers to a forked structure called the *furcula* on the fourth abdominal sternum that is held in place by a clasplike structure called a *retinaculum* on the third abdominal sternum of most members of this order. When the insect jumps, the forked structure is released with sufficient force against the surface of the ground to propel the animal into the air.

Collembola are cosmopolitan, small-to-minute, soft-bodied, wingless hexapods. About 2,000 species have been described. The oldest fossil collembolan, *Rhyniella praecursor* (Hirst & Maulik), was extracted from Devonian beds of Rhynie chert in England. The compound eyes of Collembola are reduced to a few facets, and the antenna contains four to six segments. The mouthparts are reduced and concealed within the head (entognathous). The thoracic segmentation is not always clearly defined, and the legs contain four segments (a coxa, a trochanter, a femur, and a fused tibiotarsus). The abdomen consists of six segments. Metamorphosis is simple. Three commonly collected families are the Poduridae, Entomobryidae, and Sminthuridae. Some species are abundant on the surface of snow and have been given the name *snow fleas*. Most species feed in moist habitats such as decaying vegetation and rotting wood, under bark, in leaf debris, on the surface of ponds and streams, in soil, and on algae, fungi, and pollen. Some species feed on dead or moribund invertebrates in the soil. Cannibalism and predation have been documented in some groups. Parthenogenesis has been recorded in a few species. Collembola do not engage in internal fertilization. Males produce and deposit a spermatophore on the substrate. The female straddles the spermatophore, and sperm move into her genital aperture.

MICROCORYPHIA [BRISTLETAILS, ARCHAEOGNATHA, FIG. 4.5]

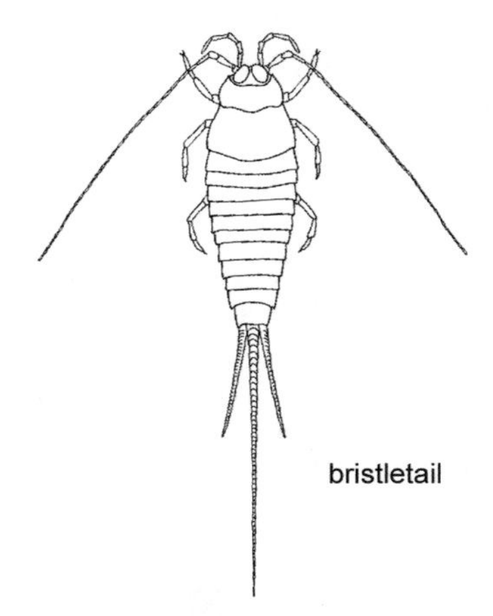

FIGURE 4.5 Microcoryphia.

Microcoryphia represent a primitive order of Ectognatha whose members were assigned to the Thysanura. The order name is derived from Greek (micro = small; corypha = head) and refers to the small head. The common name *bristletail* refers to the long cerci and appendix dorsalis. Two modern families are known (Meinertellidae and Machilidae), and about 450 species have been described. The earliest fossil record of Microcoryphia comes from Triassic deposits in Russia (Triassomachilidae). The bristletail body is moderate-sized and laterally compressed, with the thorax arched. The compound eyes are well developed, and ocelli are present. The mandible is monocondylic. The coxae of some legs display styli, and the tarsi of all legs are three-segmented. All species lack wings, and their ancestors never had wings. The abdomen is two-segmented, with styli on some segments; the appendix dorsalis is longer than the cerci. Bristletails are found under bark and stones and in grass or leaf litter. They feed on algae, moss, and other plant material. Bristletails are typically nocturnal and capable of jumping by rapid flexion of the abdomen.

THYSANURA [SILVERFISH, FIREBRATS, FIG. 4.6]

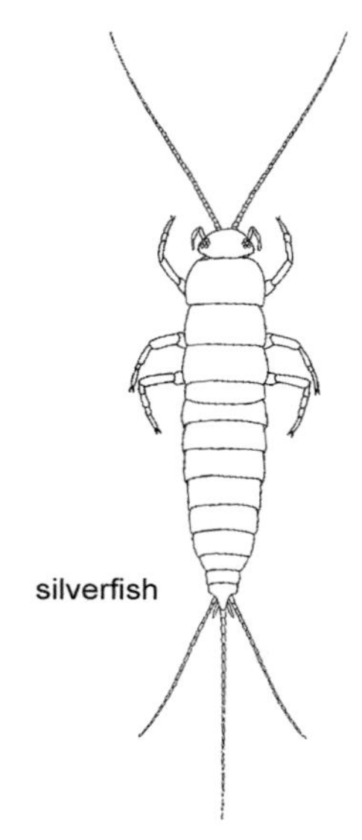

FIGURE 4.6 Thysanura.

In older classifications, the name Thysanura was applied to include the Microcoryphia (Archaeognatha in part). The name Thysanura is derived from Greek (thysanos = fringe; oura = tail) and refers to an older common name, bristletails, which is more appropriately applied to the Microcoryphia. Thysanura are called *silverfish* because as the insects move quickly, the scales of their body shimmer like those of a silver fish. Thysanura represent an order of small to moderate-sized, fusiform, dorsoventrally compressed insects. The body is usually covered with scales, which are flattened setae. The mouthparts are adapted for chewing; the mandible displays two points of articulation (dicondylic) with the head, and the compound eye is small or absent; ocelli are sometimes present. The thoracic segments are similar in size and shape; coxal styli are absent. The abdomen is two-segmented, and some sterna have styli; the appendix dorsalis and cerci are present and similar in size. Metamorphosis is simple. The most common family is the Lepismatidae. Some species, such as *Lepisma saccharinum* Linnaeus and *Thermobia domestica* (Packard), are found in domestic situations feeding on bookbindings, curtains, wallpaper paste, paper, clothing, and similar articles. Most thysanurans occur outdoors under bark and stones, in grass and leaf litter, or in rotting wood or other debris.

EPHEMEROPTERA [MAYFLIES, FIG. 4.7]

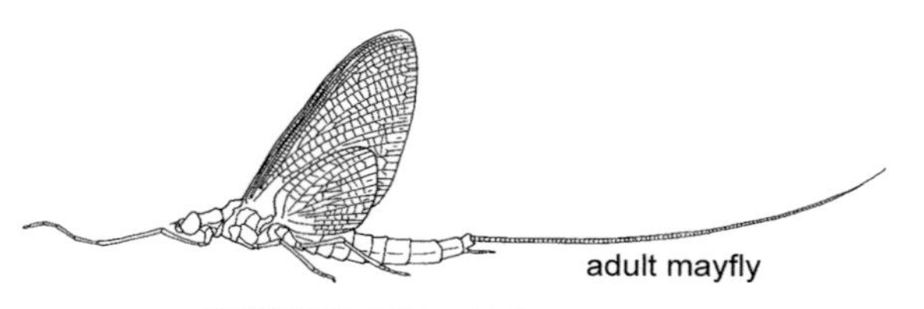

FIGURE 4.7A Ephemeroptera.

The order name is derived from Greek and refers to the short life (ephemeros) as a winged form (pteron). Mayflies are a small order of Paleoptera consisting of soft-bodied, elongate insects with at least one pair of membranous wings and two or three long, slender tails. Adults do not feed, and their mandibulate mouthparts are not strongly sclerotized. Mouthparts of the nymphal instars are adapted for chewing. The nymphs are aquatic, with gills along the sides of their abdomen. Metamorphosis is simple. Mayflies possess a unique developmental stage called a *subimago*, which is the initial winged form. This is not the adult stage; the subimago undergoes one molt before the insect is transformed into an adult. Three families are commonly collected: Ephemeridae, Heptageniidae, and Baetidae. Nymphs are found in a variety of aquatic habitats, from fast-flowing streams to the still waters of ponds, where some species burrow into the muck at the bottom. Adults are usually seen near water on vegetation

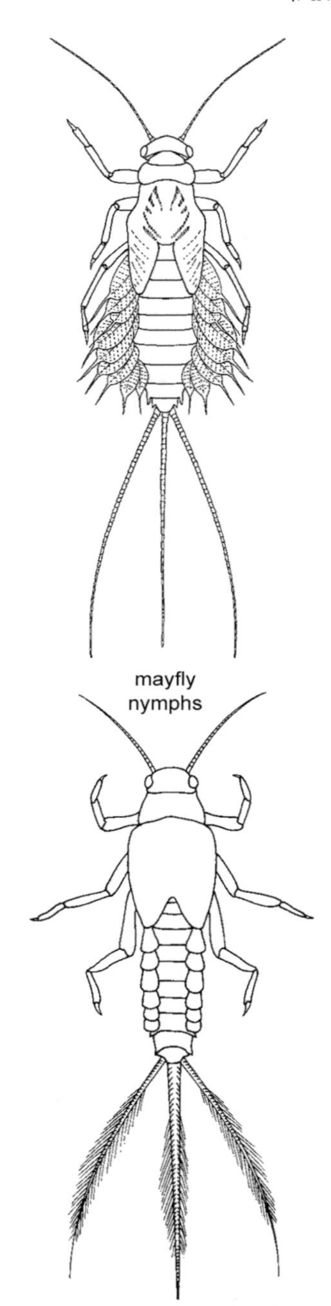

FIGURE 4.7B Ephermeroptera.

and other objects, but at times they are attracted in large numbers to lights.

Adult mayflies rarely live for more than a few days; nymphs often require an entire year to develop. Almost all mayfly nymphs are plant feeders.

ODONATA [DRAGONFLIES, DAMSELFLIES, FIG. 4.8]

The order name is derived from the Greek and means tooth, referring to the strongly toothed mandibles. The Odonata are ancient, with fossils in North America and Russia dating to the Permian. One fossil species living at that time had a wingspan of more than 70 cm. The Odonata are divided into three suborders: Anisoptera (dragonflies), Zygoptera (damselflies), and Anisozygoptera. The Anisozygoptera consist of one small family found in the Himalayas and Japan. This group is considered a transitional element between the dragonflies and the damselflies. The adult body and immature characters resemble those of the Anisoptera, and wings are suggestive of the Zygoptera.

All Odonata are predaceous in the immature and adult stages. The dragonflies are generally large-bodied insects that usually keep their wings outstretched when at rest. The damselflies are generally smaller, more delicate insects that usually fold their wings rearward over the abdomen when at rest (Corbet et al. 1960). Dragonflies and damselflies are comparatively large and have two pairs of many-veined wings. The posterior wings are as large as or larger than the front pair. The antennae are short and bristlelike, the abdomen is long and slender, the mouthparts are adapted for chewing, and metamorphosis is simple. The aquatic nymphs respire with the aid of gills. Damselflies show their gills as leaflike structures at the apex of the abdomen. Dragonflies hide their gills in the form of ridges on the rectum. Oxygen is extracted from water drawn into the

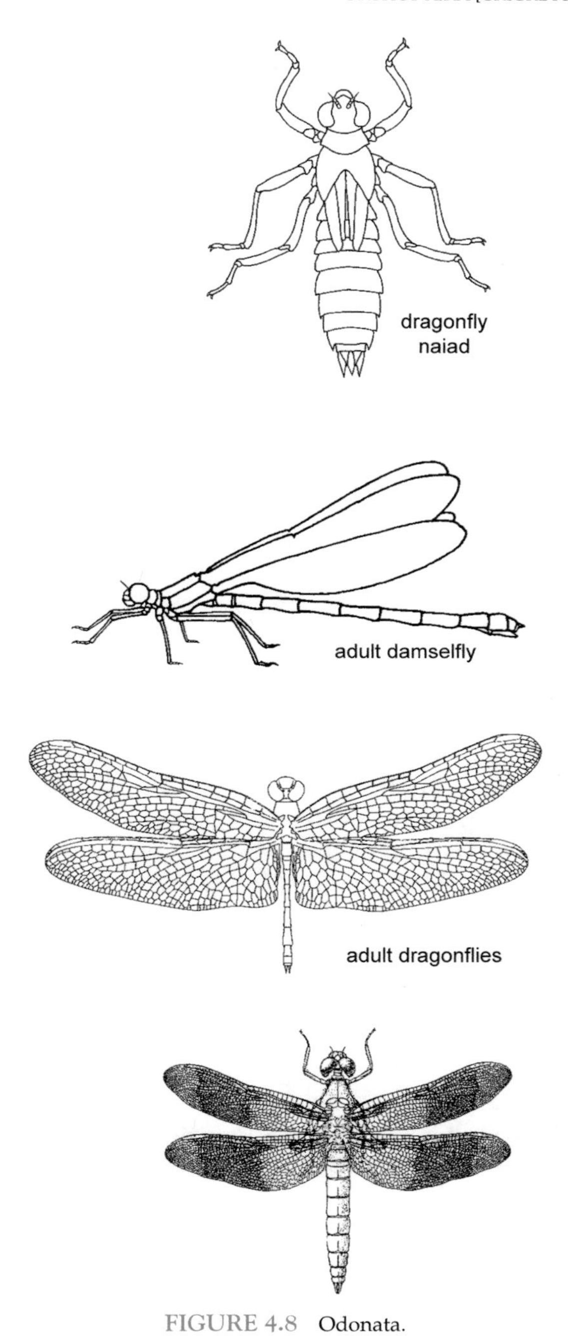

FIGURE 4.8 Odonata.

rectum. Nymphs display an elongate, scoop-like, prehensile labium that can be extended forward rapidly to grasp prey.

The nymphs may be found in most freshwater habitats and occasionally in brackish water. Some nymphs can tolerate exposure to humid air, and members of at least one genus in Hawaii are terrestrial. Nymphs may be found clinging to aquatic vegetation or in the muck at the bottom of streams or ponds. They feed on a diet of young tadpoles, aquatic insects, or other small aquatic animals. Metamorphosis usually occurs at night or early in the morning. Teneral adults avoid water.

Dragonflies are typically diurnal, sometimes crepuscular, and occasionally nocturnal. Adults are strong fliers and use sight to locate prey. Prey are captured in flight. Individuals of some species have favorite resting places and fly regular routes.

ORTHOPTERA [CRICKETS, GRASSHOPPERS, KATYDIDS, FIG. 4.9]

The order name Orthoptera is derived from the Greek and refers to straight (orthos) wings (pteron). The Orthoptera are a cosmopolitan order of terrestrial insects with more than 20,000 described species. Members of the Orthoptera are anatomically diverse. The mandibles are well developed, and the mouthparts are adapted for chewing. The antennae are multisegmented and sometimes longer than the body. The pronotum is large. The forewings (called *tegmina)* are thickened and are characterized by numerous veins; the hind wings also contain many veins but are membranous, fanlike, and folded when in repose. Most Orthoptera display hind legs enlarged and adapted for jumping. The abdomen has 11 segments, and the cerci are well developed. Metamorphosis is simple, and the nymphs resemble the adults. Many Orthoptera communicate acoustically; in this respect they differ from the mantids and cockroaches. Older classifications of the Orthoptera included groups such as the cockroaches, mantids, and

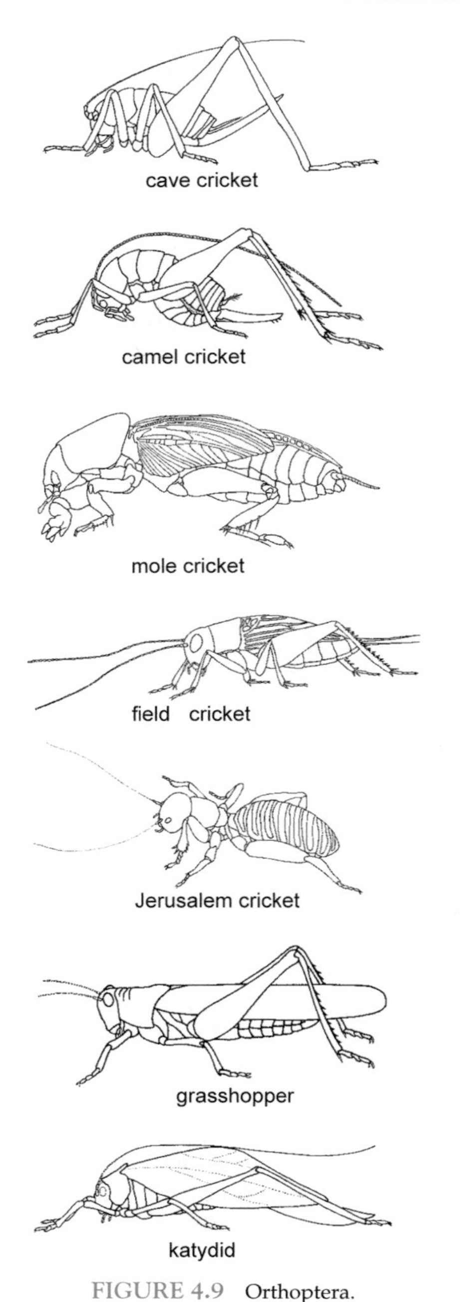

FIGURE 4.9 Orthoptera.

walkingsticks. Currently, these groups are separated, and the Orthoptera consist of about 15 families.

Some crickets are found in domestic situations, but most orthopterans inhabit vegetation such as grass and shrubs upon which they feed. Grasshoppers usually move by walking or jumping, but most adults can fly, often exposing brightly colored hindwings.

Most species of katydids are nocturnal. Many are attracted to light and food.

BLATTODEA (BLATTARIA) [COCKROACHES, FIG. 4.10], AND TERMITES, FIG. 4.11

The Blattodea (Latin, blatt = cockroach) are dorsoventrally compressed insects with the head concealed by the pronotum when viewed from the dorsal aspect. The mouthparts are adapted for chewing, and the antennae are filiform and typically display more than 30 segments. Blattodea are cursorial insects, with the hind legs similar in shape and size to the middle legs; many species are extremely rapid runners. Some species lack wings; when wings are present, the forewings are modified into moderately sclerotized tegmina that protect the membranous hind wings. The abdomen contains 10 apparent segments; the cerci usually display many segments. A few species of cockroaches are parthenogenetic. The female manufactures an ootheca or egg case. In species that deposit the ootheca on the substrate, the egg case is rather hard. Some species are ovoviviparous, with a membranous ootheca containing eggs retained within the female's brood pouch. A few species are viviparous, with the ootheca incompletely developed. The metamorphosis of cockroaches is simple, and nymphs and adults are found in the same habitats. Nymphs resemble the adult in shape but can differ considerably in coloration.

Cockroaches are omnivorous insects that feed upon plant and animal material. More than 50 species of cockroaches live in North America, of which only four are common indoors. These domestic cockroaches are nocturnal and usually avoid light. Some outdoor species may be

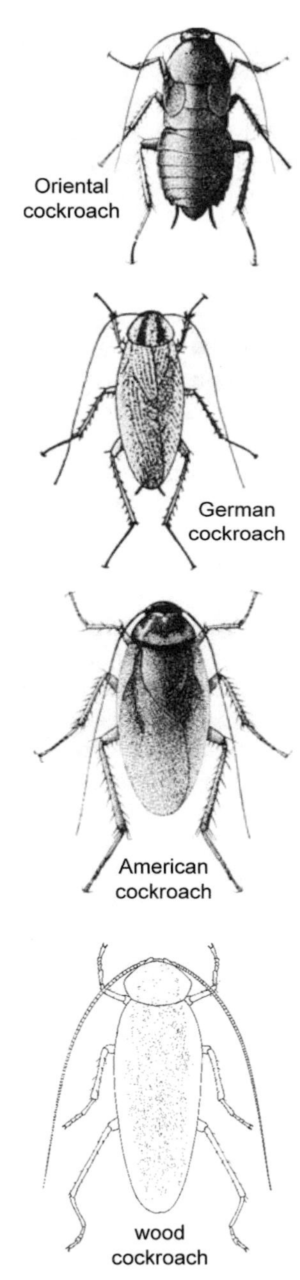

FIGURE 4.10 Blattodea.

attracted to light at night or to certain foods (Hubbell, 1956). During the day, most cockroaches hide under the loose bark of trees, beneath rotting logs, or in similar fairly moist habitats. Some species of cockroaches can be found in leaf litter.

Termites recently have been added to the order Blattodea. The common name *termite* is derived from Latin *termes,* which was the medieval name for wood worms. Termites are also called *white ants,* but they may be distinguished from true ants because termites have a broad waist and beaded antennae; winged forms have both pairs of wings the same size and shape. Termites are a cosmopolitan order including about 2,000 species. They are most abundant in tropical and subtropical regions. Termites are small, soft-bodied, polymorphic, cryptozoic insects. The antennae are moniliform or filiform; the mouthparts are mandibulate. Primary reproductives possess four membranous, similar-sized, net-veined wings that are shed; nymphs, soldiers, and workers lack wings. The abdomen shows 10 apparent segments. The cerci are one- to five-segmented, and sclerotized external genitalia are absent from most species. Metamorphosis is incomplete in termites, but each species displays several morphs.

The primary food of termites is cellulose, which is broken down by symbiotic bacteria or flagellate protozoans. Termites feed incidentally on exuviae, dead nestmates, and excrement. The association with symbionts is vital. If these microorganisms are removed from the gut of the termite, the animal will eventually die of starvation. These microorganisms live in the hindgut of the termite. This relationship is precarious because the hindgut and the microorganisms are shed during the molting process. Fortunately, the newly molted individuals receive a new colony of microorganisms by feeding at the hindgut of other termites. This process is called *proctodaeal feeding* or *trophallaxis.*

Termites display a highly developed social system. Castes include primary reproductives (a king and a queen), secondary reproductives (neotenics), soldiers (male and female), and workers (male and female). Reproductives swarm periodically in great numbers. Bonding occurs in the swarm, with a royal pair (king and queen) then excavating a nuptial chamber. Copulation occurs

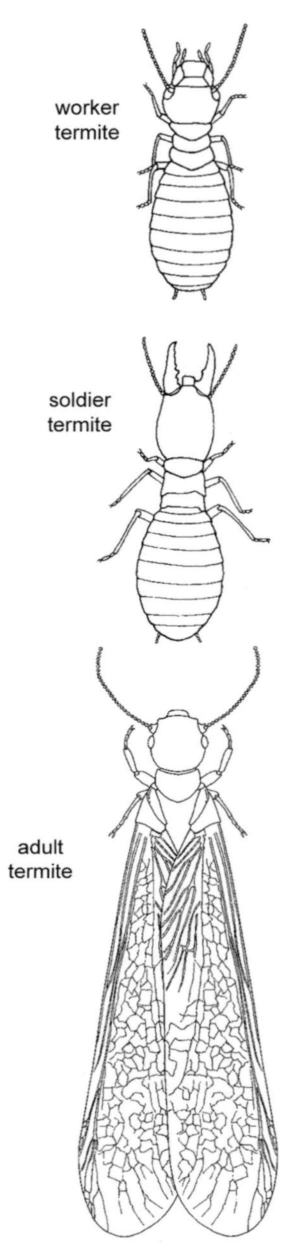

FIGURE 4.11 Isoptera.

within the chamber, and a new colony is established. After the dispersal flight from the parent colony, the wings of the adults are usually shed, leaving only short stubs that are visible under low magnification. The royal pair is long-lived, perhaps as long as 50 years.

Wingless forms are sterile adults of the worker and soldier castes. Workers provide food for the colony, construct new tunnels and chambers, and care for the egg-laying queen. In many groups of termites, the queen's abdomen is so swollen that she is immobile. This condition, called *physogastry,* is typical in some groups. Soldiers defend the colony against attack. Their head is greatly enlarged and usually displays powerful biting mandibles. In some species of termites the mandibles are supplemented with a beak through which a fluid may be ejected to repel enemies. Soldiers of some species display highly sclerotized heads that are modified to block passage through tunnels of the colony. Such heads are termed *phragmotic.*

Termites are responsible for the economically significant destruction of wood and wood products. Based on the habitat, termites construct several types of homes, including mounds (termitaria), arboreal nests, and subterranean nests. They are a common landscape scene in some tropical and subtropical areas of the world. Termitaria are constructed of earth mixed with salivary-gland secretions and excrement and are striking in their architectural diversity and range of size. Some species of termites construct termitaria nearly 7 m tall. Termitaria are exceedingly hard and must be penetrated with iron or steel implements. Arboreal termites include species that live in dead trees, wooden buildings, or furniture not in contact with soil. Subterranean termites live in the soil, sometimes creating mounds protruding a few meters above the surface. Colonies of subterranean termites may be located by digging into these mounds, by prying apart rotting stumps, or by turningover rotting logs and searching in the soil beneath.

Colonies of arboreal termites are more difficult to find, because often there is no external evidence of an infestation. Careful examination may show the entrance holes made by the reproductive adults as they entered the wood, but these holes usually are sealed with cementlike plugs secreted by the termites. If the termites are working close to the outer edge of the wood, surface blisters in the paint or a flaking of the surface of unpainted wood may be a clue to the presence of a colony.

For quantitative studies of termites, Pearce (1990) suggests using a cardboard coaster sandwiched between two glass plates held together with a clip. The cardboard is moistened and the entire trap buried in the soil. The termites feed on the moistened cardboard, and the entire trap can be lifted from the soil to assess damage and to collect termites without disturbing the trap.

MANTODEA [MANTIDS, FIG. 4.12]

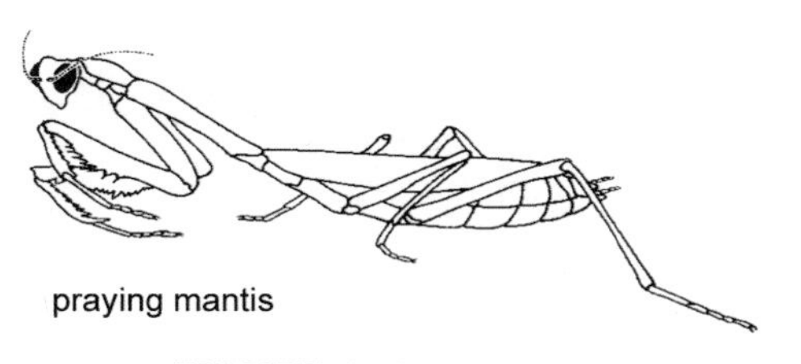

FIGURE 4.12 Mantodea.

Mantids are sometimes called *praying mantids* or *soothsayers* (Greek, manti = soothsayer) because their forelegs are held in a supplicatory position resembling prayer. Nearly 2,000 species have been described. The group is cosmopolitan in distribution but predominantly tropical or subtropical. The mantid body size is typically moderate to large. The head rotates, and live specimens can give a turn to the head that seems quizzical. The pronotum is elongate, narrow, and maneuverable on the mesothorax. The forelegs are raptorial, with the forefemur and tibia extended forward to capture prey. The prey is held between the femur and tibia with opposable spines on the venter of the femur and tibia when the forelegs are retracted toward the head. The middle and hind legs are slender and are not modified for running or jumping. The abdomen is 11-segmented, and the cerci consist of many segments. Eggs are protected within an ootheca.

All mantids are predaceous as nymphs and adults. Mantids are usually found on vegetation, where they prey on other insects. During the autumn and winter, mantid oothecae can be found attached to shrubs or grass stems. Adult mantids usually die in the fall soon after the eggs are laid, but during the summer mantids are voracious feeders, which suggests that they can be important biological control agents in the garden. Indeed, the nymphs and adults consume large numbers of insects, but only those readily visible on plant foliage. Soil-dwelling or stalk-boring insects are not attacked by mantids.

PHASMATODEA (PHASMIDA) [WALKINGSTICKS, LEAF INSECTS, FIG. 4.13]

Phasmids are called *walkingsticks* because they are slow-moving insects with elongate bodies that resemble twigs or sticks (Greek, phasm = phantom). Some tropical species are called *leaf insects* because their bodies are dorsoventrally flattened and expanded laterally so that the insect resembles a leaf. Phasmids are cosmopolitan in distribution, and more than 2,500 species have been described. Several species are more than 30 cm long. The phasmid's body is often spinose, and the head is typically prognathous. Ocelli are usually absent, and the mouthparts are of the chewing type (mandibulate). The prothorax of phasmids resembles that of mantids, in that it is movable on the mesothorax. However, wings are absent from most species of phasmids. Winged species are most common in tropical areas; when present, the forewings are modified into tegmina and the hind wings are broad with a uniform branching pattern. The legs are gressorial, with all legs similar in appearance and adapted for walking; the coxae are small and separated. The abdomen is 11-segmented; cerci are not segmented, but sometimes they are long and

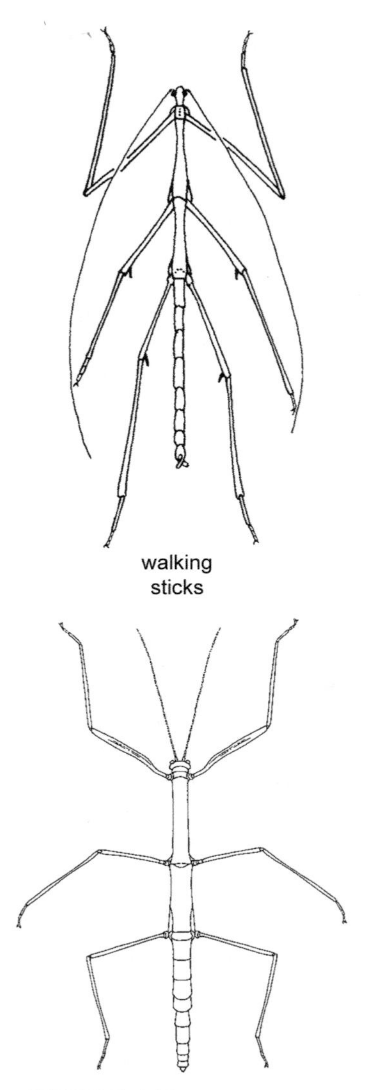

FIGURE 4.13 Phasmatodea.

clasperlike in males. Metamorphosis is simple, and the nymphs resemble the adults. Eggs are not enveloped by an ootheca. Instead, they are scattered individually on the ground. Phasmids may remain in the egg stage for more than a year.

The biology and ecology of Phasmatodea have been reviewed (Bedford, 1978; Carlberg, 1986). All species of phasmids are phytophagous and may be located on trees or shrubs upon which they feed. Their slow movements and resemblance to twigs or leaves and cryptic coloration (usually green or brown) makes these insects difficult to detect. Sweeping or beating trees and shrubs may yield some individuals. In spring or early summer, newly hatched nymphs may be found, which in the most common species, *Diapheromera femorata* (Say), are only about 0.5 cm long. The nymphs mature in about 6 weeks, attaining a length of 8–10 cm. Usually phasmids are leaf-feeding insects. However, in captivity the newly molted nymphs have been observed eating their cast exuviae.

GRYLLOBLATTODEA (GRYLLOBLATTARIA) [ROCK CRAWLERS, FIG. 4.14]

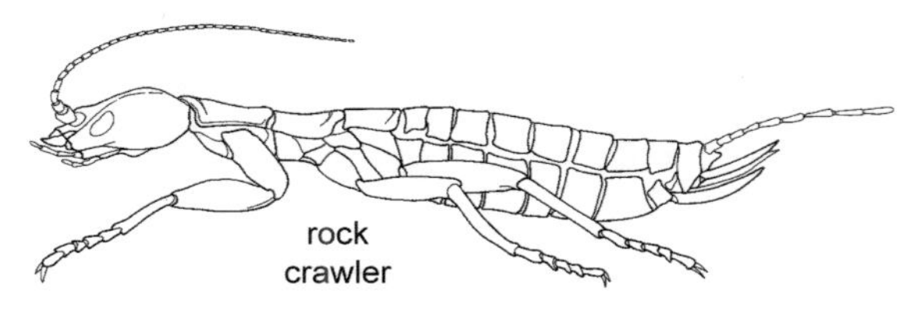

FIGURE 4.14 Grylloblattodea.

Grylloblattodea are rare insects, found along the edges of glaciers, have features resembling both crickets (Latin, gryll = cricket) and cockroaches (Latin, blatt = cockroaches). They derive their common name from their habit of moving about slowly on rocks in cool habitats. The order consists of one family and about 20 cryptozoic species found in cold, wet habitats of North America, Japan, and Russia. Rock crawlers are soft-bodied, elongate, slender, mandibulate insects from 15 to 30 mm long. They are characterized by a prognathous head with compound eyes that are reduced or absent; the ocelli are absent and the antennae are long with many segments. The prothorax is large, wings are absent, and the legs are cursorial and adapted for running with large coxae. The cerci are long and segmented; the ovipositor is strongly exserted.

Rock crawlers are typically found in rotting logs, beneath stones, and on talus slopes near snow. They are active at low temperatures. Their diet consists of moss and other insects. The life cycle requires several years to complete. Grylloblattoids are of no economic importance, but they are interesting because they are unusual and rarely collected.

DERMAPTERA [EARWIGS, FIG. 4.15]

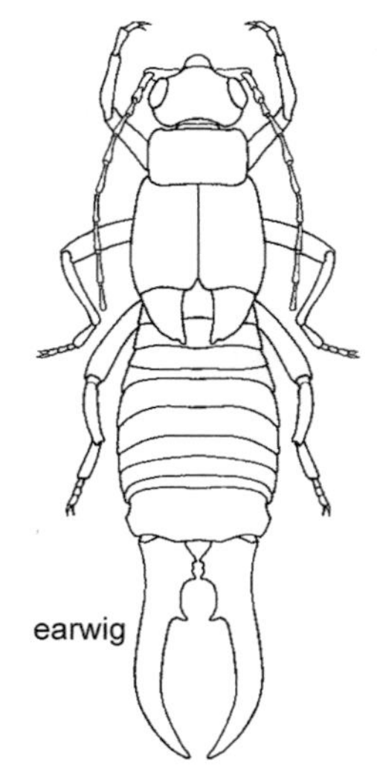

FIGURE 4.15 Dermaptera.

The order name Dermaptera is derived from Greek (derma = skin; pteron = wing) and refers to the skinlike appearance of the forewings. The common name is derived from the erroneous superstition that earwigs crawl into the ears of sleeping people. The order Dermaptera is cosmopolitan and currently consists of about 1,200 named species. Dermaptera are most common in tropical and warm temperate regions. Dermaptera represent a distinct group that is probably related to the Grylloblattodea. Dermaptera are elongate insects that measure up to 50 mm long. The head is prognathous, mouthparts are of the chewing type, and the prothorax remains free from the mesothorax. The adults usually have two pairs of wings. The forewings are short, leathery, and elytriform or modified into tegmina. The hind wings are large, membranous, fanlike or circular, and folded under the forewings when the insect is at rest. The legs are cursorial, with three segments forming the tarsus. The abdomen is often telescopic, and the cerci are modified into forceps that can pinch if the earwig is handled. Metamorphosis is incomplete, and the nymphs resemble the adults.

Earwigs are typically nocturnal insects that are sometimes collected at lights. During the day, earwigs usually hide in cracks and crevices, under the bark of trees, or in rubbish on the ground. One species, *Anisolabis maritima* (Géné), is commonly collected under stones or driftwood along the Atlantic and Pacific coasts. The eggs are laid in a burrow in the ground, and the female cares for them until they hatch.

Earwigs feed as scavengers on decaying organic matter, but some species may also feed upon the living tissues of flowers, ripening fruits, and vegetables. Earwigs occasionally feed upon aphids and other small insects. Earwigs living in cooler habitats are predominantly herbivorous; those living in warm temperate and tropical regions are predominantly predaceous. Predatory species feed on a variety of insects but seem to prefer soft-bodied larvae.

EMBIIDINA (EMBIOPTERA) [WEBSPINNERS, FOOTSPINNERS, FIG 4.16]

These insects have been called Embioptera, but the correct name is Embiidina (Greek, embi = lively), probably due to the fact that they can run backward very rapidly. The common names stem from their ability to spin silk with glands on the feet. Embiidina constitute a widespread group of about 200 named species and an indeterminate

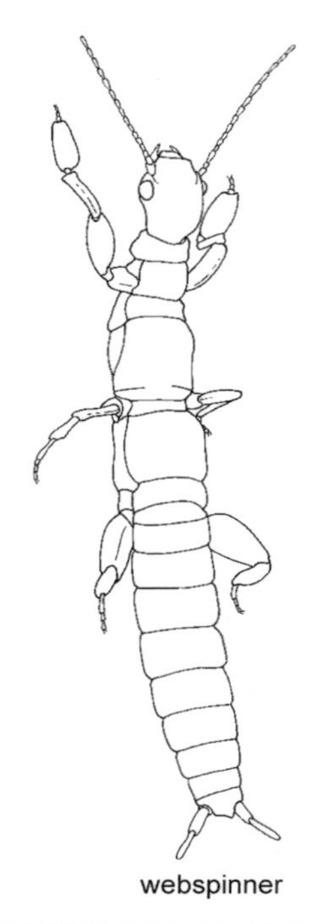

FIGURE 4.16 Embiidina.

number of undescribed species. Embiidina are well represented in tropical and warm temperate regions. Webspinners are elongate, mostly small insects with a prognathous head and chewing mouthparts. The antennae are filiform, and ocelli are absent. Females are wingless; males are winged or wingless, and sometimes both conditions occur in males of one species. When wings are present, the wing venation is reduced. Legs display three tarsomeres. Males, females, and all instars of nymphs spin silk from glands located on the forebasitarsus. The silk issues from hollow, tubular structures on the ventral surface of the basitarsus and second tarsomere. The abdomen appears 10-segmented, but the 11th segment is reduced; the cerci are two-segmented and tactile. All species are gregarious, with nymphs and adults occupying silken galleries. Rapid rearward movement of webspinners is enabled through large depressor muscles in the hind tibiae. Species feed on bark, moss, lichen, dead leaves, and other plant material. Cannibalism may occur, and males are sometimes eaten by females after copulation. Eggs and early instar nymphs are guarded by the female.

Male webspinners are sometimes seen at lights during the night, but females are wingless and are found in the galleries. Galleries may be located among debris and in cracks and fissures or on the surface of bark.

PLECOPTERA [STONEFLIES, PLAITED-WINGED INSECTS, FIG. 4.17]

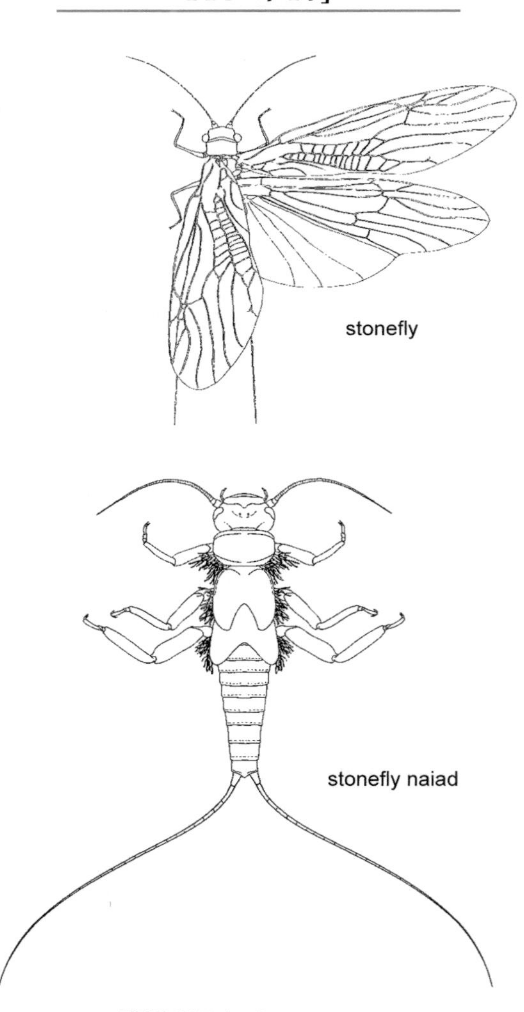

FIGURE 4.17 Plecoptera.

The scientific name of the order Plecoptera is derived from Greek (plekein = to fold; pteron = wing) and refers to the habit of fanlike folding of the hindwing beneath the forewing in repose. The common name *stonefly* refers to the habitat in which specimens are often collected. The Plecoptera are a cosmopolitan order containing about 1,000 named species. Most genera of stoneflies are found in the Holarctic, where the order radiated or numerically expanded. Adult stoneflies are soft-bodied, moderate-sized, and dorsoventrally compressed, with a body seemingly loosely jointed. The head shows two to three ocelli, and the antennae are long, many-segmented, and slender. The mouthparts are of the chewing type but are sometimes reduced. The prothorax is large and mobile; the mesothorax and metathorax are smaller and subequal in size. Wings are membranous, with numerous veins and crossveins; a few species are brachypterous or apterous (usually males). Legs bear three tarsomeres. The abdomen has 10 segments; the cerci contain one to many segments. Metamorphosis is incomplete.

Most species of stoneflies are diurnal as adults and found near water. Immatures live in water, and the nymphs are called *naiads.* A few species found in the Southern Hemisphere apparently exist in damp terrestrial situations. Plecoptera naiads typically prefer clean, cold, moving freshwater. They demonstrate an intolerance of pollution and narrow tolerances of water temperature and related parameters. The life cycle of most species lasts one year, but a few species live longer. Some species apparently modify their life cycle in response to changing environmental conditions. Adults are short-lived. Males die soon after copulation; females die soon after egg laying. Evidence suggests that females must feed as adults for their eggs to develop. Females deposit eggs on the surface of water during flight when the apex of the abdomen touches the water surface. Some species broadcast or deposit eggs beneath the surface of the water or upon submerged objects. Stonefly eggs are small, with an adhesive coating activated by water. Females may lay more than one batch of eggs; total egg production may reach 1,000.

Stonefly naiads possess long antennae and cerci; their mouthparts are adapted for chewing. Some North American species display branched gills on the thorax and at the base of the legs. Naiads of species found elsewhere may display respiratory gills on the mouthparts, cervix, thorax, or abdomen or within the anus. In North America, stonefly naiads resemble mayfly naiads. However, stonefly naiads differ in the location of the gills and possess two tarsal claws instead of one. Stonefly naiads are found in running water, under stones, or clinging to submerged piles of drifted leaves and other debris. Specimens can be collected with an aquatic net, or a dipper or by hand and dropped into vials of alcohol or held for rearing. To rear Plecoptera, the food preferences of the naiads must be determined. The species most commonly collected are plant feeders, but some species are carnivorous. To reduce the time needed for rearing, some collectors schedule their collecting trips to coincide with the season of adult emergence for a particular species. The adults almost always emerge at night or early in the morning. By visiting streams at night, mature naiads can be collected as they crawl out of the water just before their final molt. These specimens should be held in temporary cages until the teneral adults expand their wings and dry.

Most insects hibernate or become inactive in cold weather, but adults of many species of Plecoptera emerge during the winter. Species that mature in spring or summer often fly to lights at night. During the day, adults are usually found resting on vegetation or foliage. Most stoneflies are poor fliers but agile runners.

PSOCOPTERA [BOOKLICE, BARKLICE, FIG. 4.18]

The name Psocoptera is nonsensical because it is derived from Greek and means "chewing wings." The name stems from the attempts of Shipley (1904) to standardize the endings of names of insect orders. The common names of Psocoptera are more descriptive. The terms *booklice* and *barklice* refer to the habitats of some frequently

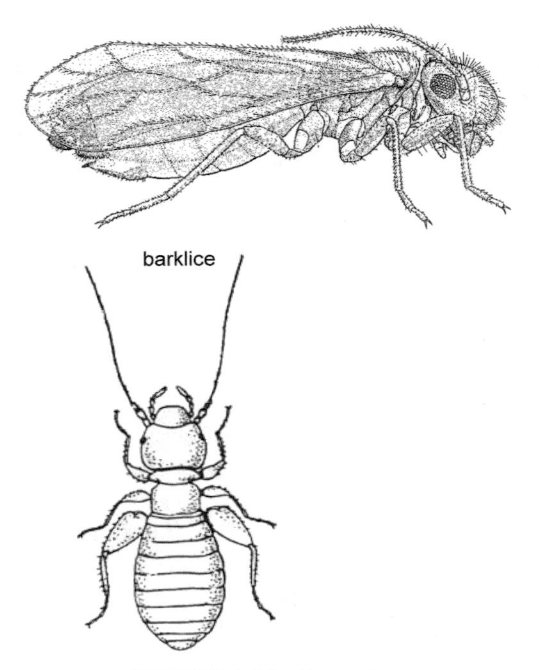

FIGURE 4.18 Psocoptera.

encountered species. The Psocoptera form a cosmopolitan order of small to moderate-sized, soft-bodied insects currently containing about 1,800 species. The head is large and movable, the antennae are long and filiform, and the mouthparts are mandibulate; ocelli are present in winged forms but absent in wingless forms. The prothorax is reduced in winged forms; the meso- and metathorax are fused in wingless forms. The wings are membranous with venation reduced, and the pairs are held together when active; at rest the wings are held obliquely over the body. Legs display two or three tarsomeres. The abdomen consists of nine segments, and cerci are absent. Members of the Psocoptera resemble psyllids or aphids but differ from them in having chewing mouthparts. Psocoptera undergo incomplete metamorphosis.

Some usually wingless species found on old books or papers have received the common name *booklice.* In damp locations in houses and granaries, booklice may attain colossal numbers. Most species of Psocoptera occur outdoors on the trunks and leaves of trees and shrubs, on lichen-covered stones, and on fences. Psocids feed on mold, lichen, pollen, cereals, and starchy materials.

ZORAPTERA [ZORAPTERANS, FIG. 4.19]

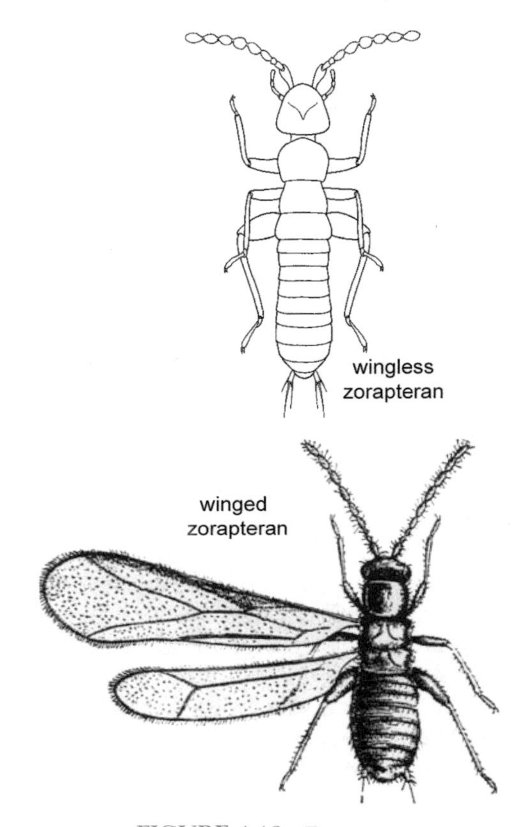

FIGURE 4.19 Zoraptera.

Zoraptera constitute one of the smallest orders of Insecta, consisting of one family and about 30 named species. Geographically, zorapterans are widespread. The scientific name is derived from Greek (zoros = pure; a = without; pteron = wing) and is somewhat misleading. When the group name was proposed, entomologists believed that zorapterans were all wingless. Subsequently, it has been learned that some zorapterans are winged and that winged adults cast off their wings. Adults are minute to small-bodied, with a hypognathous head and chewing mouthparts. The adult antenna is moniliform with nine segments; the nymph displays eight antennal segments. The prothorax is well developed in all specimens, but the mesothorax and metathorax are not differentiated in primitively wingless forms. The wings of zorapterans are membranous with reduced venation and they are shed at the base. Wingless forms lack compound eyes or ocelli; winged forms possess compound eyes and ocelli. The legs display tarsi with two segments. The abdomen is composed of 11 segments, the ovipositor is vestigial or absent, and cerci are present but not segmented. Development is gregarious, usually under planks, in piles of old sawdust, in rotting logs, under bark, or in association with termites. Zorapterans are apparently fungivorous or necrophilous or both. Only two species are found in North America; they occur from Pennsylvania and Iowa to Texas and eastward but have no economic importance.

PHTHIRAPTERA [CHEWING LICE, WOOL-EATERS, FIG. 4.20, SUCKING LICE FIG. 4.21]

The name Phthiraptera is derived from Greek (phthir = louse and; aptera = without wings). Lice are external parasites of birds and mammals, although mostly of birds. Chewing lice are not

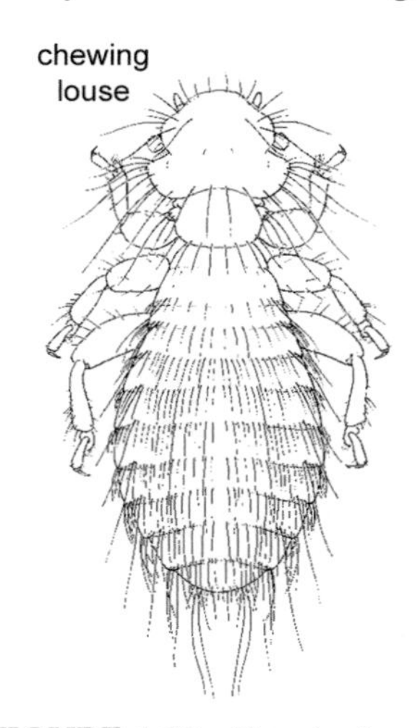

FIGURE 4.20 Chewing louse.

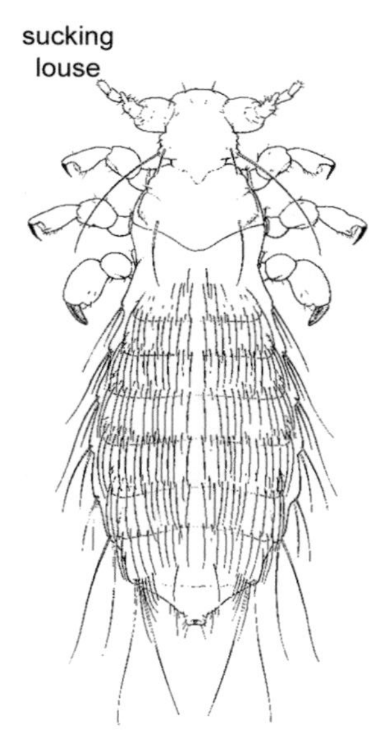

FIGURE 4.21 Sucking louse.

known to attack humans. About 3,000 species of chewing lice have been described. Morphologically, chewing lice are small-bodied and dorsoventrally compressed or flattened. The mouthparts are adapted for chewing, and all species are wingless. Metamorphosis is simple, and the immatures resemble the adults. All stages occur on the host, and chewing lice are rarely found away from the host. Six families are commonly found in North America. Of these, the Philopteridae (on poultry) and the Trichodectidae (on cattle, horses, and dogs) are perhaps the most often collected.

Anatomically, sucking lice resemble chewing lice, in that both groups are small-bodied and dorsoventrally flattened. The heads of sucking lice, however, are somewhat conical, and the compound eyes are reduced to a facet or absent. Mouthparts consist of three sclerotized piercing stylets that are concealed when not in use; the maxillary and labial palpi are reduced or absent. All of the thoracic segments are fused, and wings are never present. The tarsi are one-segmented, each with a large claw or thumblike process on the apex of the tibia. This structure serves to grasp hairs. The abdomen is nine-segmented. The female lacks an ovipositor, and cerci are absent from both sexes.

Metamorphosis in lice is simple, and the nymphs resemble adults. All stages are found on the host, including eggs that are glued to hair.

Sucking lice suck blood and are parasites of mammals only. Species that attack humans include the crab louse, *Pthirus pubis* (Linnaeus), the head louse, *Pediculus humanus capitis* De Geer; and the body louse, *P. humanus humanus* Linnaeus. The body louse is an important vector of typhus and other diseases. Echinophthiridae feed on marine mammals and Haematopinidae feed on horses, cattle, sheep, hogs, and other animals.

Techniques for mounting lice were described by Kim et al. (1986).

THYSANOPTERA [THRIPS, FRINGE-WINGED INSECTS, FIG. 4. 22]

The order name Thysanoptera is derived from Greek (thyasnos = fringe; pteron = wing) and alludes to the long marginal fringe found in winged specimens. The common name *thrips* is singular and plural; we do not refer to a "thrip." Thrips are a cosmopolitan order consisting of about 4,000 named species. They are related to Hemiptera. Thrips are small and slender, but some species may attain a length of 12 mm. The compound eyes usually have large facets; ocelli are present in winged adults only. Another curious feature of thrips is that the right mandible is absent. The left mandible is adapted for rasping; the labrum and labium form a cone and the maxillae are adapted for piercing. Wings, when present, are narrow with reduced venation and display a long marginal fringe; wings are held over the abdomen in repose, but they are not folded. The tarsi are one- or two-segmented, and the pretarsus forms an eversible bladder that is used for adhesion. The abdomen of thrips is 10-segmented; an ovipositor is present or absent, but cerci are always absent. Metamorphosis is complex among thrips and shows a transition between simple and complete metamorphosis. Thus, nymphal instars 1–2 are active, feeding, and wingless or wing buds are internal; instar 3 (prepupa) and instar 4 (pupa) are quiescent and nonfeeding, with external wing buds.

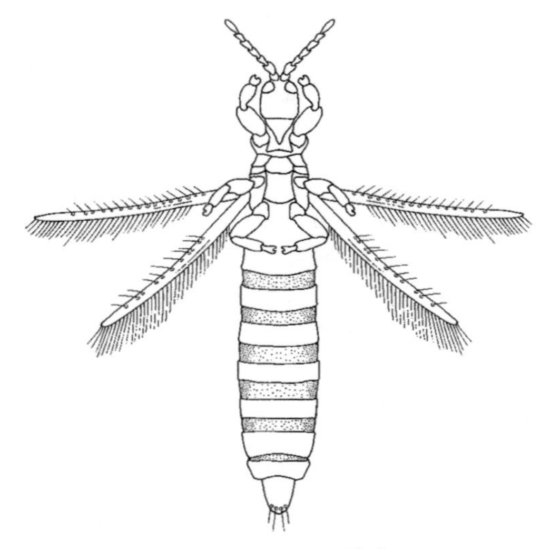

FIGURE 4.22 Thysanoptera.

Parthenogenesis and viviparity are recognized in some species. Thrips are predominantly phytophagous, a few species are fungivorous, and a few species are predaceous. Eggs are inserted into plant tissue, under bark, or into crevices. Thrips are typically multivoltine. Most thrips feed on flowers, pollen, leaves, buds, twigs, and other parts of plants. Thrips infestation symptoms on plants includes curled leaves or deformed buds.

HEMIPTERA [TRUE BUGS, FIG. 4.23, as well as two suborders that make up the group formerly called Homoptera, scale insects, mealybugs, whiteflies, leafhoppers, aphids, FIG. 4.24]

The order name Hemiptera is derived from Greek (hemi = half; pteron = wing) and alludes to the condition of the forewing in which the basal part is thickened and the distal part is membranous. However, due to reclassification, other members of this order possess wings that are uniform throughout. The etymology of the word *bug* is uncertain. The term *bug* is used commonly to refer to all insects. However, in the strict sense only species of the Hemiptera are the "true bugs."

The hemipteran head is usually prognathous, the antenna displays four or five segments, and the beak typically displays three or four segments. Forewings are modified into hemelytra, and the tarsi hold two or three segments with two apical claws. Members of the order Hemiptera are diverse in habits and habitats. Most species are terrestrial, but many Hemiptera live in or on the water. Most Hemiptera are plant feeders, but some species

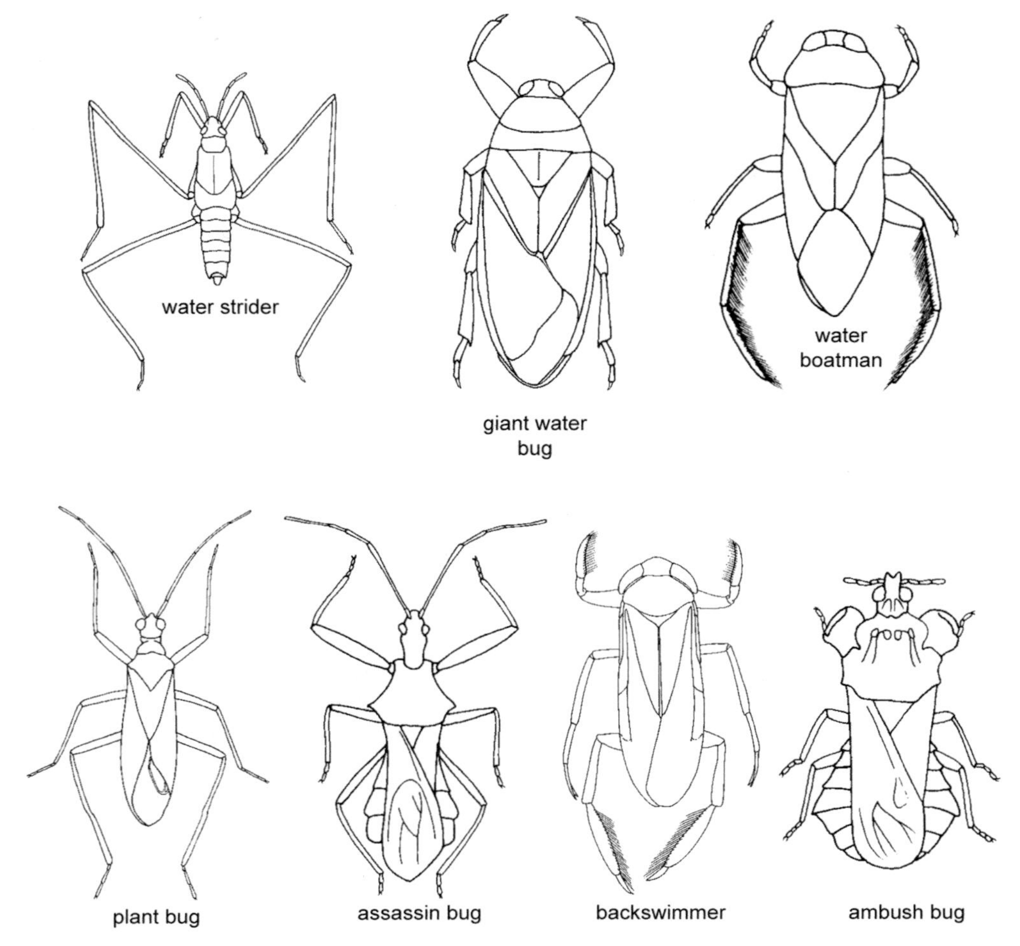

FIGURE 4.23 Hemiptera.

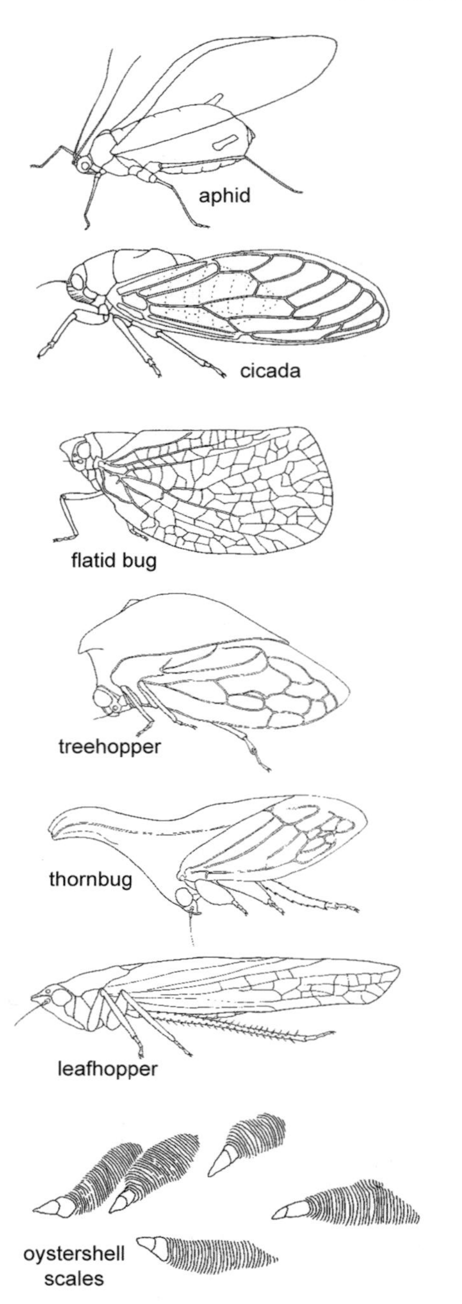

FIGURE 4.24 Several examples of Homoptera.

are predaceous on other insects. A few species imbibe the blood of humans and other vertebrate animals and are important vectors of diseases in the tropics.

Some species are attracted to lights. Sifting or using a Berlese funnel on leaf litter or on soil around plant roots usually will yield some Hemiptera. Some species are found in association with ants and may mimic the ants in appearance. Species of Cimicidae (bed bugs) are ectoparasites of birds, bats, and other animals, including humans. Cimicids usually feed on their host at night and hide in cracks and crevices during the day. Cimicids are virtually wingless and can be collected easily with small forceps.

Hemipteran insects lack cerci. Metamorphosis is characterized as simple, and the immatures resemble the adults in most species. However, more complex patterns of development are seen in some scale-like insects. Aleyrodidae pass through a pupal stage; males of scale insects pass through a prepupal or pupal stage (or both); the life stages of the Phylloxeridae are even more complex (Stoetzel, 1985).

Anatomically, Hemipterans are exceedingly diverse. Wingless species, such as scale insects, are sometimes difficult to recognize as insects.

All Hemiptera are terrestrial and most feed on plants. Some (such as whiteflies, psyllids, and leafhoppers) transmit plant diseases and are regarded as serious economic pests. Others (such as soft scales, armored scales, mealybugs, and aphids) can develop into numerically large, highly dense populations. Such Hemipterans cause problems through the accumulation of honeydew, which provides a substrate for the development of fungi. Cicadas can reach epidemic proportions and cause damage to trees and create intolerable levels of sound.

Scale insects and mealybugs are particularly susceptible to attack by chalcidoid wasps (Tachikawa, 1981; Viggiani, 1984); leafhoppers are attacked by dryinid wasps (Olmi, 1984a, b); the egg stage of many Auchenorrhyncha is attacked by species of mymarid wasps (Huber, 1986). In some instances, hyperparasites can be reared from Hemiptera (Viggiani, 1990). Accurate host associations are very important in our understanding of parasitic wasps and their importance in biological control.

COLEOPTERA [BEETLES, FIG. 4.25]

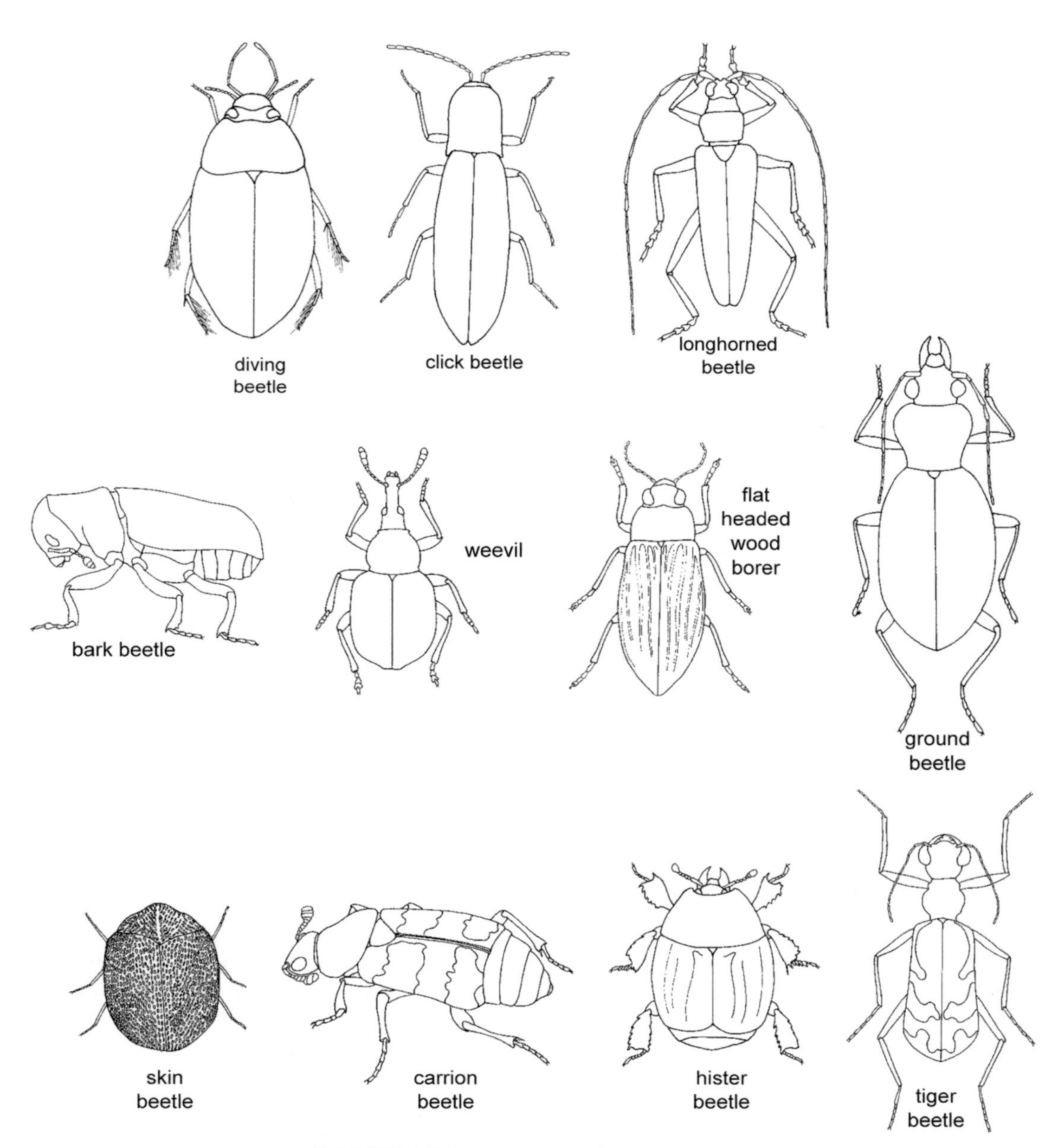

FIGURE 4.25 Several examples of Coleoptera.

The order name Coleoptera is derived from Greek (koleos = sheath; pteron = wing) and alludes to the thickened forewings that protect the hind wings. This order contains about 300,000 described species of insects and is the largest of the insect orders. More than 25,000 described species occur in the United States, with many undescribed species. Approximately 120 families are recognized, which are grouped into three suborders. This order is a favorite among collectors, and many amateurs specialize in collecting select families of beetles.

Beetles have mouthparts adapted for chewing and antennae that are exceedingly variable in segmentation and shape. The prothorax is well developed; the mesothorax is generally reduced; the abdomen is broadly joined to the thorax. Beetles usually have two pairs of wings, although some beetles lack them and other species have highly modified wings. The forewings are called *elytra* (Greek, elytron = cover, sheath). Elytra are thickened, usually hard or leathery, lack veins, and often are sculptured or display pits and grooves. Elytra usually meet in a straight line dorsally along the middle of the back. The hindwings are membranous and are folded under the forewings when the insect is at rest. Several anatomical types of larvae occur within the order.

Beetles have invaded almost every conceivable aquatic and terrestrial habitat. They feed on plants and fungi. Plant-feeding beetles may be found in association with every part of a plant, from flowers to roots. Some beetles are external feeders; other species mine leaves or bore into the stalk. Many phytophagous species are serious economic pests. In contrast, some beetles are predators of other insects and are viewed as beneficial insects. Predaceous species of Coccinellidae are used in biological control. Some species of beetles develop as parasites, and many species of beetles are scavengers. Some beetles are inquilines (welcome or unwelcome guests) in nests of social insects (termites and ants). Dermestid beetles feed on dead insects and can be a problem in stored products unless protective measures are taken.

Coleoptera include a very diverse group of insects, representing many interesting behavioral differences (Bell, 1990; Chandler, 1990; Dybas, 1990; Newton, 1990; Peck, 1990; Peck & Davies, 1980; Tashiro, 1990).

STREPSIPTERA [TWISTED-WINGED PARASITES, FIG. 4.26]

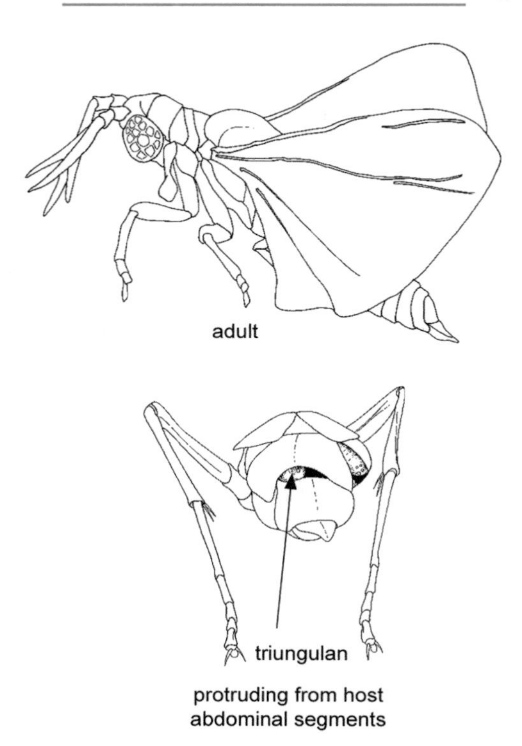

FIGURE 4.26 Strepsiptera.

The order name Strepsiptera is derived from Greek (strepsis = turning; pteron = wing) and refers to the twisted-wing condition found in males. The Strepsiptera form a widespread order of about 300 species. Adults exhibit strong sexual dimorphism. Females of most Strepsiptera lack eyes, antennae, and legs; the head and thorax are fused, and the general body shape is larviform. Females of some free-living species (some Mengeidae) display compound eyes, a developed head, antennae, and chewing mouthparts. All males resemble a generalized insect, the forewings are reduced (elytraform), and the hind wings are enlarged and membranous with radial venation. Strepsiptera are morphologically unique among insects in lacking a trochanter in the adult leg.

The immature stages of Strepsiptera are parasitic on Thysanura, Blattodea, Mantodea, Orthoptera, Hemiptera, Diptera, and Hymenoptera. Postembryonic development of parasitic species is hypermetamorphic: the first-instar larva is called a *triungulinid*. The free-living first instar resembles the triungulin of Coleoptera but differs in lacking a trochanter in all legs, and the antennae and mandibles are not well formed. The triungulinid is very active and searches for a host. After a host is found, the triungulinid enters the host's body and then molts into a legless, wormlike form that feeds and pupates within the integument of the host. The adult male is winged and leaves the host, but the female remains in the host with its body protruding from between the host's abdominal segments. After producing large numbers of the tiny triungulinids, the female dies.

Strepsiptera also parasitize several other insects (bees, wasps, flies, bugs). Hosts may be recognized by the small saclike females protruding from the often-distorted abdomens of their hosts (stylopized) (MacSwain, 1949).

MECOPTERA [SCORPIONFLIES, HANGINGFLIES, FIG. 4.27]

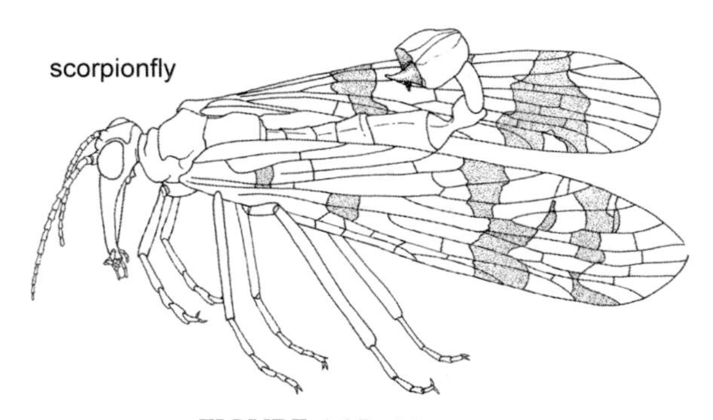

FIGURE 4.27 Mecoptera.

The order name Mecoptera is derived from Greek (meco = long; ptera = wing). Some Mecoptera are called *scorpionflies* because the upturned genitalia of some males resemble the tail of a scorpion. The name *hangingflies* pertains to one family (the Bittacidae), which resemble craneflies and hang suspended from vegetation by their front and middle legs. Mecoptera are moderate-sized (to 25 mm long), slender-bodied insects with relatively long wings. The head is prolonged into a rostrum beneath compound eyes, and the mouthparts are mandibulate and positioned at the end of the rostrum. Most Mecoptera are winged; some are brachypterous and a few are apterous. Winged species have four wings similar in size, shape, and venation; the wings are not folded when in repose. The metathorax is fused with the first abdominal tergum, but the first sternum is free. The abdomen is 11-segmented; cerci are two-segmented in the female and one-segmented in the male (or rarely absent). Metamorphosis is complete in Mecoptera. The larvae of some species possess compound eyes; the pupa is decticous and exarate.

Scorpionfly adults and larvae feed primarily on living and dead insects, but some adults are attracted to nectar and fermenting fruit. Some species apparently feed on moss. Males are predaceous, but females do not take living prey. Larvae of scorpionflies resemble the larvae of sawflies and some Lepidoptera but differ from the latter by the absence of crochets (tiny hooks) on the prolegs and from the former by having seven or more ocelli.

The larvae of most scorpionflies are found in the soil or in leaf litter and most adults are found in heavily wooded areas on vegetation. Adults are not strong fliers. Adults of the so-called snow scorpionflies of the family Boreidae emerge during winter and often can be picked from the surface of snow during winter months. Members of the Panorpidae (common scorpionflies) are active only during the daytime.

NEUROPTERA [ALDERFLIES, ANTLIONS, DOBSONFLIES, FISHFLIES, LACEWINGS, OWLFLIES, SNAKEFLIES, FIG. 4.28]

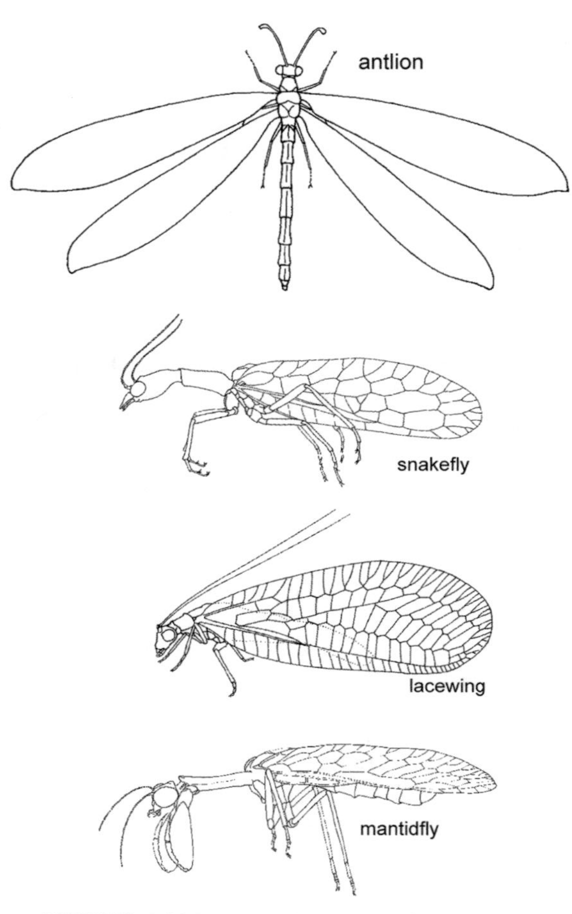

FIGURE 4.28A Several examples of Neuroptera.

The order name Neuroptera is derived from Greek (neuro = nerve; pteron = wing) and refers to the reticulate nature of the wing venation. The common names are generally applied to specific families or groups of Neuroptera based on their appearance or habits. The Neuroptera are a relatively small, cosmopolitan order of endopterygote, neopterous insects that are best represented in tropical regions. Specimens vary in size from small to very large, with a wingspan of more than 100 mm. Neuroptera are soft-bodied; the compound eyes are well developed, but ocelli are usually absent. The antennae are long, multisegmented, and sometimes display a club-shaped enlargement at the apex. Mouthparts are of the mandibulate biting type. The prothorax is movable, and the entire thorax is loosely organized. Neuroptera usually display four large, membranous wings that are subequal in size; venation is abundant and netlike. The wing coupling mechanism is simple; when the insect is at rest, the wings are held rooflike over the back. The abdomen is 10-segmented (except in Chrysopidae); cerci are absent. The immatures do not resemble the adults, and pupation usually occurs inside a silken cocoon. The pupa is decticous and exarate; pupal mandibles are well developed.

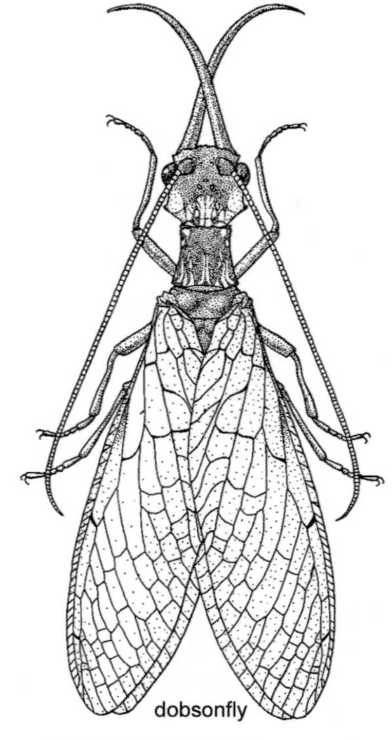

FIGURE 4.28B Neuroptera.

Classification of Neuroptera varies among entomologists. Here we recognize three suborders and 14 families. Neuroptera here are considered in the broadest sense to include the dobsonflies (Megaloptera) and snakeflies

(Rhaphidioidea). Neuropterans such as dobsonflies and fishflies (Corydalidae), alderflies (Sialidae), brown lacewings (Hemoerobiidae), green lacewings (Chrysopidae), and antlions (Myrmeleontidae) are frequently collected.

Neuroptera are predaceous as larvae and adults. As a consequence they are highly beneficial to human agricultural efforts. In fact, green lacewings are sold commercially for the control of various pests. Many Neuroptera will bite if handled incautiously when collected. The larvae of several members of the order are aquatic and commonly concealed under stones in streams. Adults of aquatic immatures usually remain on vegetation near water. They are relatively poor fliers. Many adults of terrestrial and aquatic immatures are attracted to lights. Immature myrmeleontids (called antlions) are found partially buried at the bottom of small pits that they dig in sand or dust to trap ants or other insects. The rare mantispid larvae are parasitic on spider eggs.

TRICHOPTERA [CADDISFLIES, FIG. 4.29]

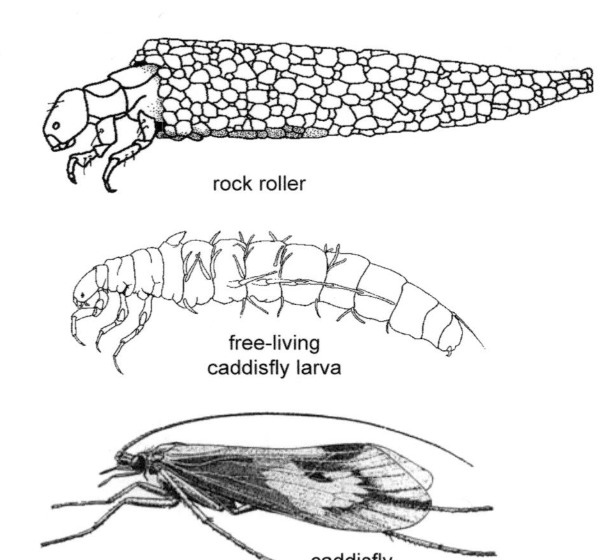

FIGURE 4.29 Immature and adult Trichoptera.

The order name Trichoptera is derived from Greek (trich = hair; pteron = wing) and refers to the hairy appearance of the wing. Caddisflies are soft-bodied insects with long, slender antennae and two pairs of membranous wings. The wings are clothed with setae and held rooflike over the back when the insect is at rest. The larval mouthparts are adapted for chewing; the adult mouthparts are adapted for feeding on liquids. About 5,000 species in 20 families are known for the order.

Trichoptera larvae are aquatic and display a pair of hooklike appendages at the apex of the abdomen. Many larvae live in characteristic cases constructed of pebbles, sand grains, twigs, or other materials found in ponds and streams. The abdominal hooks are used to drag the case about as the larva feeds. When the larva is ready to pupate, the animal attaches the case to a rock or other fixed object in the water. Case-making larvae are mostly detritus or plant feeders. Some caddisfly larvae do not make cases but instead spin silken webs, which are used to capture food drifting in the stream. A few species do not build cases or webs but are free-living and predaceous. When collecting immature Trichoptera, Preserve the case in alcohol along with the larva, pupa, or both.

Adult caddisflies are weak, usually crepuscular (dawn or dusk) fliers. They are generally found during the day resting on vegetation, bridges, or other objects near ponds and streams.

LEPIDOPTERA [BUTTERFLIES, SKIPPERS, MOTHS, FIG. 4.30]

The order name is derived from Greek (lepis = scale; ptera = wing) and refers to the wings usually covered with scales (flattened setae). The common name *butterfly* applies to most of the diurnal Lepidoptera with clubbed antennae; the common name *skipper* refers to diurnal Lepidoptera of the family Hesperiidae, with fast, erratic flight and whose antennae are hooked at the apex; the common name *moth* is applied to nocturnal Lepidoptera whose antennae are not clubbed. The Lepidoptera are a cosmopolitan

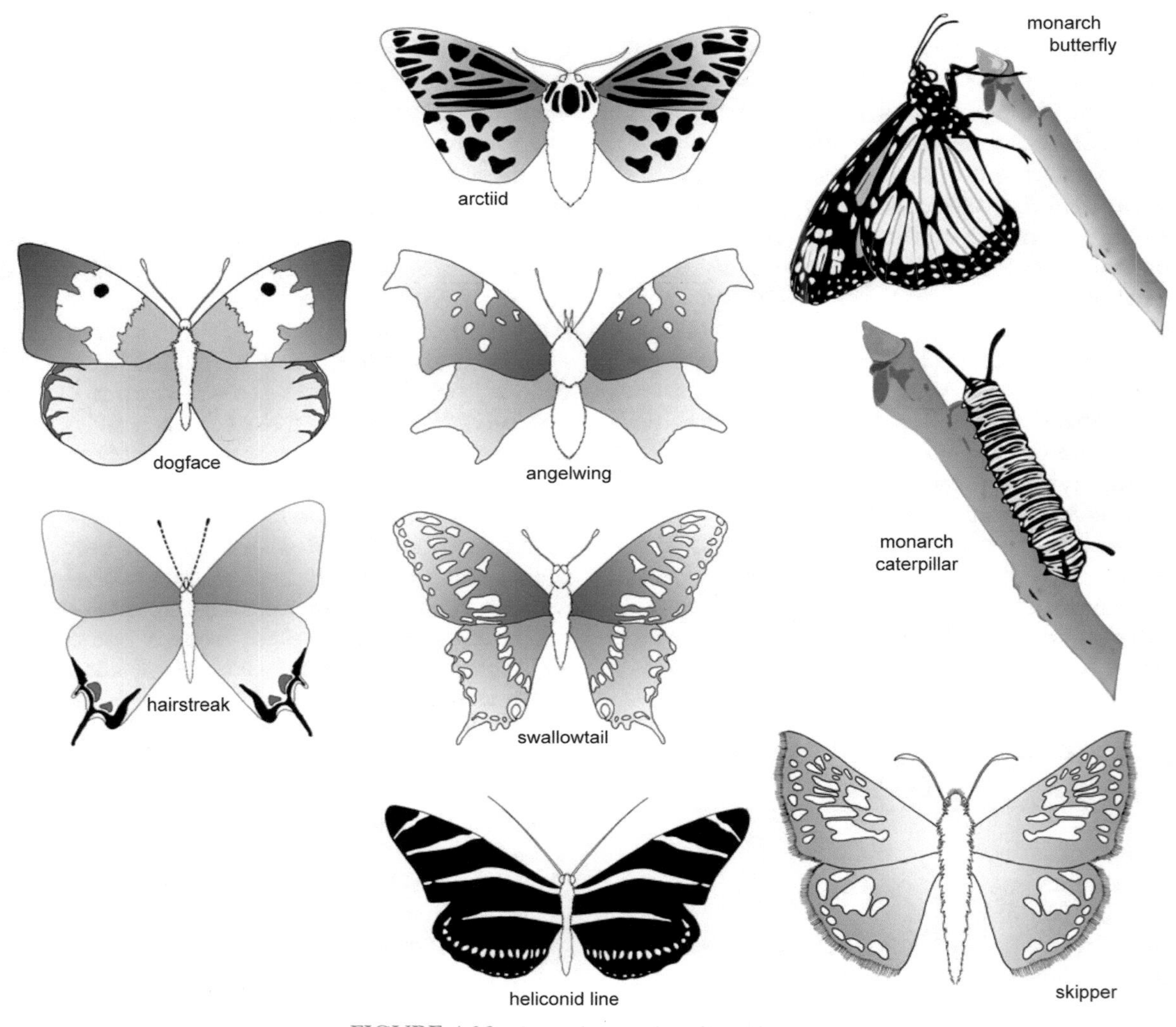

FIGURE 4.30 Several examples of Lepidoptera.

order of Holometabola consisting of about 150,000 described species placed in about 80 families. More than 11,000 species have been described in North America, but many species await description.

Lepidoptera are among the most popular insects collected by amateur entomologists. The mouthparts of adult Lepidoptera are adapted for imbibing nectar and water; a few primitive Lepidoptera have vestigial mandibles. The mouthparts of most Lepidoptera are formed into a long, spirally coiled haustellum or proboscis. The coil actually is derived from elongate maxillary galeae; the maxillary palpi are small or absent, but the labial palpi are well developed. Adult Lepidoptera all possess large, multifaceted compound eyes; some Lepidoptera have two ocelli, but many lack ocelli. (Sometimes scales must be removed from the head to find the ocelli.) The antennae of adult Lepidoptera are elongate and multisegmented but otherwise diverse in appearance. The antennae may be covered with scales or display featherlike branches. The adult head is freely movable; the prothorax is small and membranous and often displays a pair of articulated sclerites called *patagia*. Most Lepidoptera have four membranous, scale-covered wings; the wing-coupling mechanisms are diverse. Legs are usually long and tapered, but the posterior or anterior pair may be reduced in some groups; the tarsi display five segments. The adult abdomen contains 10 segments, and cerci are not present.

The larvae of Lepidoptera are commonly called *caterpillars.* The name *caterpillar* is derived from Latin and literally means "hairy cat," probably in reference to the many caterpillars that are covered with setae. Caterpillars usually have a well-differentiated, strongly sclerotized head and mandibles adapted for chewing. The thorax consists of three thoracic segments, each bearing a pair of legs with five segments; each leg terminates in a single claw. The larval abdomen also consists of 10 segments, and there are "prolegs," which usually are displayed on segments 3–6 and 10. The prolegs are apically truncate, with the truncate surface called the *planta.* Minute, sclerotized hooks called *crochets* are found along the margin of the planta. These hooks are used as holdfast structures and are diagnostically important. Lepidoptera pupae are decticous and exarate or adecticous and obtect.

Most lepidopterous larvae feed on live plants. Some species feed on or in the leaves, and other species bore into the stems, seeds, or other parts of the plant. The larvae of many species are host-specific and will starve rather than feed on a plant other than the preferred species. Larvae of some Lepidoptera are predaceous (Montgomery, 1982, 1983). A few larvae are scavengers on dead plant or animal matter, including woolen clothing. The larvae of many Lycaenidae are associated with ants and can be found in ant nests (Hinton, 1951).

Most adult Lepidoptera fly actively, but a few (such as female bagworms) are wingless or possess wings too short for use in flight. Most species of butterflies are diurnal, and are brightly colored. Whereas, most moths are nocturnal and less colorful.

DIPTERA ["TRUE" FLIES, FIG. 4.31]

The order name Diptera is derived from Greek (dis = twice; pteron = wing) and alludes to the presence of two membranous wings used in flight. Perhaps more than any other order of insects, families of Diptera are called by common names: mosquitoes, gnats, midges, punkies, no-see-ums, horse flies, bat flies, snipe flies, robber flies, house flies, bottle flies, and so forth. This, no doubt, is a consequence of the long-term intimate relationship between flies and humans. Incidentally, by convention we separate the word "fly" from its descriptor when referring to dipterous insects. We combine the word "fly" with its descriptor when referring to nondipterous insects whose names contain that word (e.g., scorpionflies, sawflies). Names notwithstanding, the Diptera constitute a cosmopolitan order of holometabolous insects that consists of about 150,000 described species. The order contains about 120 families, and in North America over 18,000 species have been described. Flies are found in virtually all habitats of all zoogeographical realms; they are omnipresent!

Adult Diptera are relatively soft-bodied and recognized by a typically large, manipulable head with large and multifaceted compound eyes. Most flies have three ocelli, some species have two ocelli, and a few species are anocellate. The antennae of flies are highly variable and range from moniliform and multisegmented to three-segmented and aristate. The mouthparts of Diptera are adapted for sucking

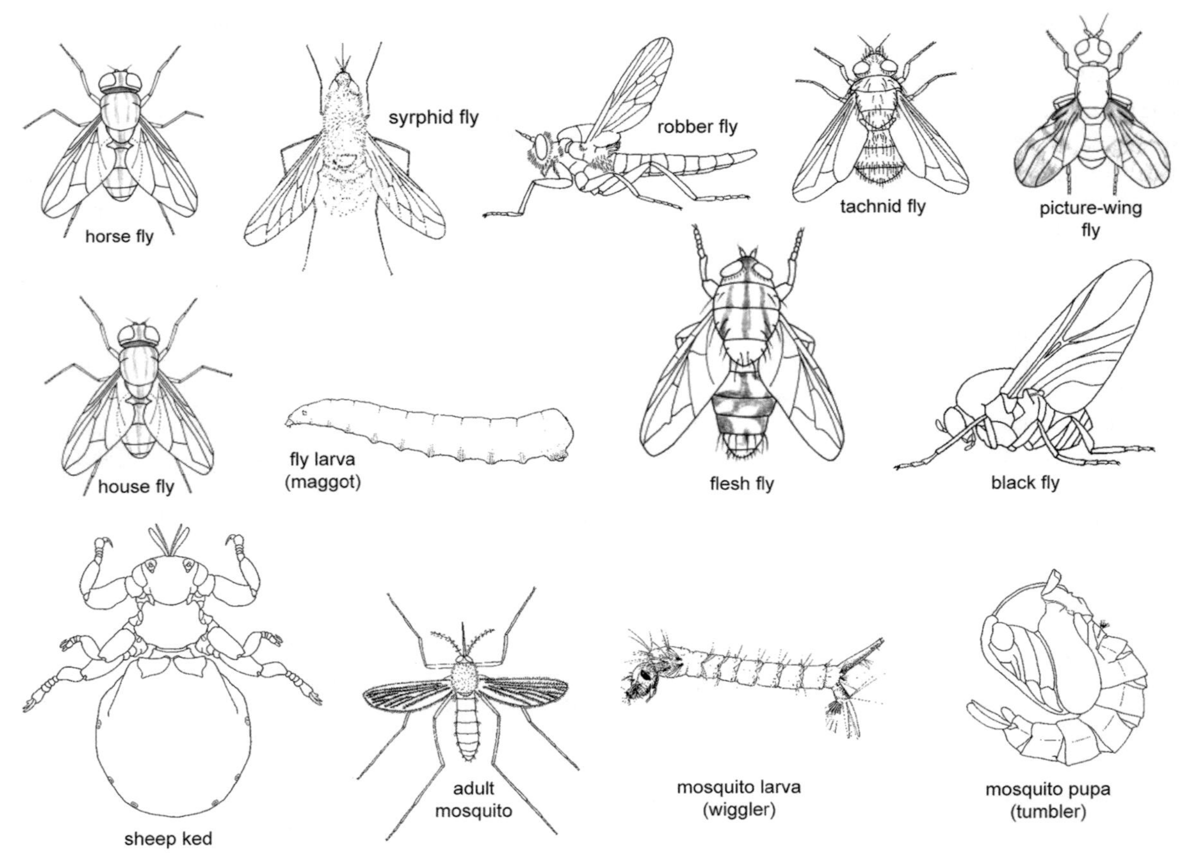

FIGURE 4.31 Examples of various Diptera.

and form a proboscis or rostrum. A piercing-type proboscis is found in predatory and blood-sucking species; a sponging-type proboscis is found in house flies; a few primitive flies have nonfunctional mandibles. The prothorax and metathorax of Diptera are reduced; the mesothoracic wings are membranous and used in flight; the metathoracic wings have become modified into clublike halteres that are used as balancing organs. Other orders may have species with only one pair of wings, but no other insects have knobbed halteres. (The halteres or hamulohalteres of winged male scale insects are not knobbed and usually are tipped with one or more hooked setae. Most winged male scale insects also differ from flies in having a single, long, stylelike process at the apex of the abdomen.) A few species of flies lack wings. The legs of flies are variable in shape and structure. The abdomen has 11 segments in the ancestral or primitive condition and 10 segments in the derived condition (10 and 11 fuse to form the proctiger). The distal segments of "higher" Diptera are often modified into a telescopic "postabdomen." Cerci are present at the apex of segment 10; the female cerci are primitively two-segmented (most Nematocera, Brachycera) or reduced to one segment (throughout Diptera); male cerci consist of one segment on the proctiger. Male genitalia are complex.

The larvae of higher Diptera are frequently called *maggots.* All larvae are apodous, but some species have one or more pairs of prolegs. In some primitive families, the larvae possess a distinct head capsule and the pupae are free-living. In more advanced flies, as in the muscoid families, a head is not apparent on the larvae. Larvae often pupate inside the last larval exuviae, which is called a *puparium.* The pupae of Diptera are adecticous and exarate or obtect.

The biological importance of Diptera cannot be overemphasized: bloodsucking species transfer many diseases, including malaria; house flies transfer enteric diseases; the nuisance value of biting flies cannot be measured. The larvae of some species cause myiasis. In terms of agriculture, the impact of some Diptera is notable. Tephritid fruit flies are serious pests of fruit. Quarantine and eradication programs in some areas cost tens of millions of dollars yearly in response to the threat posed to agriculture by fruit flies. Leaf- mining and gall-forming Diptera cause conspicuous damage to foliage and other plant parts, but their economic impact is questionable. In contrast to these negative impacts on humans and their activities, many species of Diptera are beneficial as pollinators; other species are predators or parasites of insect pests. Most species of flies probably are neither beneficial nor harmful to humans, but often are collected and studied because of the unusual ecological habitats they occupy.

Habitats of Diptera are numerous and varied, but most dipterous larvae live in aquatic or semiaquatic habitats. Perhaps the most unusual habitats involve the seeps of crude petroleum in which larvae of some ephydrid larvae (shore flies) develop or the hot springs in which some stratiomyid larvae are found. A rain- filled can by the side of the road may harbor mosquito larvae (Culicidae), and stones in a fast-flowing stream may hold black fly larvae (Simuliidae). A dipper or small aquatic net is useful in collecting such larvae. A sweep net or aerial net can be used effectively to collect the adults, which generally remain near the water. Marsh fly adults (Sciomyzidae) are slow-flying and rest on emergent vegetation along the margins of freshwater. They often perch with their head directed downward. Marsh fly larvae are obligate feeders on freshwater snails, fingernail clams, land snails, and slugs.

Some Diptera larvae live in plant tissues, where they mine leaves, form galls, or feed in the stems or roots. Some species pupate in the soil.

SIPHONAPTERA [FLEAS, FIG. 4.32]

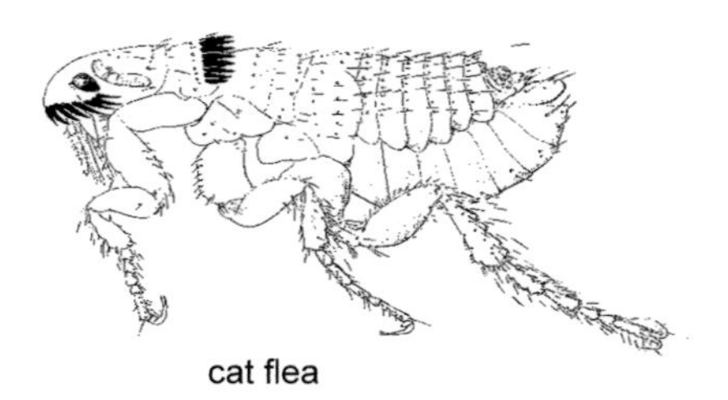

FIGURE 4.32 Siphonaptera.

The order name Siphonaptera is derived from Greek (siphon = tube; aptera = wingless) and refers to fleas as wingless siphons. This name seems appropriate for reasons explained later. Fleas represent an order of about 1.400 species of holometabolous, endopterygote Neoptera, divided into about 20 families. Adults may be recognized by the body, which is wingless and typically less than 6 mm long. Fleas are laterally compressed and display bristles and cuticular projections that are directed rearward. These structures are sometimes modified into combs called *ctenidia* that may be found on the gena, prothorax, or metathorax. The compound eyes are absent or replaced by one large lateral ocellus on either side of the head. Mandibles are not present, and the mouthparts are of the piercing-sucking type. The piercing elements are derived from the sclerotized epipharnyx and the laciniae of the maxilla; the maxillary and labial palpi are well developed in adult fleas. The antennae are short, typically with three segments. The antennae are held in depressions called on the side of the head. All fleas are apterous. The legs are spinose, with large coxae, and the hind legs are adapted for jumping. The abdomen contains 10 segments, with tergum I reduced and sternum I absent. A complex sensory apparatus called the *sensillum* is positioned at the apex of the abdomen. Larval fleas display 13 body segments, lack appendages, and show a wormlike appearance. The head capsule is well sclerotized but lacks eyes. The mandibles are brushlike and the antennae are one-segmented. The terminal body segment displays anal struts. The larva undergoes three instars before constructing a pupal cocoon. The pupa is adecticous and exarate; pupal wing buds are visible in some species, which suggests that fleas had winged ancestors.

Adults of all flea species are external parasites of mammals and birds. Fleas such as the cat flea, *Ctenocephalides felis* (Bouche), and the dog flea, *C. cans* (Curtis), are well known as household pests. However, fleas are most important to humans because some species can transmit bubonic or sylvatic plague, endemic typhus, and other serious diseases. Some species serve as intermediate hosts of tapeworms.

Adult fleas reflect many features that are adaptations for a parasitic existence on the bodies of their hosts. Adult fleas have lost the need for wings because they live in the fur and feathers of their hosts. Moving around on the body of a host with fur and feathers is difficult, so the flea's body has become greatly compressed laterally to facilitate movement between hairs and feathers. To protect the flea from damage, its body has become heavily sclerotized. The flea is armed with combs or bristles that engage fur and resist dislodgment from the host during grooming. The antennae are short and fitted into grooves for protection from damage. The piercing-sucking type mouthparts are an adaptation for feeding on warm-blooded vertebrates.

Flea larvae live in the nests of their hosts. The larvae do not suck blood because their mouthparts are not adapted for that kind of feeding. Instead, flea larvae feed on organic debris (including dried blood) within the nest or habitation of the host. Fleas commonly infest animals that live in nests or burrows such as birds and rodents but seldom are found on cattle, deer, or other hoofed animals.

HYMENOPTERA [SAWFLIES, ANTS, WASPS, BEES, FIG. 4.33]

The order name Hymenoptera is derived from Greek (hymen = membrane; pteron = wing) and refers to the four membranous wings. The Hymenoptera are a cosmopolitan order with about 125,000 described species. The Hymenoptera are subdivided into two suborders, the Symphyta (sawflies) and Apocrita (bees, wasps, ants, parasitic Hymenoptera). The adult head is mobile, usually hypognathous but sometimes prognathous; the head is not fused with the pronotum. The compound eyes typically are large and multifaceted; the eyes are reduced to a few ommatidia in some ants and parasitic species. Three ocelli usually are present, but these are absent from some species. Antennal segmentation and shape are variable, but the antenna typically is *geniculate* (elbowed). The antenna frequently shows sexual dimorphism, with the funicle sometimes *ramose* (branched) in the male and apical segments sometimes differentiated into a club in the female. Mandibles always are present in adult Hymenoptera, and the mouthparts usually are adapted for biting- chewing; some bees show adaptations for

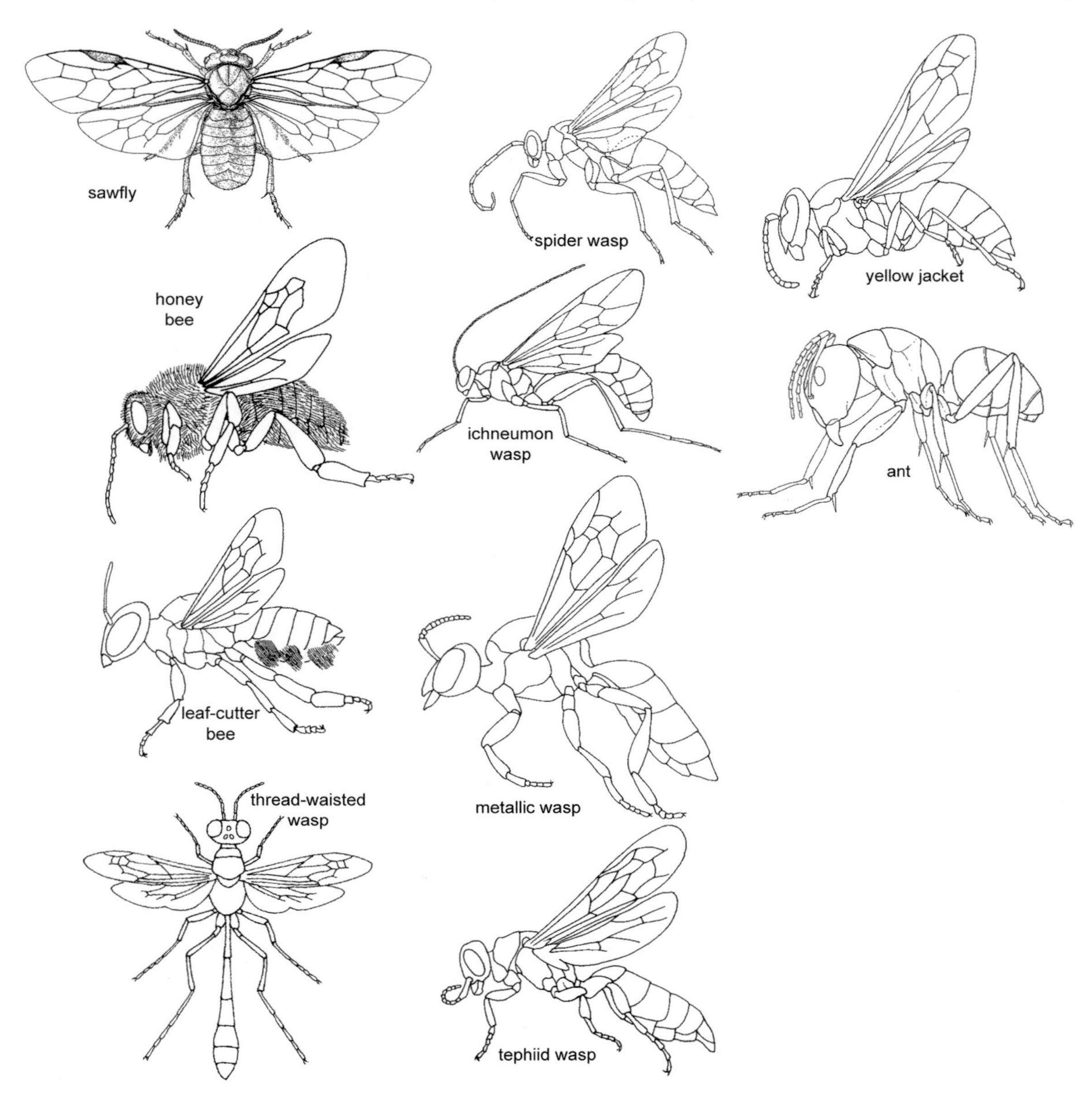

FIGURE 4.33 Several examples of Hymenoptera.

chewing-sucking. Most Hymenoptera have four membranous wings, with the forewing usually substantially larger than the hind wing; the forewing and hind wing are connected during flight by hooks called *hamuli*. The forewing venation is complex and extensive in large- bodied species; it is drastically reduced in small to minute parasitic species. The hind wing venation forms a few cells at most; otherwise the venation is reduced. Apterous species are distributed throughout the

Hymenoptera. Reproductive ants develop wings but shed their wings after nuptial flight; the worker and soldier castes of ants are wingless. Among the Apocrita the first abdominal segment is incorporated into the thoracic region and called a *propodeum*. The propodeum and thorax collectively are called the *mesosoma* or alitrank (in ants). The propodeum usually is separated from the remainder of the abdomen by a constriction of the second abdominal segment (called the *petiole)* or segments two and three (called the *postpetiole* in ants). The abdomen behind the petiole is called the *metasoma* or *gaster* (in ants). The abdomen of Hymenoptera has 10 segments in the ancestral condition; the first segment (propodeum) and second segment (petiole) typically lack the sternal elements. Female Hymenoptera display a sclerotized, elongate, tubular ovipositor (sawflies, parasitic Hymenoptera) or sting *(aculeata)* the cerci are one-segmented or modified into two setose patches *(pygostyli)*. The male genitalia are complex and sometimes diagnostically important.

Hymenoptera larvae are extremely disparate in terms of their anatomy. The larvae of Symphyta are caterpillarlike, with thoracic legs and prolegs. The prolegs lack crochets, a character that may be used to separate these larvae from the larvae of Lepidoptera. The larvae of Apocrita are apodous or vermiform. Some species display hypermetamorphic development. That is, the first-instar larva does not resemble subsequent instars. The number of instars among Hymenoptera is variable. Some egg parasites may have only two instars; some sawflies mayhave seven instars. The pupae of Hymenoptera are adecticous and usually exarate, but some species display an obtect condition.

The biology of Hymenoptera is exceedingly diverse. The kinds and numbers of Hymenoptera that occur all around us are staggering (Bentley, 1992; Darling & Packer, 1988). Species are phytophagous, parasitic, predaceous, and gall-forming. Among the apocritous larvae, the gut is not connected to the anus until feeding has been completed. Polyembryonic development has been shown in some species, and eusociality has evolved several times among the bees and ants. Parthenogenesis is universal within the order; females possess the diploid *(2n)* complement of chromosomes and males possess the haploid *(n)* complement of chromosomes.

The Hymenoptera have beneficial and noxious species. Hymenoptera are highly beneficial in terms of crop pollination. Many bees are important pollinators of plants, and some are valued for the honey they produce. The parasitic species are of ever-increasing importance as biological control agents of agricultural pests. The plant-feeding forms include some of the most destructive defoliators of forest trees.

The parasitic Hymenoptera represent a significant portion of the Hymenoptera, yet they are poorly understood. More than 50,000 species have been described, but several times that number remain undescribed. These insects are multitudinous and attack all stages of other insects. On the other hand, some hymenoptera can be extremely damaging. In fact, the plant-feeding forms include some of the most destructive defoliators of forest trees. In addition, females of the social wasps and bees are well known for their ability to sting and as a result, many people are hospitalized each year.

SUMMARY

Additional information can be found by searching the following, as well as other published studies.

Insect Orders:

- Essig, 1942, 1958
- Jacques, 1947
- Peterson, 1948
- Chu, 1949

- Matheson, 1951
- Brues et al., 1954
- Usinger, 1956
- Borror & White, 1970
- Richards & Davies, 1977
- Hollis, 1980
- Parker, 1982
- Ross et al., 1982
- Arnett, 1985
- Scott, 1986
- Borror et al., 1989

Arachnids:

- Rohlf (1957)
- Evans et al. (1964)
- Cooke (1969)

Ticks and mites:

- Lipovsky, 1951, 1953
- Strandtmann & Wharton, 1958
- Furumizo, 1975
- Jeppson & Kiefer, 1975
- Krantz, 1978
- McDaniel, 1979
- Evans, 1992

Insects have been on earth for an estimated 280 million years, mites even longer. Considering both numbers of individuals and numbers of species, insects and mites represent a dominant form of life. Clearly, there are more insects and mites alive than any other group of metazoan animals. In terms of biological diversity, insects occupy more habitats and adopt more lifestyles than any other group of organisms; mites rank second. Insects represent humankind's principal competitors for food and fiber; mites often cause considerable damage to plants and food. Insects and mites both directly and significantly impact the health and welfare of humans and other animals. They are at once both beneficial as scavengers, biological control agents, and environmental indicators and detrimental as humans' most damaging pests. Considering their geological age, numbers, biological diversity, and economic importance, the study of insects and mites is compelling.

The ability to identify specimens to the species level is critical in biological sciences (Gotelli, 2004) and is basic to pest management because the identity of the pest dictates what controls should be considered. Arthropods that become the subject of legal investigations or forensic examinations also must be identified to the species level. Similarly, legislation covering quarantines, conservation surveys and international transfer of specimens (Gottelli, 2004) relies on specific identification (Walter & Winterton, 2007).

Dichotomous Keys

A dichotomous key is a tool that allows the user to determine the identity of arthropods based primarily upon appearance (its morphology). 'Dichotomous' means 'divided into two parts.' Therefore, dichotomous keys normally offer two mutually exclusive choices, depending upon what the specimen looks like. The key will eventually lead the user to its correct name. Once specimen is identified, one may assume that much of its biology, distribution, and key behaviors are similar to others in that group.

Experience in working through the keys will help tremendously. Sometimes particular specimens are difficult to classify, even with a key. On occasion, a rare specimen may not easily key out at all. Do not be discouraged. Consult with a person who is well trained in classification or contact another professional entomologist, preferably a specialist in that particular group, to assist.

The Illustrated Dichotomous Key to Arthropod Classes and the Illustrated Dichotomous Key to the Orders of Insects, that follow provide another example of how keys work and should help beginning diagnosticians classify specimens to the class and order level. Thousands of similar keys exist for family, genera, and species separation, but are far too many in number to include in this book. Diagnosticians accumulate many textbooks and independently published keys as part of their work and will refer to them as needed.

Remember that the following keys are designed for adult specimens only. They will not work for immature insects. Other dichotomous keys are available for immature life stages. Keys may exist for specific groups of arthropods, such as for selected pests of certain crops, pests generally found in homes, or insects occurring in a particular region or state. Keys to exclusive groups of insects, such as butterflies or beetles, or even more select groups – such as keys to the biting ticks east of the Mississippi – may exist, making for a very large number of keys available. No diagnostician possesses all keys, but rather collects those specifically needed and searches for others when the need arises. This strategy is best because keys change over time with the discovery of new insects or the rearrangement of groups of insects.

Beginning diagnosticians should become familiar with the use of keys so as to avoid misidentifying arthropods that look alike. In addition, expertise in working through dichotomous keys is a foundation for more in-depth identification (to the family, genus, or species level). The following illustrated key can be used for most living classes of Arthropods. In this key, classes are separated based on differences in external morphology.

ILLUSTRATED DICHOTOMOUS KEY TO CLASSES OF ARTHROPODA

1a. Antennae absent (in immature stages and some adult insects, the antennae are considerably reduced or absent; such forms will not key well here)................................2

1b. Antennae present 4

2a. Body usually with seven pairs of appendages, including five pairs of legs; abdomen rudimentary (sea spiders).................**PYCNOGONIDA**

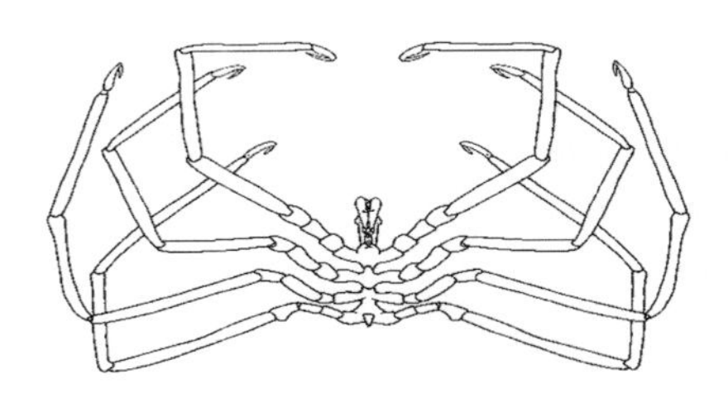

2b. Body with six pairs of appendages (rarely fewer), including four pairs of legs (rarely five); abdomen usually well developed but sometimes fused with cephalothorax 3

3a. Abdomen with booklike gills on ventral surface; large animals up to 50 cm long, with hard, expanded shell and long, spinelike tail (horseshoe crabs) Class MEROSTOMATA, order or subclass ....**XIPHOSURA**

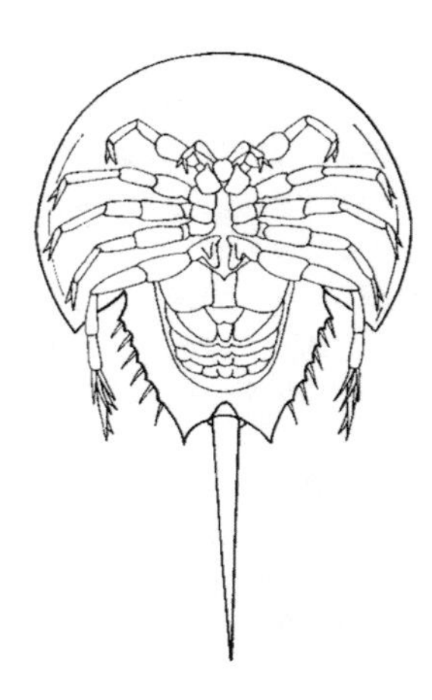

3b. Abdomen without booklike gills; smaller forms rarely over 7 cm long, body not as above (spiders, scorpions, mites, etc.)...**ARACHNIDA**

4a. With two pairs of antennae (one pair may be rudimentary in sowbugs); head and thorax fused to form cephalothorax; breathing by gills (crabs, lobsters, shrimps, sowbugs, etc.)..**CRUSTACEA**

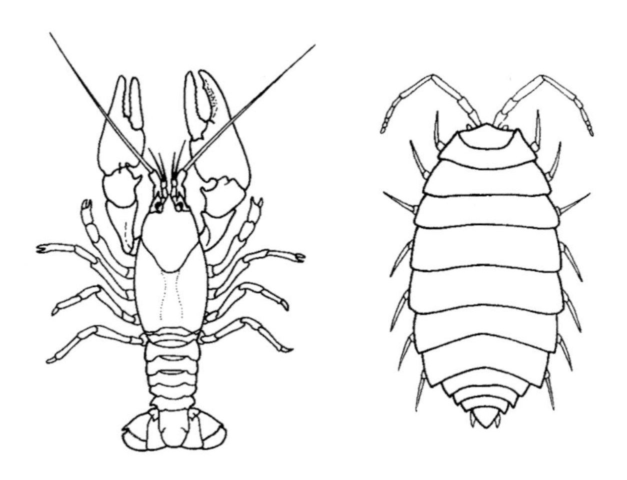

4b. With one pair of antennae; head and thorax separate; breathing by tracheae..................5

5a. Body with head, thorax, and abdomen; thorax with three pairs of legs at some stage in life cycle; abdomen sometimes with appendages that resemble thoracic legs; wings present in most species (all insects).....**HEXAPODA (INSECTA)**

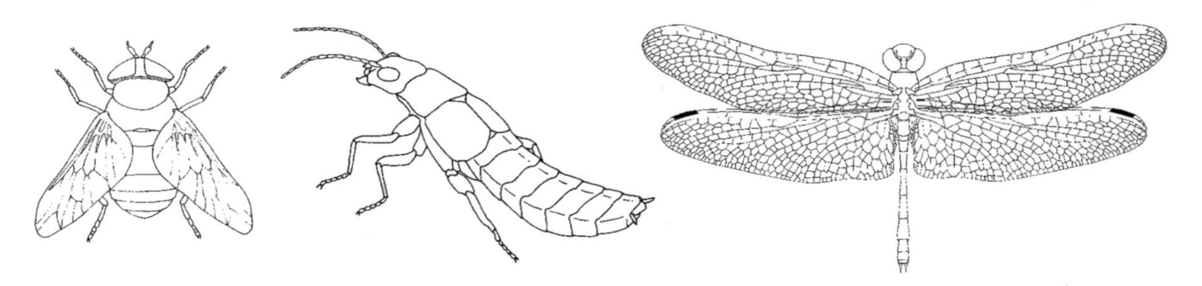

5b. Body more or less wormlike; most body segments behind head with legs; body with nine or more pairs of legs; wings absent (myriapodan classes) 6

6a. Most body segments each with two pairs of legs; slow-moving animals (millipedes) ....**DIPLOPODA**

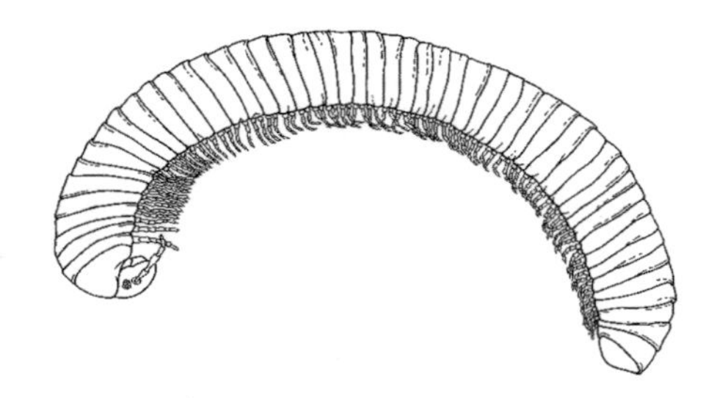

6b. Most segments of body with at most one pair of legs; speed variable..........................7

7a. Body rather large, more or less flattened, with 15 or more pairs of legs; often reddish brown, rapidly moving animals.................... ....**CHILOPODA**

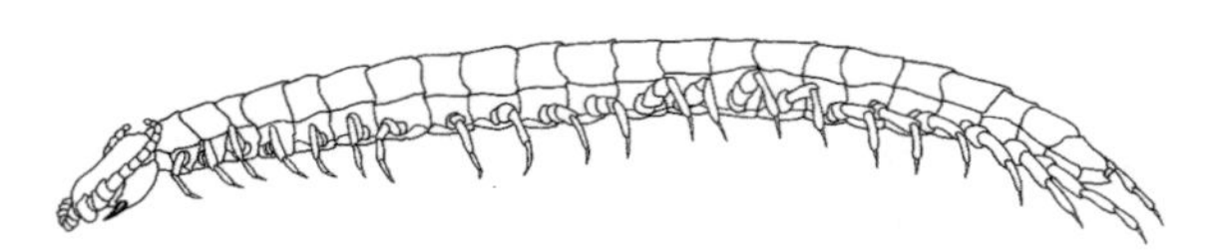

7b. Body small to minute (not over 8 mm long), usually cylindrical, with 9–12 pairs of legs, whitish or pale-colored 8

8a. Antennae branched; nine pairs of legs; body minute (1–1.5 mm long); found in leaf litter etc. (pauropods).................**PAUROPODA**

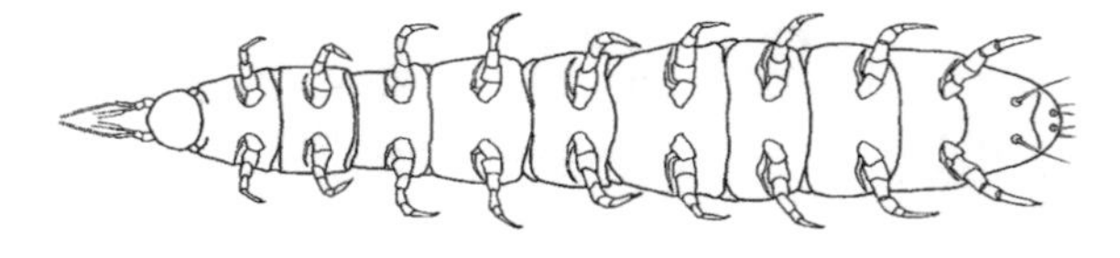

8b. Antennae not branched; 10–12 pairs of legs; body to 8 mm long, cylindrical, centipede-like; found in moist habitats (symphylans).........**SYMPHYLA**

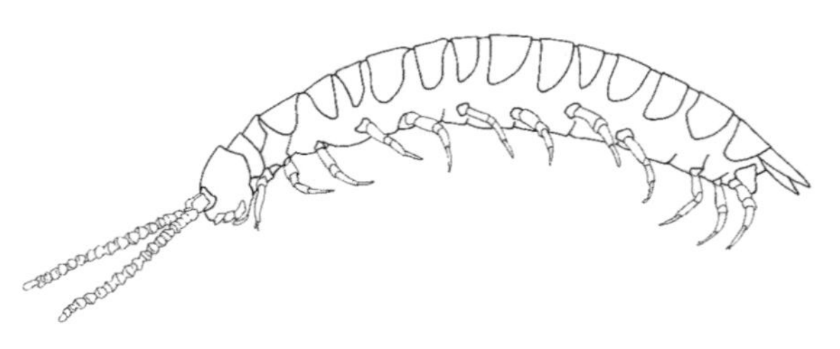

CLASS ARACHNIDA

The Arachnida are the second largest class of arthropods in terms of number of species and species important to agriculture. Most arachnids, including all spiders, are predaceous. Some arachnids are parasitic on animals, including humans. Many mites (subclass Acari) are important pests of plants.

Key to Orders of Arachnida

1a. Abdomen not segmented, or if segmented (rarely) then with distinct sclerites (as in Asiatic family Liphistiidae); spinning organs on ventral side of abdomen.............2

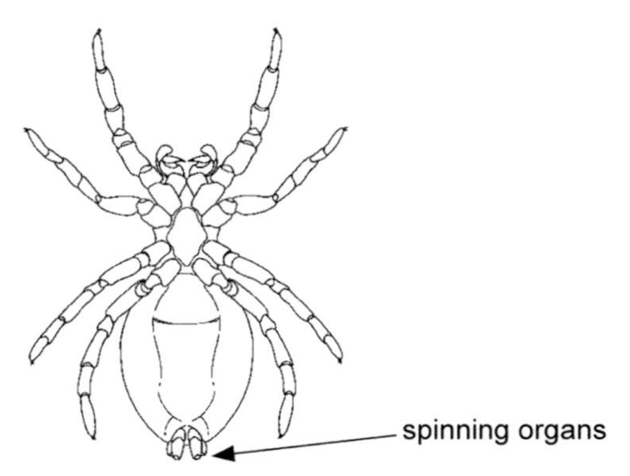

Ventral view of spider.

1b. Abdomen distinctly segmented; without silk-spinning organs on abdomen 3

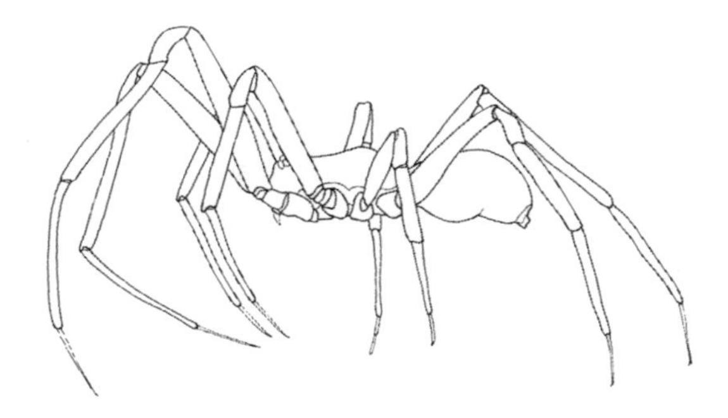

2a. Abdomen joined to cephalothorax by narrow, short stalk; abdomen usually soft and weakly sclerotized, abdomen with spinning organs (spiders; not further considered here).......................**ARANEIDA**

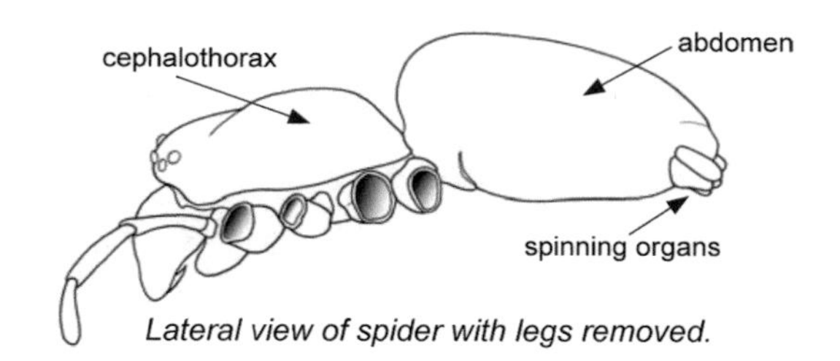

Lateral view of spider with legs removed.

2b. Abdomen broadly joined to cephalothorax; abdomen comparatively tough or comparatively strongly sclerotized; abdomen without spinning organs (some mites spin silk from palpi or mouthparts (mites and ticks) ticks............**Subclass ACARI**

3a. Abdomen with posterior segments forming a long, tail-like projection.............................4

3b. Abdomen without a tail-like projection.....6

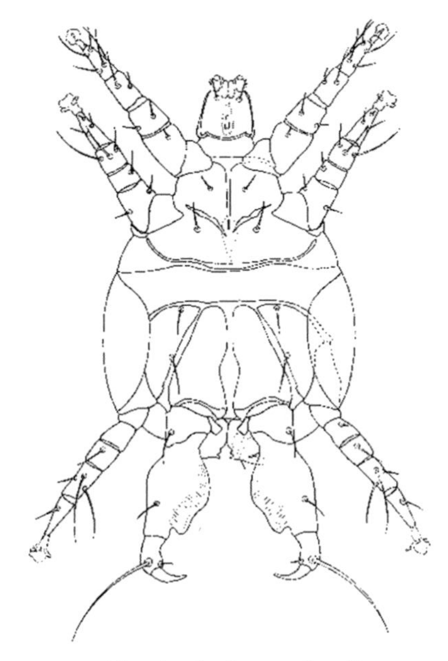

Ventral view of mite.

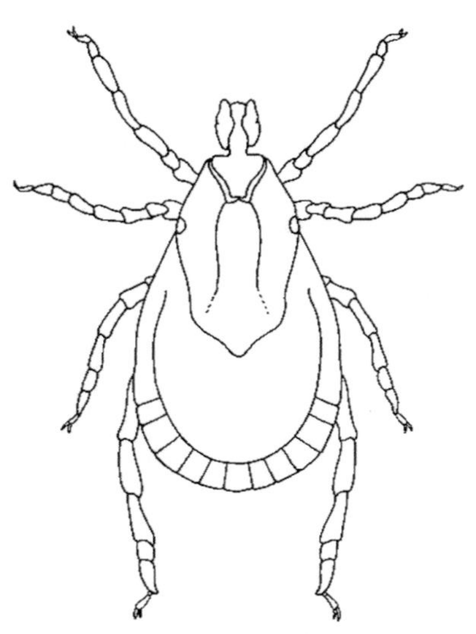

Dorsal view of tick.

4a. Tail-like projection of abdomen six-segmented, ending in bulbous, clawlike sting; abdomen broadly joined to nonsegmented cephalothorax; venter of second abdominal segment with a pair of comblike organs, the pectines (scorpions; widespread in warm, dry areas; not further considered here)................**SCORPIONIDA**

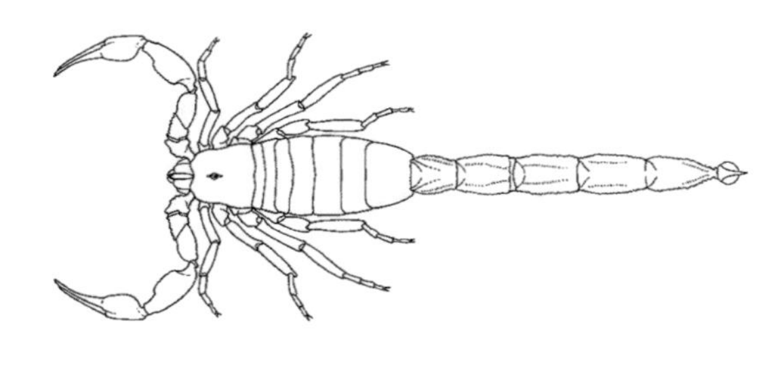

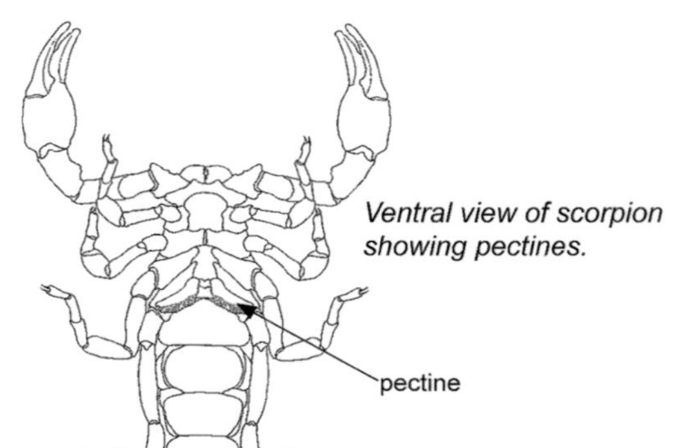

Ventral view of scorpion showing pectines.

4b. Tail-like projection of abdomen very slender, many-segmented but not ending in sting; abdomen narrowed to base, without ventral comblike organs...............................5

5a. Anteriormost leglike appendages (pedipalpi) slender; body minute, not more than 2 mm long; found in warm areas (not further considered here).........**PALPIGRADI (MICROTHELYPHONIDA)**

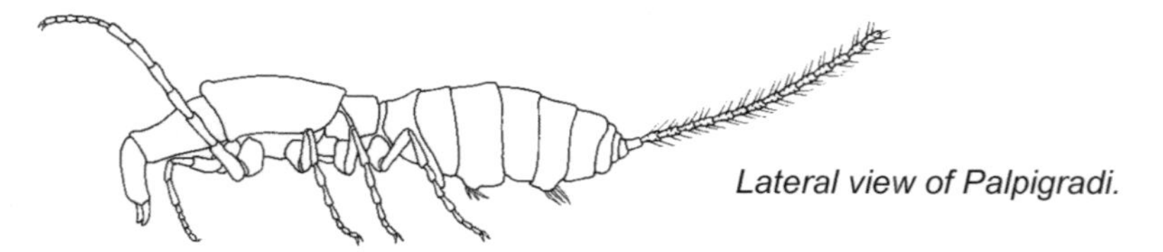

Lateral view of Palpigradi.

5b. Anteriormost leglike appendages (pedipalpi) very stout, contrasting with very long first pair of legs; body 2–65 mm long (whip scorpions, vinegaroons; not further considered here) ....**UROPYGI (THELYPHONIDA; PEDIPALPIDA in part)**

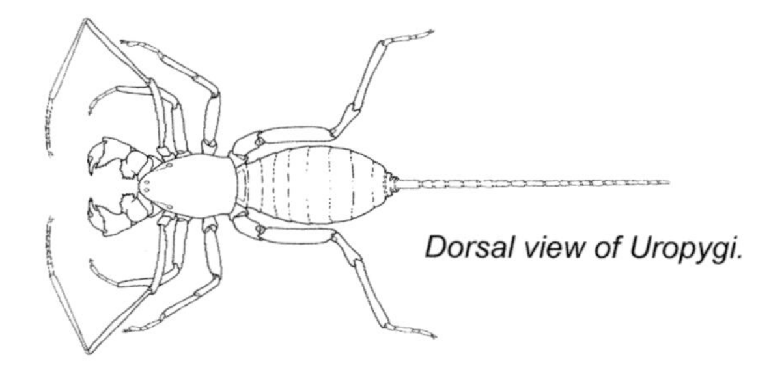

Dorsal view of Uropygi.

6a. Abdomen constricted at base; front legs very long with long tarsi; pedipalpi clawed at apex; tropical animals 4–45 mm long (tail absent, whip scorpions; not further considered here) ....**AMBLYPYGI (SCHIZOMIDA; PEDIPALPIDA in part)**

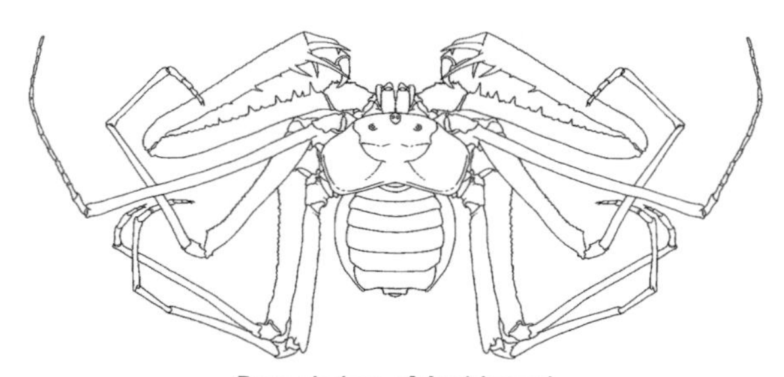

Dorsal view of Amblypygi.

6b. Abdomen broadly joined to cephalothorax; front tarsi not lengthened; other characters variable 7

7a. Pedipalpi with large, pincerlike claws (pseudoscorpions; not further considered here)..................**PSEUDOSCORPIONIDA (CHELONETHIDA)**

Dorsal view of Pseudoscorpionida.

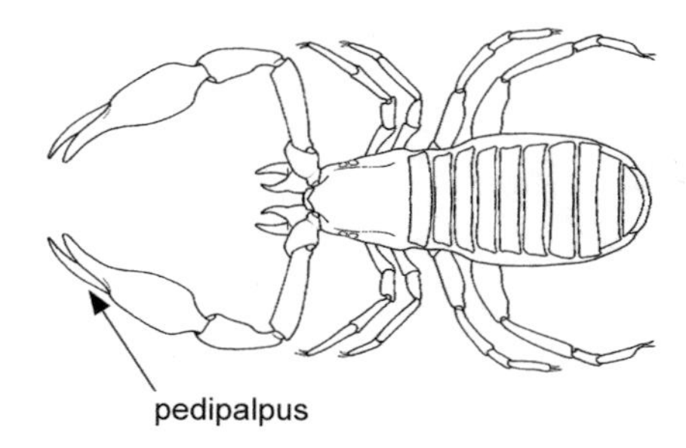

7b. Pedipalpi without pincerlike claws............8

8a. Head distinct from three-segmented thorax; chelicerae large and powerful, with pincers moving up and down; pale-colored, nocturnal, large animals (up to 7 cm long); occur in Florida and the southwest (wind scorpions; not further considered here)......... ....**SOLPUGIDA**

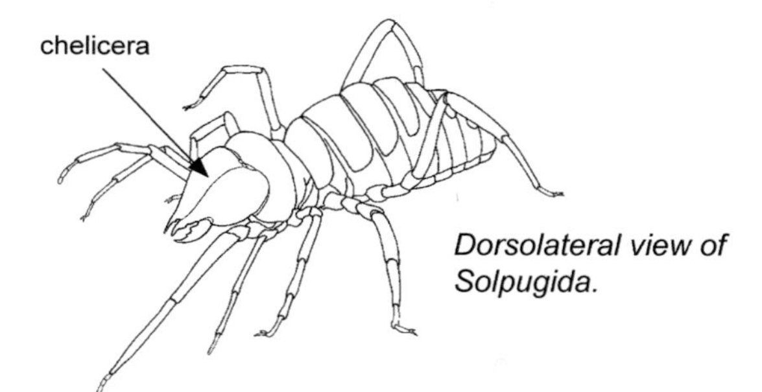

Dorsolateral view of Solpugida.

8b. Cephalothorax present, not divided into head and three segments; chelicerae usually smaller, pincers not moving up and down; other characters variable..........................9

9a. Abdomen apparently four-segmented with lateral and dorsal sclerites, and small, several-segmented endpiece; eyes absent; heavy-bodied animals 5–10 mm long, with moderately long legs; tropical to Texas (ricinuleids; not further considered here)...... ....**RICINULEI**

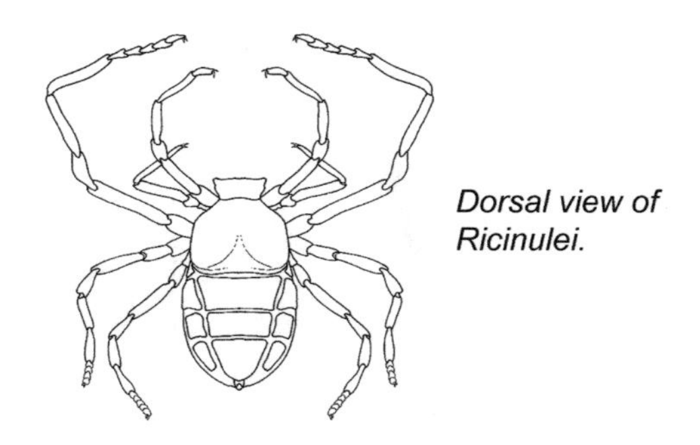

Dorsal view of Ricinulei.

9b. Abdomen usually appearing seven-segmented from above, without separate lateral sclerites; body 5–10 mm long, with legs very long; distribution widespread (daddy longlegs, harvestmen; not further considered here)........................**OPILIONES (PHALANGIDA)**

Lateral view of Opiliones.

SUBCLASS ACARI

The acarines are so varied in form that their anatomical terminology has developed along considerably different lines than that of the insects. Most acarines are very small and require special techniques for their preservation and examination. The small size, morphological plasticity, and complex terminology make acarines considerably more difficult to identify than most insects. Only recently has their higher classification attained a measure of stability.

Past acarologists considered the Acarina as an order equivalent to the other orders of the class Arachnida (such as Araneida and Scorpionida). Current students of the mites regard the Acarina as the subclass Acari without reference to the status of other orders. The classification of the subclass Acari is variously constructed. Wolley (1988) divides the sublcass Acari into the orders Astimata, Gamasida, Opilioacarda, Ixodida, Holothyrida, Oribatida (Sarcoptiformes, Oribatei, Cruyptostigmata, and Oribatoidea), and Actinedida (Trombidiformes and Prostigmata). Evans (1992) scheme lists the orders Notostigmata, Holothyrida, Ixodida, Mesostigmata, Prostigmata, Astigmata, and Oribatida. The group names used here are left without rank designation until problems with higher classification are resolved. Specific procedures for mounting mites may also be found in Appendix II.

A few definitions for the more unfamiliar words may be helpful in the following key. A *sejugal furrow* is a line of demarcation that separates the podosoma and opisthosoma; a *hypostome* is the anteroventral region of the gnathosoma (foremost part of the mouthparts); *Haller's organ* is a sensory organ found on tarsus I of ticks.

Key to Some Primary Groups of the Subclass Acari

1a. Without visible stigmata (breathing pores) posterior of coxae II; coxae not free, often fused with ventral body wall, forming coxosternal regions delimited by epimera, but sternum lacking; sejugal furrow or interval present, causing legs III to be farther from legs II than the latter are from legs I; number of legs sometimes reduced..................**Order ACARIFORMES, including PROSTIGMATA ASTIGMATA, and CRYPTOSTIGMATA**

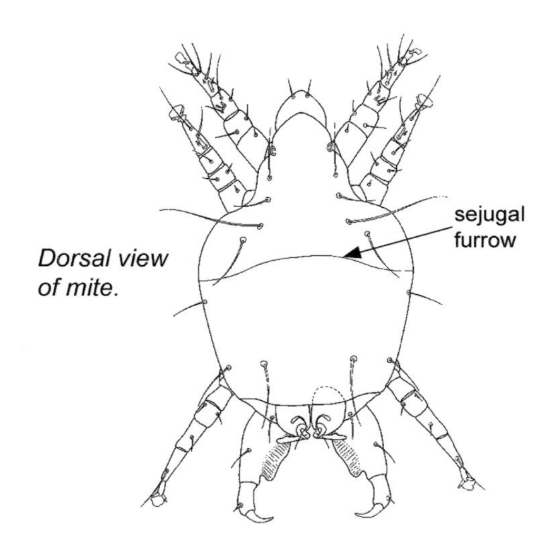

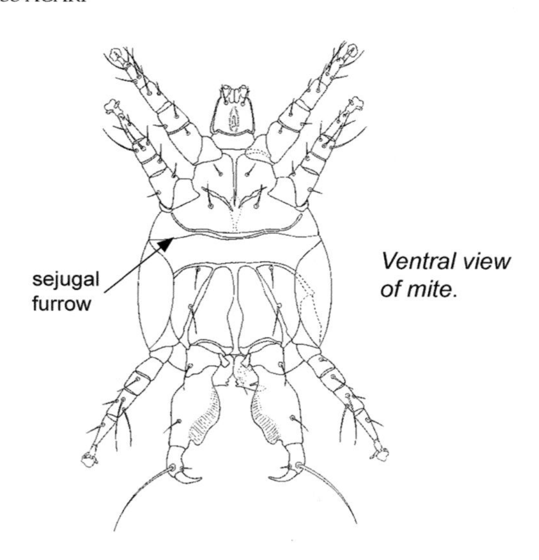

1b. With one to four pairs of dorsolateral or ventrolateral stigmata posterior of coxa II; coxae free, distinct; sternum nearly always present (lacking in Ixodida); sejugal furrow or interval lacking; distance between legs II and III not greater than between I and II and III and IV, all of which are present...**Order PARASITIFORMES 2**

2a. Pedipalpal tarsus without claws; hypostome modified into piercing organ with backward-directed teeth; stigmata present behind coxa IV or lateral above coxal intervals II to III, each surrounded by stigmal plate; sternum absent; Haller's organ present on upper side of tarsus I (large, bloodsucking acarines called ticks; length usually well over 2 mm)..**Suborder IXODIDA**

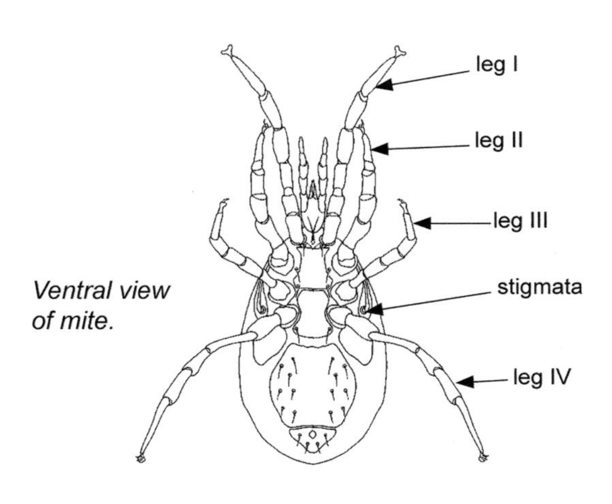

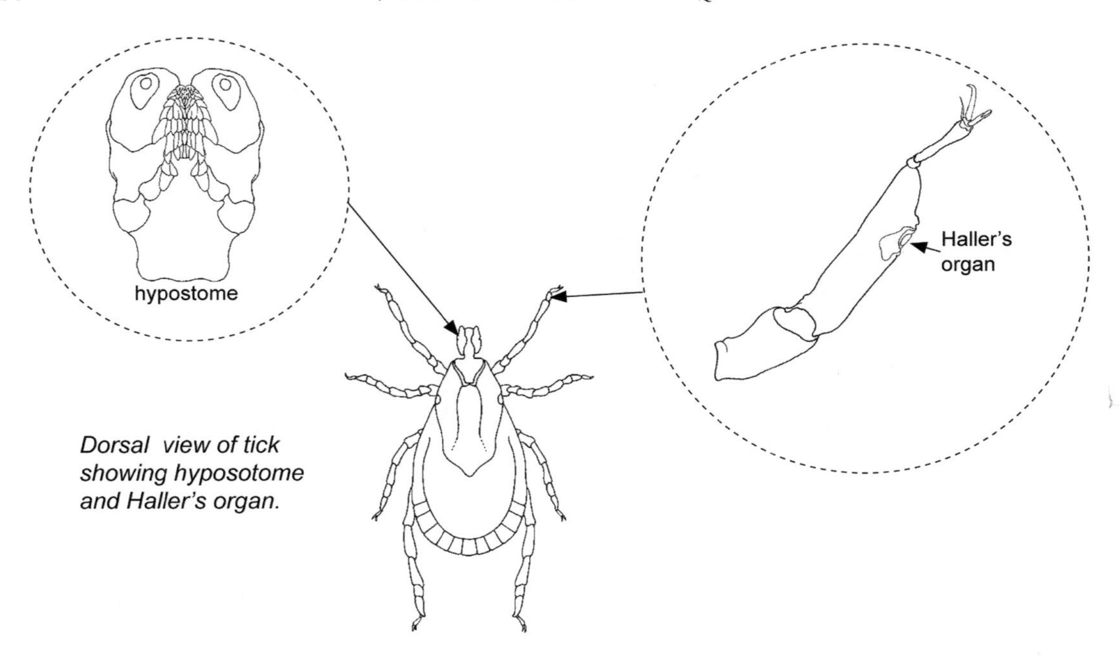

Dorsal view of tick showing hyposotome and Haller's organ.

2b. Pedipalpal tarsus with terminal, subterminal, or basal claw (simple or tined); hypostome serving only as floor of gnathosoma, without teeth; Haller's organ lacking (usually smaller mites)...........**Suborders MESOSTIGMATA, HOLOTHYRINA, and OPILIOACARIDA**

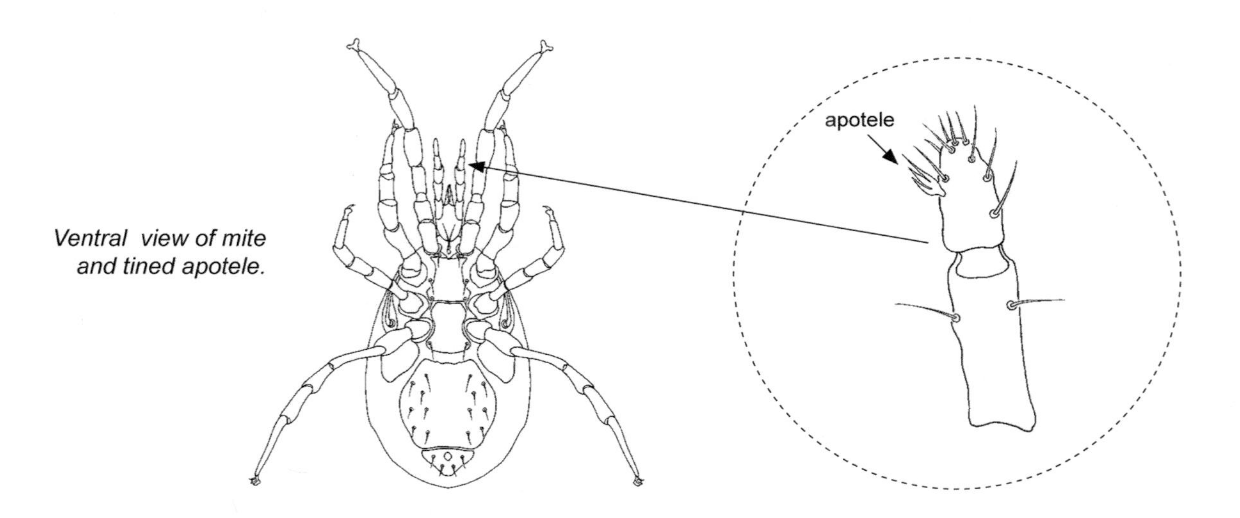

Ventral view of mite and tined apotele.

ILLUSTRATED DICHOTOMOUS KEY TO INSECT ORDERS

The following key includes suborders sometimes considered as orders. Any comprehensive survey of the larger orders of insects will include exceptions and aberrant forms that do not fit well in a brief key. Treating all aberrant forms is impractical, particularly when including known immature forms. Some orders appear more than once in the key. With specimens that do not key satisfactorily here, the reader should consult other references or an experienced systematic entomologist. This is especially true of pupal forms of Trichoptera, Mecoptera, and some Neuroptera. Interested students should obtain a copy of Borror et al. (1989) for comprehensive keys to the North American insect fauna.

1a. Wings present and well developed............2

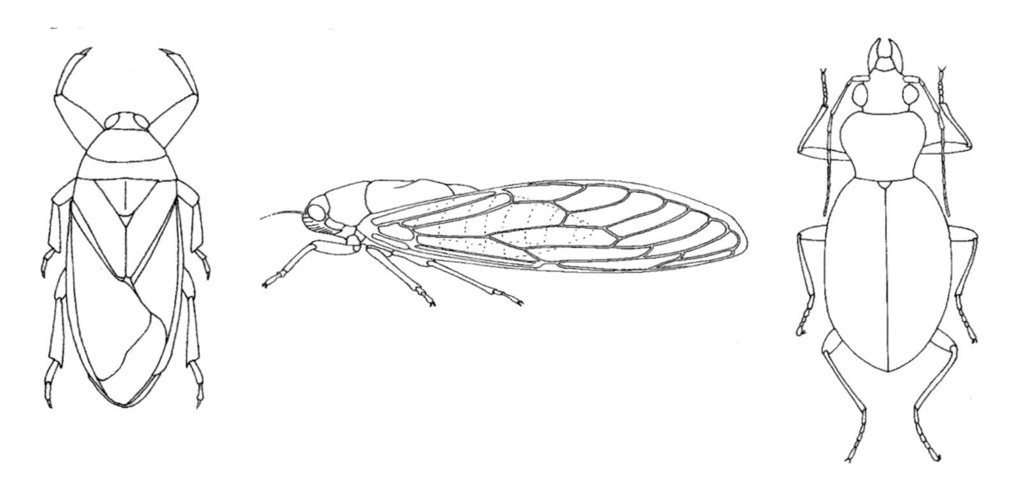

1b. Wings absent or not suitable for flight (wingless adults and immature stages) 34

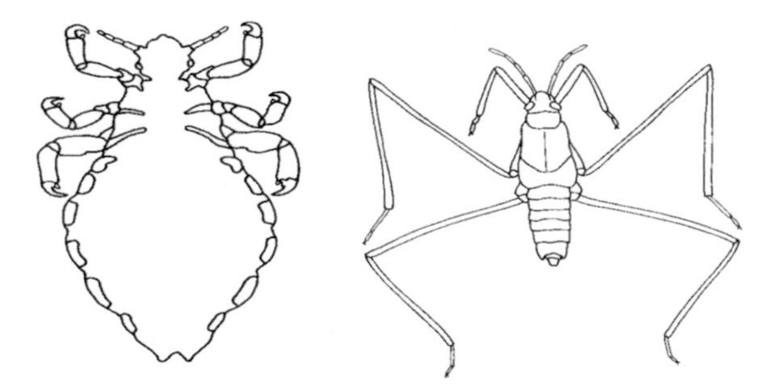

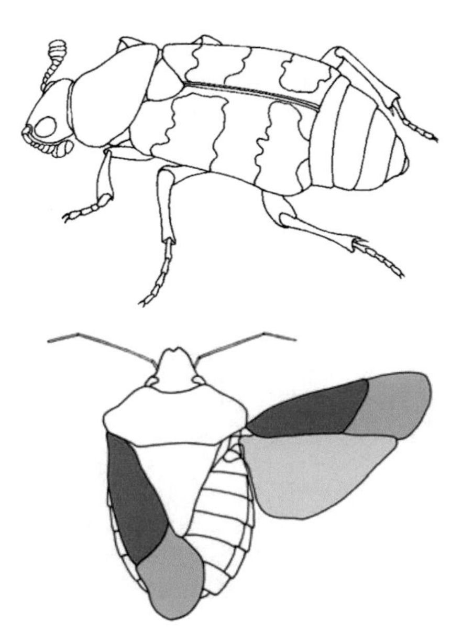

2a. Forewings at least partly horny, leathery, or strongly differing from completely membranous hindwings; hindwings sometimes absent...3

2b. Forewings completely membranous or membranous at base; hindwings variable, usually present and membranous, sometimes modified....................................12

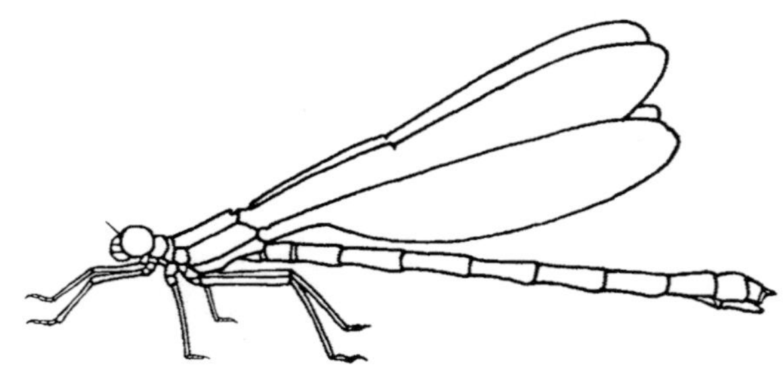

3a. Forewings (wing covers, elytra) uniformly horny or leathery and without apparent veins; hindwings, if present, folded lengthwise and crosswise, concealed beneath forewings when at rest; mouthparts including mandibles....................................4

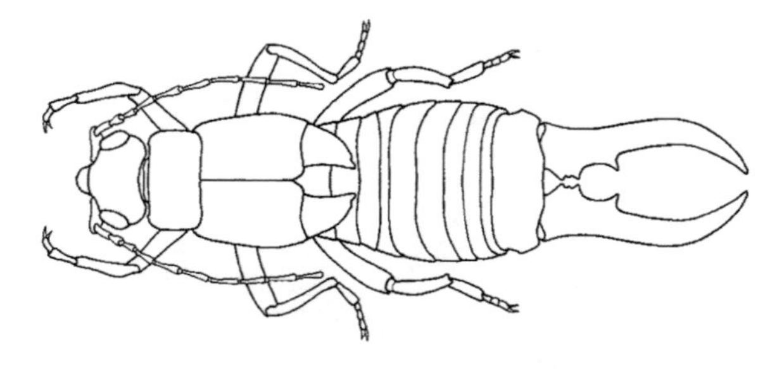

3b. Forewings (hemelytra, tegmina) with variable texture and with veins; hindwings not folded crosswise; mouthparts variable...5

4a. Apex of abdomen with heavy, forceps-like cerci; wings short and most of abdomen exposed; hindwings delicate, almost circular, radially folded......**DERMAPTERA**

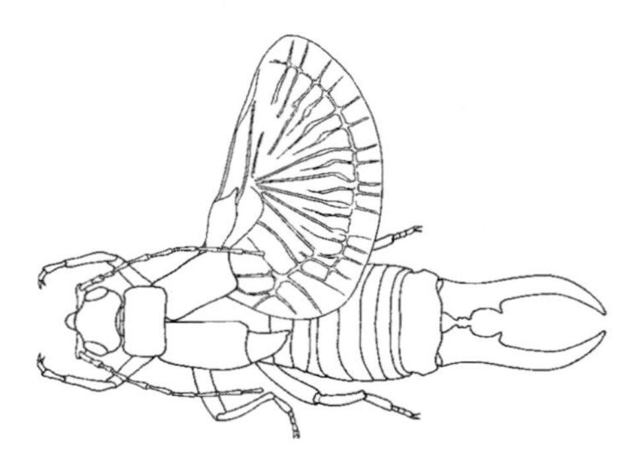

4b. Apex of abdomen without heavy, forceps-like cerci; wings usually long or covering most of abdomen; if forewings short, then hindwings elongate or absent...................
COLEOPTERA

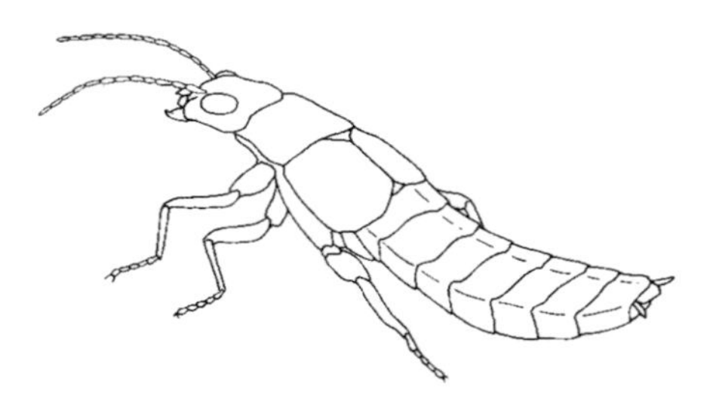

5a. Mouthparts adapted for sucking, forming a jointed or segmented beak..........................6

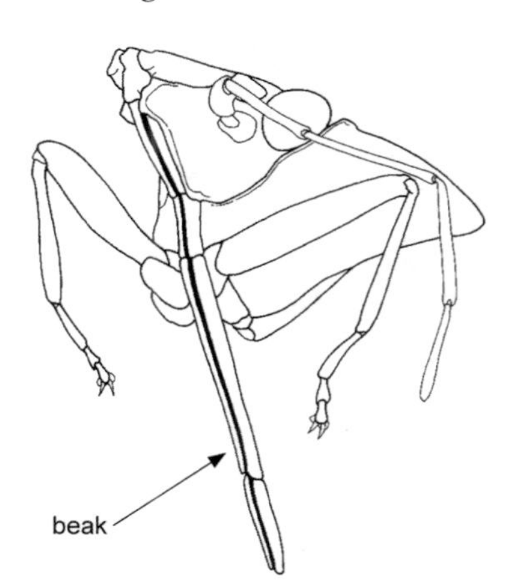

5b. Mouthparts adapted for chewing, forming mandibles that move laterally....................7

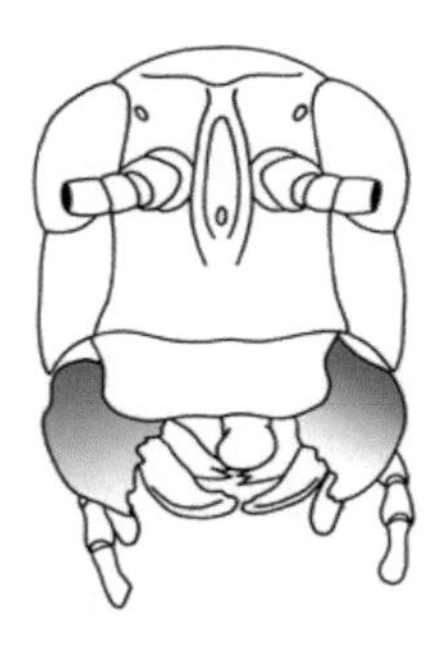

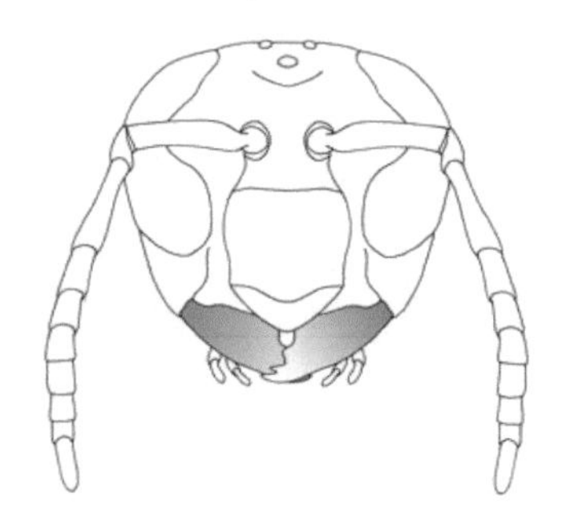

6a. Forewings hardened or leathery basally and membranous apically, usually held flat on abdomen with apices overlapping; beak arising from anterior part of head and usually capable of movement forward of head.................**HEMIPTERA**

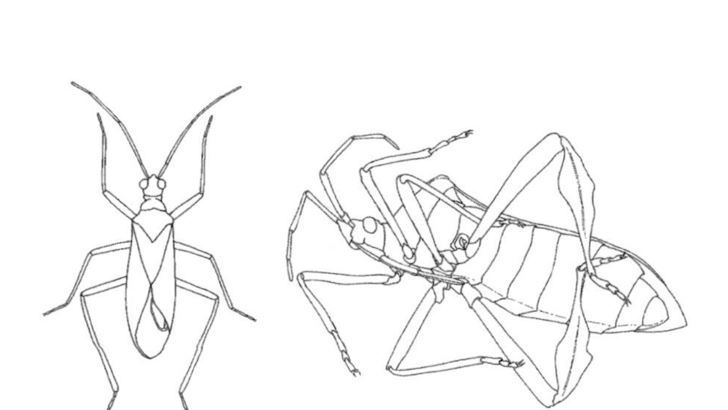

6b. Forewings completely or predominantly membranous, held more or less rooflike over abdomen; beak arising from posterior-ventral part of head and projecting downward and rearward between forelegs...............**HOMOPTERA**

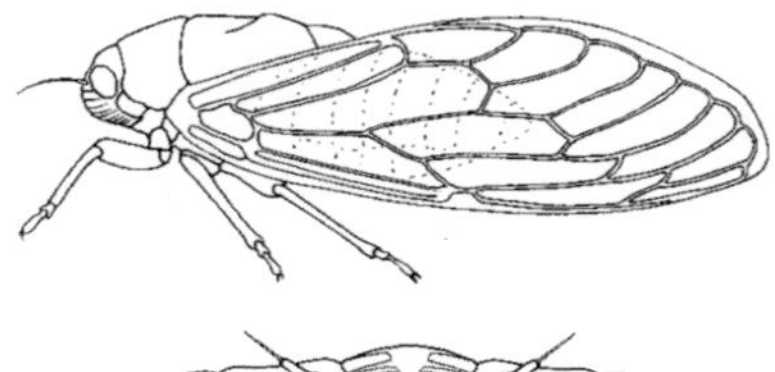

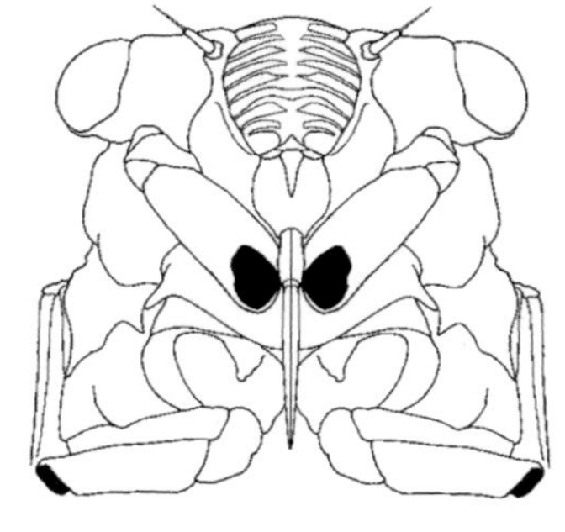

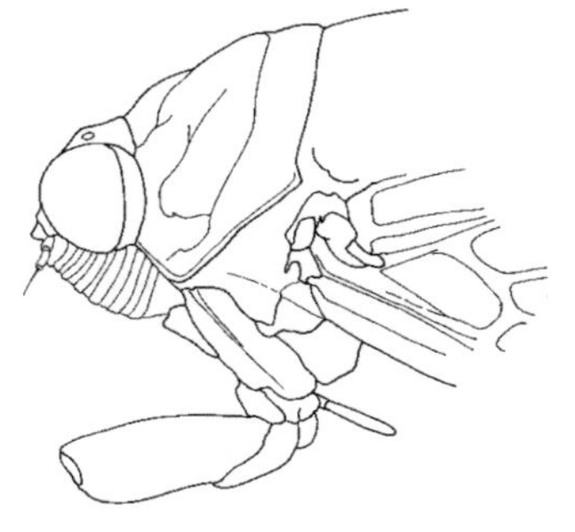

7a. Hind wings not folded at repose, similar to forewings; both pairs of wings with thickened, very short basal part separated from remainder of wing by suture; most of wing easily detached at suture; body soft and usually pale-colored; social insects living in colonies (see also couplet 32).
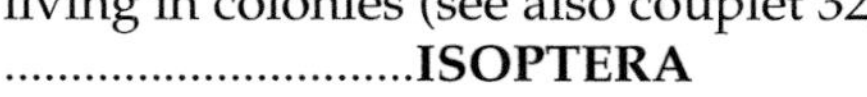
...............................**ISOPTERA**

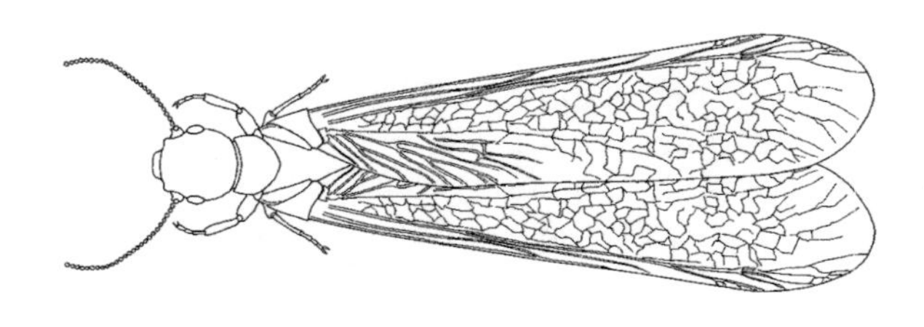

7b. Hindwings folded fanwise at repose, broader than forewings; both pairs of wings without basal suture for wing detachment; body hardness and coloration variable.......8

8a. Minute insects, usually less than 6 mm long; forewings small, clublike; antennae short, with few segments; parasites of other insects..........................**STREPSIPTERA** males

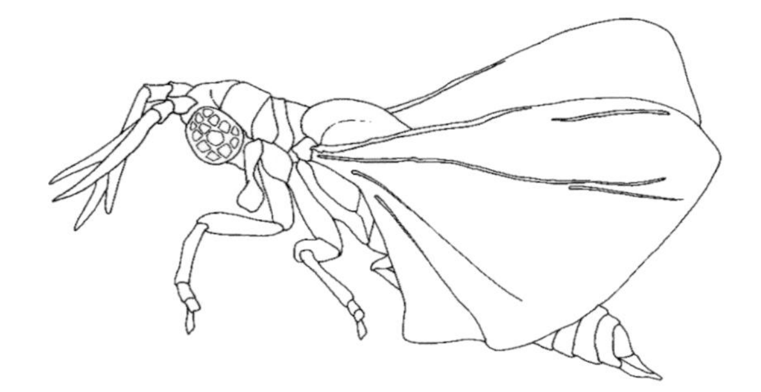

8b. Usually large or moderately large insects; forewings usually flat and long; antennae usually lengthened and slender, many-segmented (orthopteran orders).................9

9a. Hind femora enlarged, modified for jumping............**ORTHOPTERA**

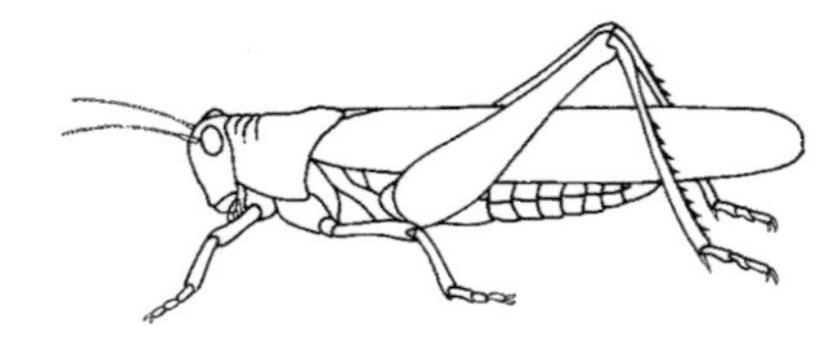

9b. Hind femora not enlarged, similar in size and shape to femora of other legs or not apparently adapted for jumping...............10

10a. Cerci short, not segmented; body usually elongate and slender, sticklike ....**PHASMIDA**

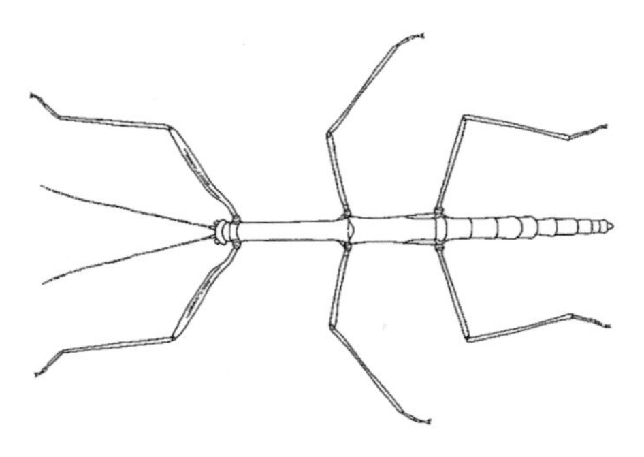

10b. Cerci long or short but segmented; body usually not elongate or sticklike..............11

11a. Forelegs raptorial with tibial and femoral spines modified for grasping prey; middle and hind legs adapted for walking ....**MANTODEA**

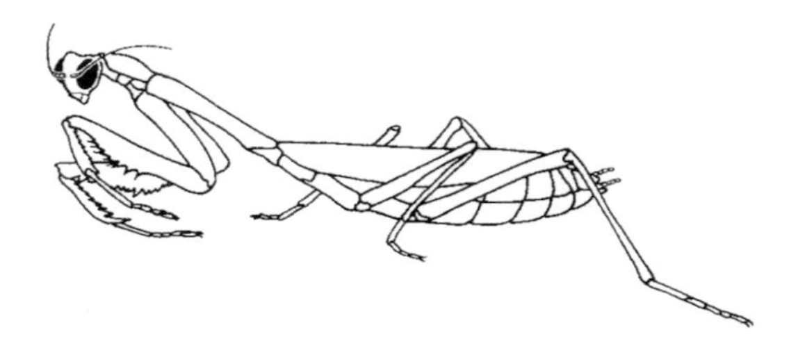

11b. All legs similar in size and shape, without spines adapted for grasping prey; all legs adapted for walking.. **BLATTODEA**, Section 6.9

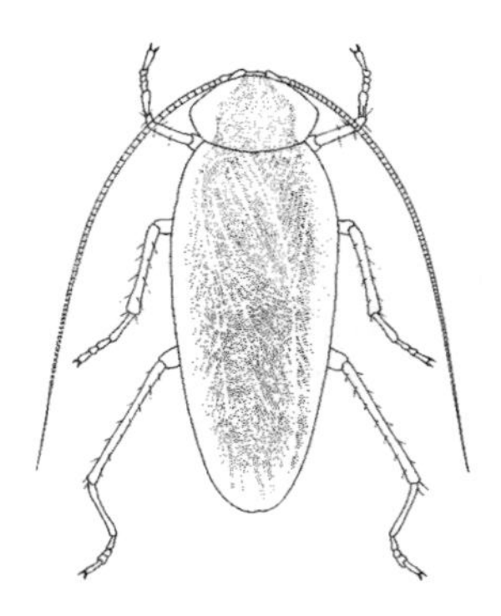

12a. Two well-developed wings present: forewings obviously used in flight; hind wings modified, sometimes small and clublike..13

12b. Four wings present: forewings obviously used in flight, hind wings sometimes small but flat or straplike and not clublike..15

13a. Mouthparts forming a piercing-sucking or lapping proboscis, rarely rudimentary or absent; hind wings replaced by clublike halteres; abdomen without tail filaments... ....**DIPTERA**

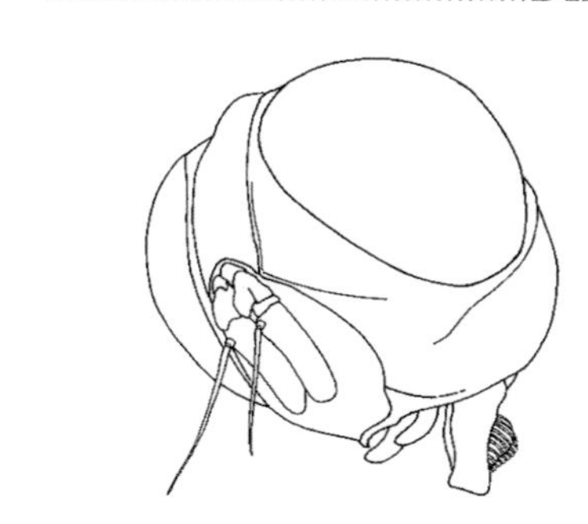

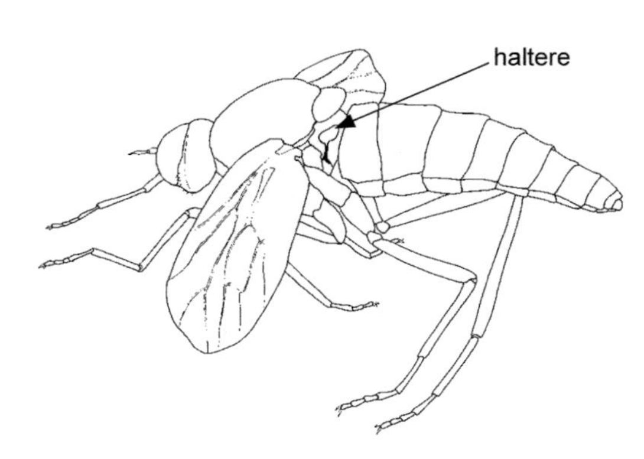

13b. Mouthparts not functional; hind wings not formed into clublike halteres; abdomen with tail filaments..................14

14a. Hind wings not halterelike; antennae inconspicuous, with small scape and pedicel, flagellum bristlelike; forewings with numerous crossveins (few mayflies)................**EPHEMEROPTERA**

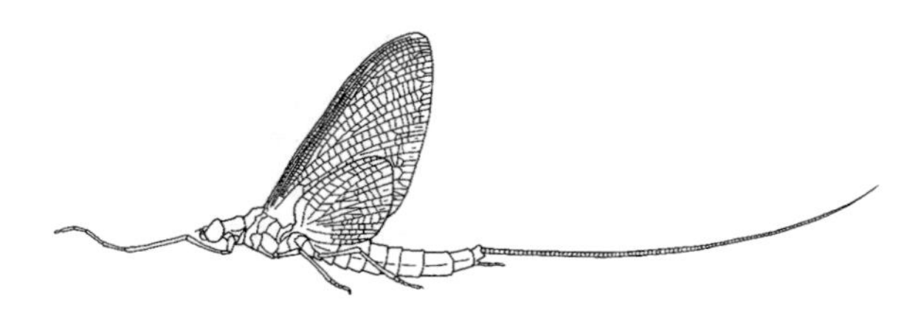

14b. Hind wings reduced to halterlike structures; antennae conspicuous, flagellum not bristlelike; forewings with venation apparently reduced to one forked vein (male scale insects) ....**HOMOPTERA**

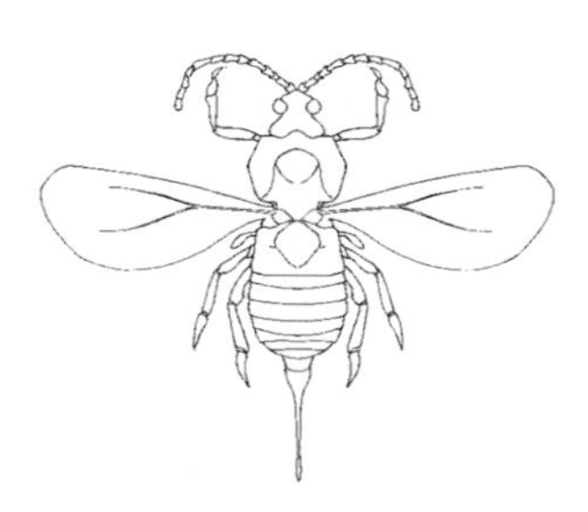

15a. Wings long, narrow, almost veinless, with long marginal fringe; tarsi one- or two-segmented, with apex swollen; mouthparts conical, adapted for piercing and sucking plant tissues (minute insects)..................................... **THYSANOPTERA**

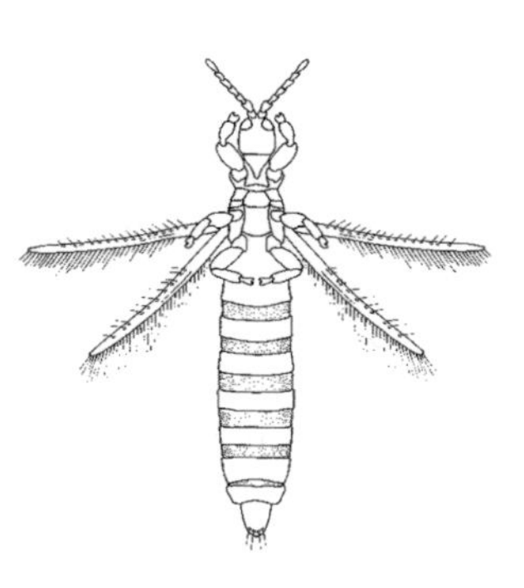

15b. Wings relatively broad, veins usually conspicuous and at least one crossvein present, marginal fringe absent or not longer than width of wing; tarsi with more than two segments and apex not swollen; mouthparts variable................ 16

16a. Wings, legs, and body at least partially covered with elongate, flattened scales (setae) and often with hairlike setae; wings hyaline (transparent) under color pattern formed by scales; mouthparts tonguelike (rarely rudimentary), forming a helically coiled tube; small mandibles present only in a few families of small moths with wingspread not over 12 mm................**LEPIDOPTERA**

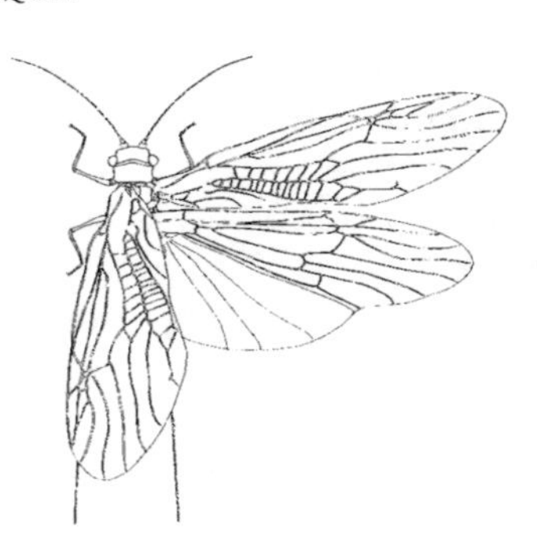

17b. Hind wings usually not larger than forewings, without plaited anal area; antennae often inconspicuous, bristlelike..20

18a. Tarsi three-segmented; cerci well developed, usually long and many-segmented.................................
PLECOPTERA

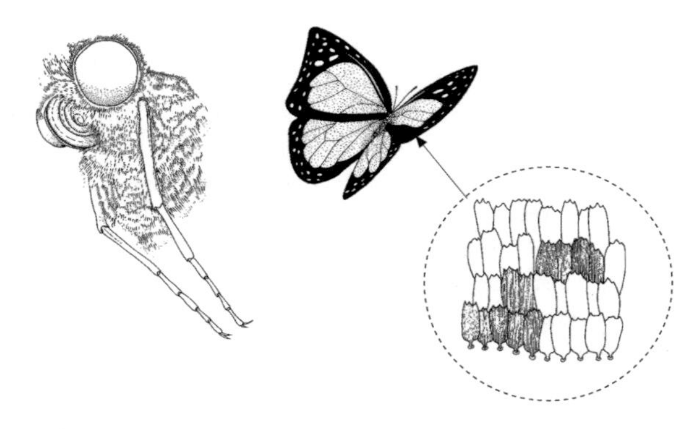

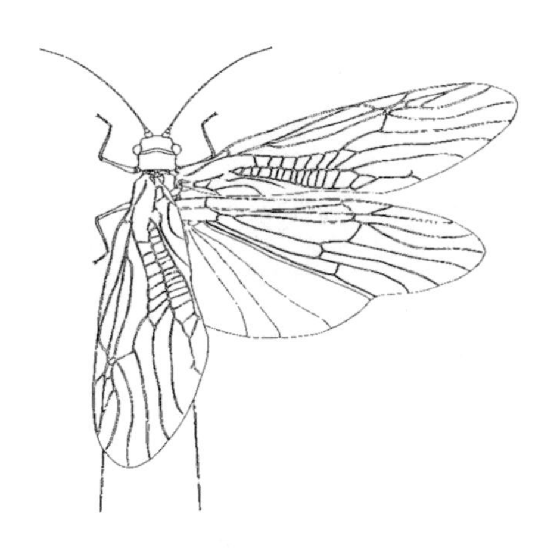

16b. Wings, legs, and body not covered with flattened scales, although a few scales sometimes present; color pattern of wing involving wing membrane or hairlike setae or both; mandibles typically present..........17

17a. Hind wings usually larger than forewings, with broad anal area, plaited when wings folded; antennae conspicuous..................18

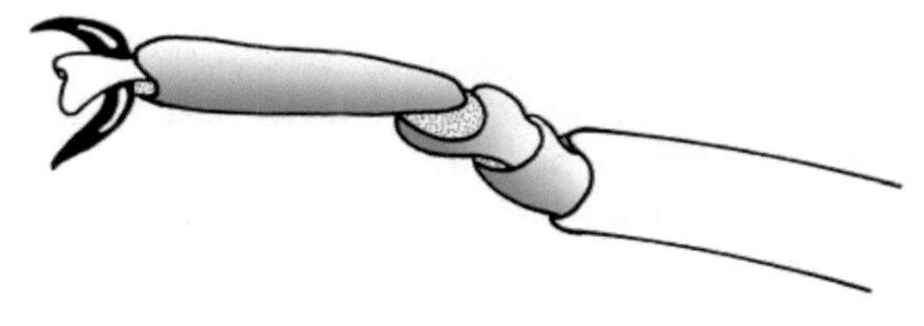

18b. Tarsi five-segmented; cerci not well developed...19

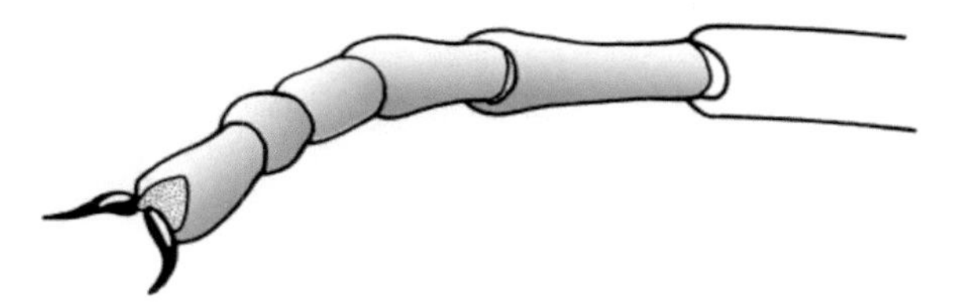

19a. Wings with several subcostal crossveins, surface without hairlike setae or scales ....**NEUROPTERA**, Suborder **MEGALOPTERA**

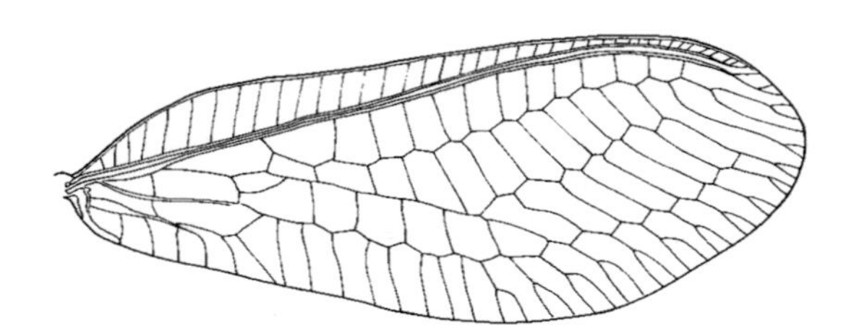

19b. Wings without subcostal crossveins, surface with hairlike setae or scales. **TRICHOPTERA**

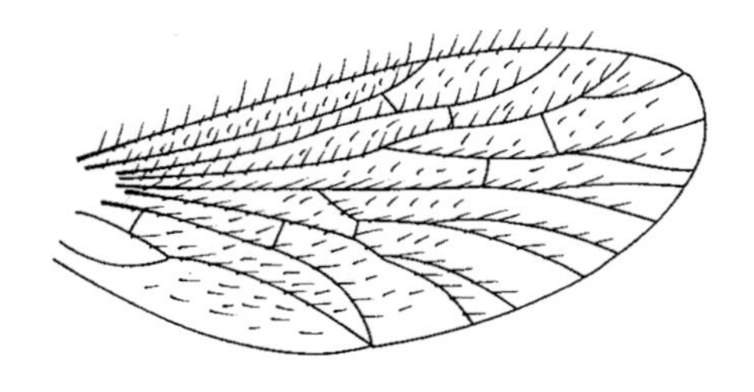

20a. Antennae short, bristlelike; wings with numerous crossveins forming a network; mouthparts with mandibles near eyes...21

20b. Antennae large or wings with a few crossveins or mouthparts near beak..........22

21a. Hind wings much smaller than forewings; abdomen with long tail filaments..................**EPHEMEROPTERA**

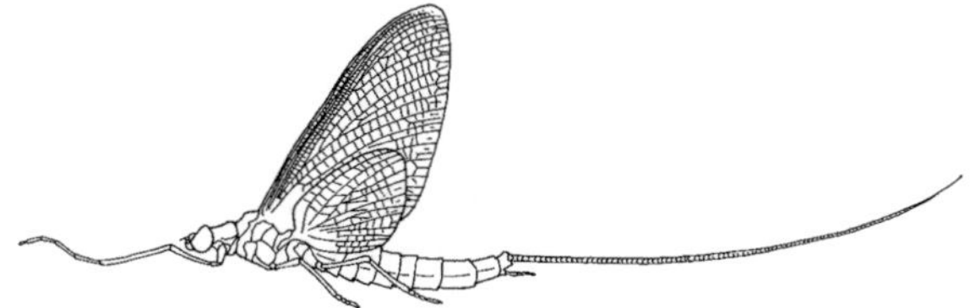

21b. Hind wings about as large as forewings; abdomen without long tail filaments...............**ODONATA**

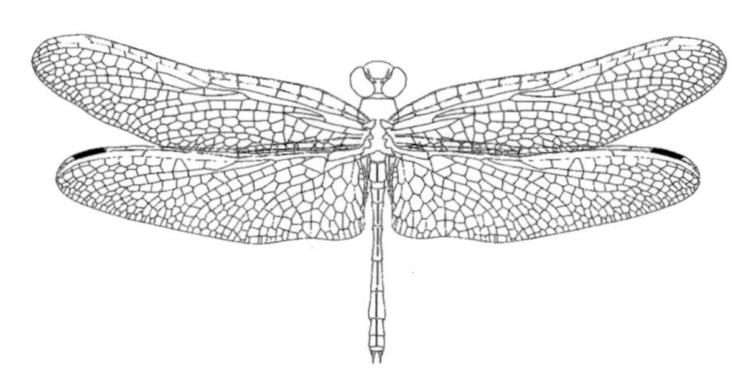

22a. Head beneath eyes beaklike with mandibles at apex; hind wings not folded; wings usually with color pattern and numerous crossveins; male genitalia usually swollen, turned forward, and with strong pair of forceps.. **MECOPTERA**

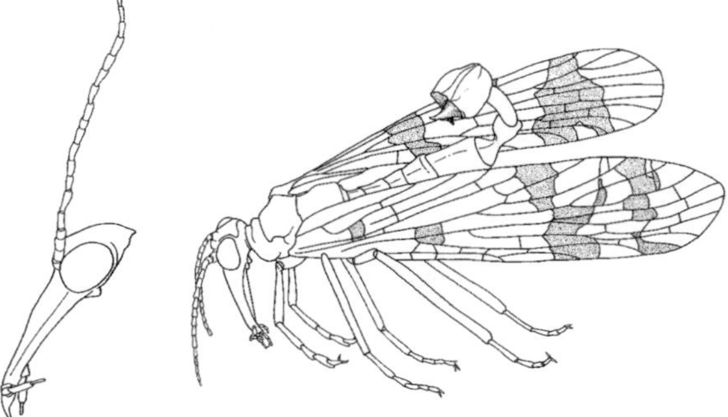

22b. Head beneath eyes not beaklike or formed into a conical tube; wing characters variable; male genitalia without conspicuous forceps................................23

23a. Mouthparts sometimes absent, when present consisting of proboscis without chewing mandibles; cerci absent; wings with few crossveins..................................24

23b. Mouthparts with mandibles adapted for chewing; cerci sometimes present; wing venation variable......................................26

24a. Wings covered with scales that form color pattern; antennae with many segments; mouthparts (when present) consisting of helically coiled haustellum (tongue)... **LEPIDOPTERA**

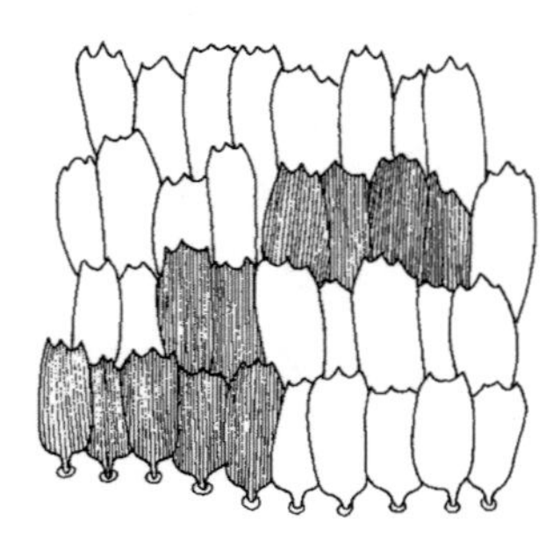

24b. Wings not covered with scales; antennae with few segments; mouthparts forming a segmented piercing beak.......................25

25a. Beak arising from anterior part of head................**HEMIPTERA**

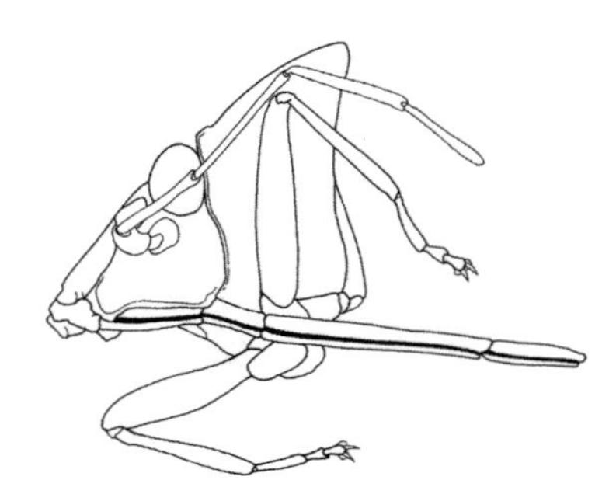

25b. Beak arising from posterior part of head, extended downward between forelegs..**HOMOPTERA**

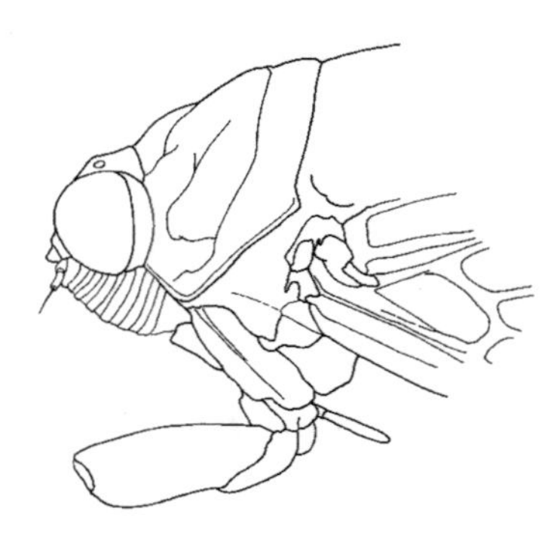

26a. Body and wings covered with whitish powder; wings bordered anteriorly by very narrow cell without row of crossveins; insects less than 5 mm long...................**NEUROPTERA, Suborder PLANIPENNIA (Coniopterygidae)**

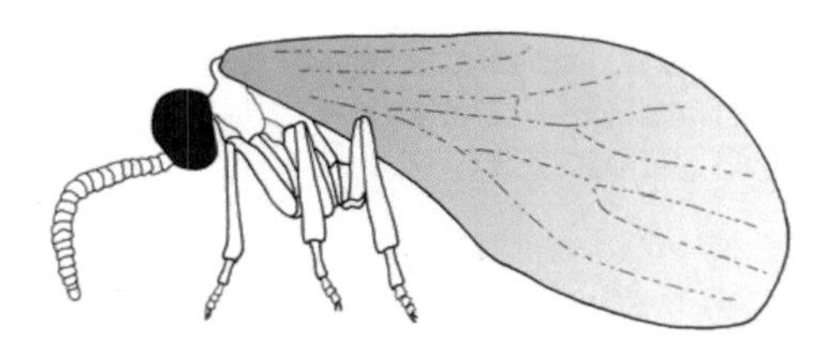

26b. Body and wings not covered with whitish powder, other characters differing..........27

27a. Tarsi five-segmented.................................28

27b. Tarsi with four or fewer segments...........31

28a. Prothorax typically fused with mesothorax; forewings with fewer than 20 cells; hind wings smaller than forewings; abdomen usually constricted at base, forming a petiole; sting or appendicular ovipositor present .... **HYMENOPTERA**

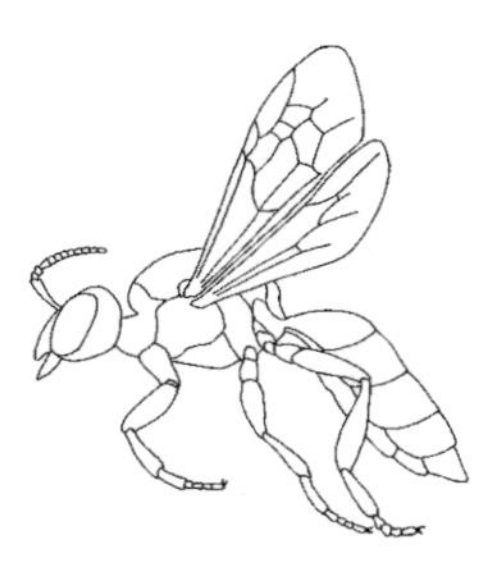

28b. Prothorax more or less free, sometimes long; forewings with more than 20 cells; forewings and hind wings approximately equal in size; abdomen not constricted to form petiole; sting or appendicular ovipositor not present...............................29

29a. Prothorax cylindrical, much longer than head; forelegs simlar to other legs, not enlarged.......................................
NEUROPTRA, Suborder RAPHIDIODEA

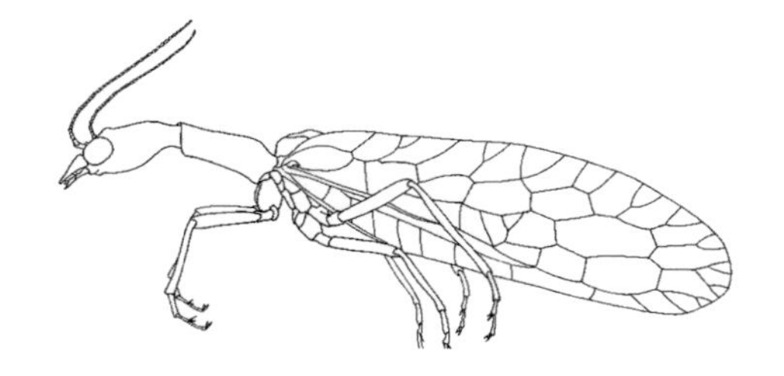

29b. Prothorax not longer than head; if longer, then forelegs enlarged and adapted for grasping prey...30

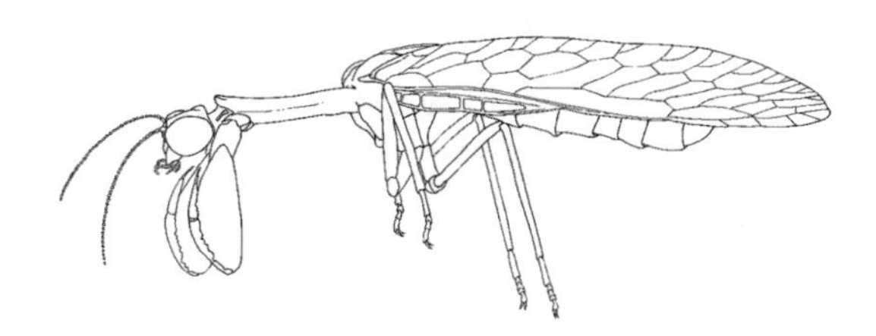

30a. Costal cell with many crossveins ....**NEUROPTERA, Suborder PLANIPENNIA**

30b. Costal cell without many crossveins ......**MECOPTERA,** Section 6.26.

31a. Wings equal in size or hind wings rarely larger; tarsi three- or four-segmented 32

31b. Hind wings smaller than forewings; tarsi two- or three-segmented........................33

32a. Forebasitarsi not swollen; wings dehiscent (see also couplet 7)................................ ....**ISOPTERA**

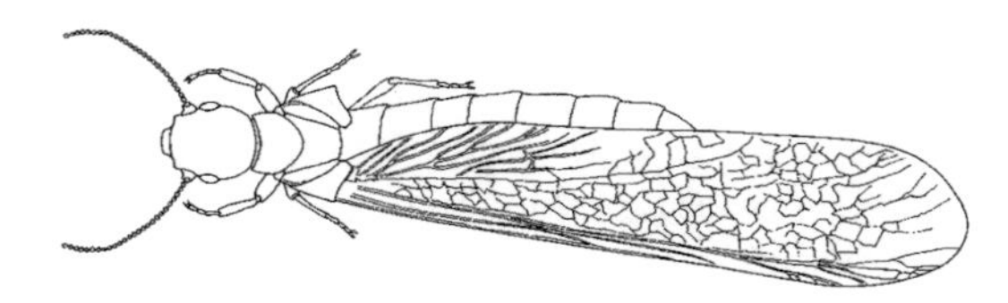

32b. Forbasitarsi swollen....................................... ....**EMBIIDINA**

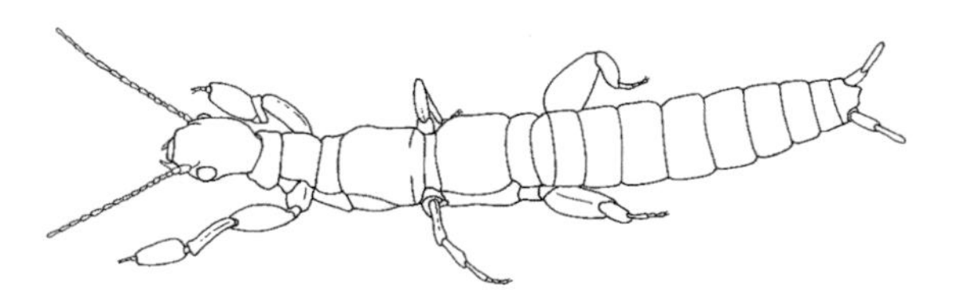

33a. Cerci absent; wings remain attached to body; antennae slender, with 13 or more segments (insects commonly collected).........................**PSOCOPTERA**

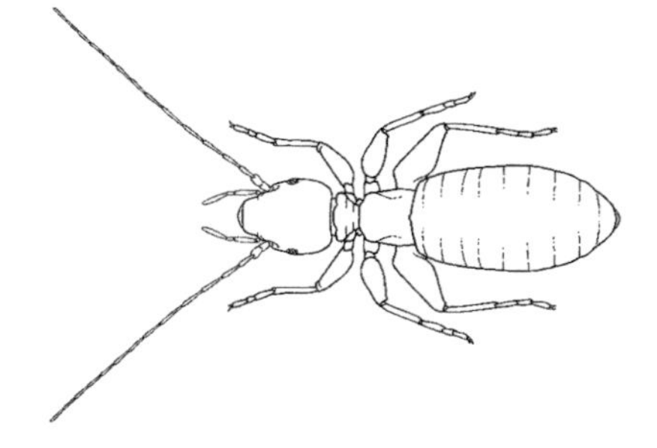

33b. Cerci present, although short, ending in bristle; wings shed; antennae with nine beadlike segments (insects seldom collected)...............**ZORAPTERA**

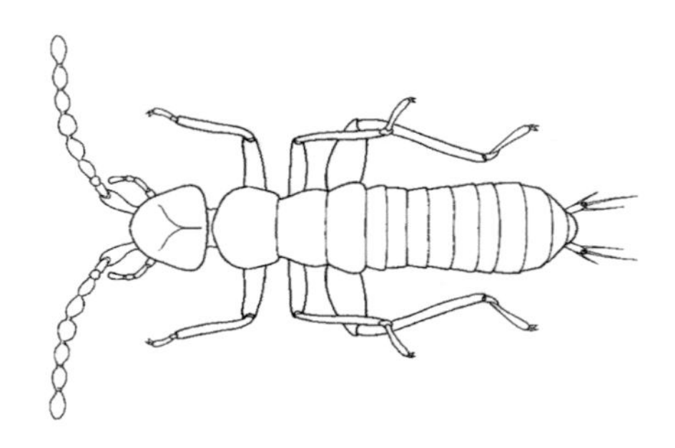

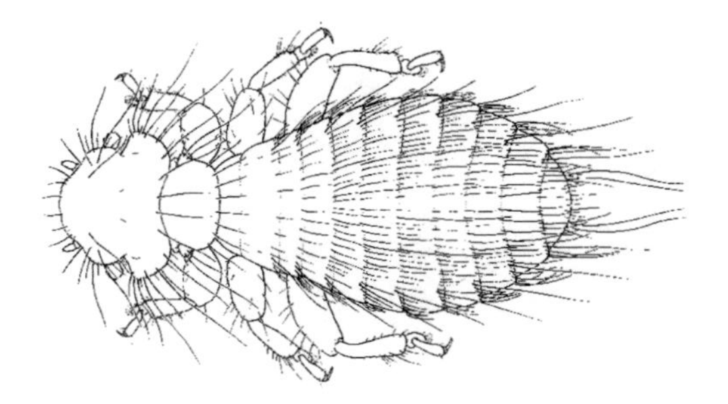

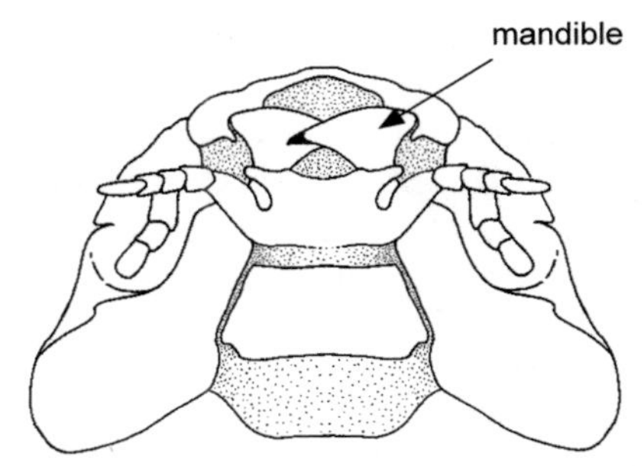

34a. Body with more or less distinct head, thorax, and abdomen; legs jointed, enabling animal to move........................35

34b. Without distinctly separate body parts, or without legs, or not able to move..........78

35a. Parasites of warm-blooded animals..36

35b. Not parasites of warm-blooded animals...40

36a. Adult body strongly compressed laterally; mouth forming a short, sharp, downward-projecting beak; powerful jumping insects; adults found on vertebrate hosts; immatures found in nest of host.................................... **SIPHONAPTERA**

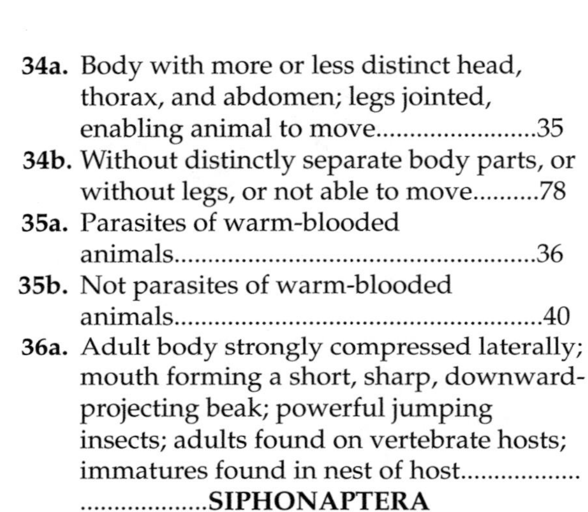

36b. Adult body not compressed laterally; mouthparts variable; not powerful jumping insects; habitats mandible variable...........37

37a. Mouthparts with mandibles adapted for chewing, directed forward; insects generally oval in outline with more or less triangular head; parasites of birds and mammals.....**MALLOPHAGA**

37b. Mouthparts forming a beak or otherwise modified; body shape variable; feeding habits diverse..38

38a. Antennae inserted in pits and not visible when viewed from above (also maggot-shaped larvae without antennae)................................**DIPTERA**

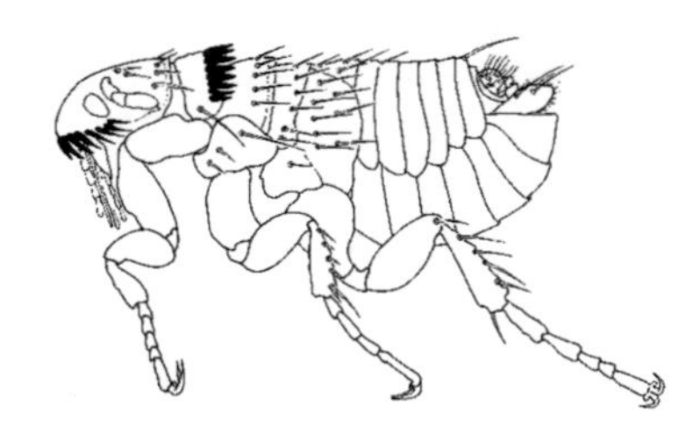

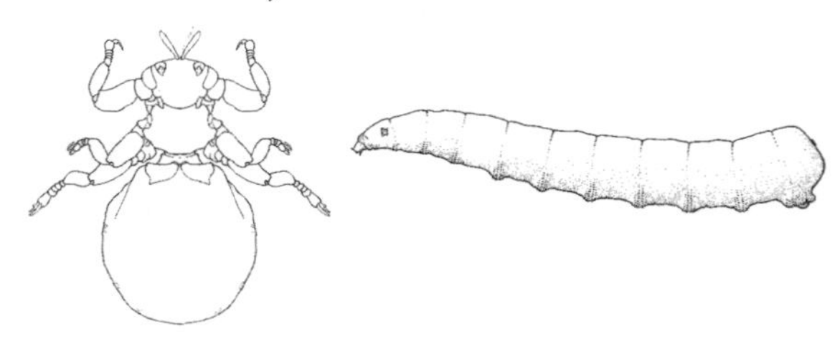

38b. Antennae present, short but not in pits..39

39a. Beak not jointed; tarsi forming a hook for grasping hairs of host; parasites remaining on host.............**ANOPLURA**

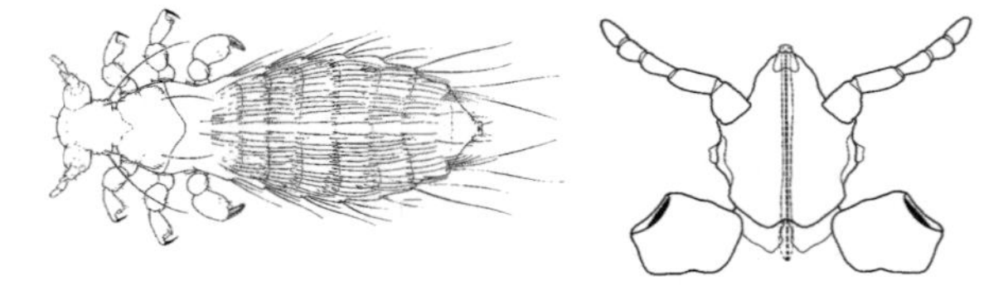

39b. Beak jointed; tarsi not hooked; parasites not remaining on host (bed bugs and related insects).............**HEMIPTERA**

40a. Aquatic insects, usually breathing by gills; larval and some pupal forms..................41

40b. Terrestrial insects, breathing by spiracles or rarely without breathing organs........49

41a. Mouth forming a strong, pointed, downward-curved beak...............Immature **HEMIPTERA**

41b. Mouth with mandibles...........................42

42a. Mandibles extending straight forward, united with maxillae to form piercing jaws...............Some larval **NEUROPTERA**

42b. Mandibles moving laterally, forming biting jaws..43

43a. Immature insects living within cases formed of sand, pebbles, leaves, twigs, etc.; usually with external tracheae serving as gills........................Some larval **TRICHOPTERA**

43b. Immature insects not living within cases... 44

44 a. Abdomen with lateral organs serving as gills (a few larval Trichoptera and Coleoptera also key here).........................45

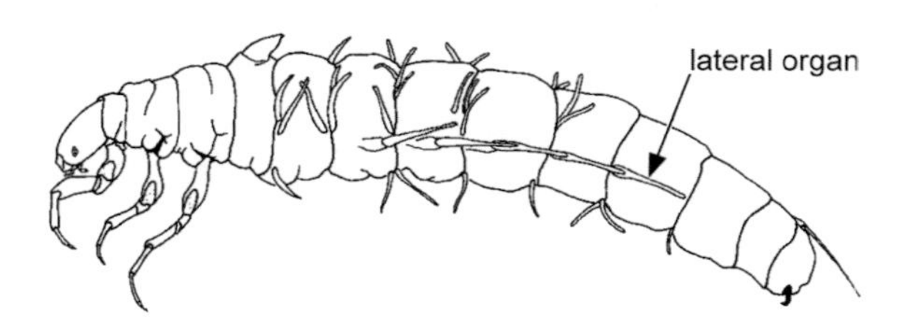

44b. Abdomen without external gills (some larval Trichoptera also key here)..............46

45a. Abdomen with two or three long tail filaments... Immature **EPHEMEROPTERA**

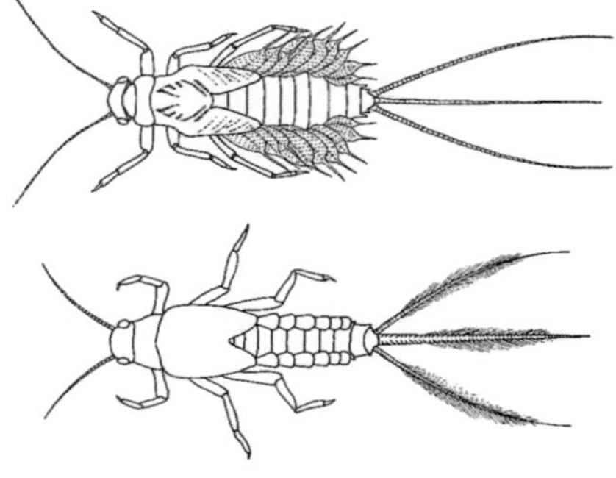

45b. Abdomen with short end processes (larvae of some Trichoptera key here) ....Larval **NEUROPTERA,** Suborder **MEGALOPTERA**

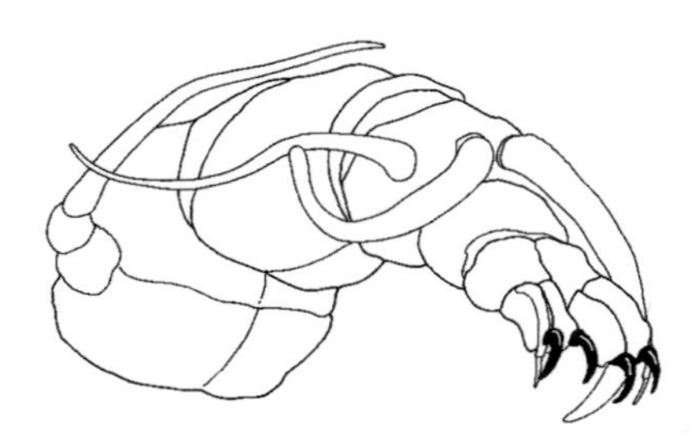

46a. Lower lip (labium) folded backward, extensible, and furnished with a pair of jawlike hooks..........Immature **ODONATA**

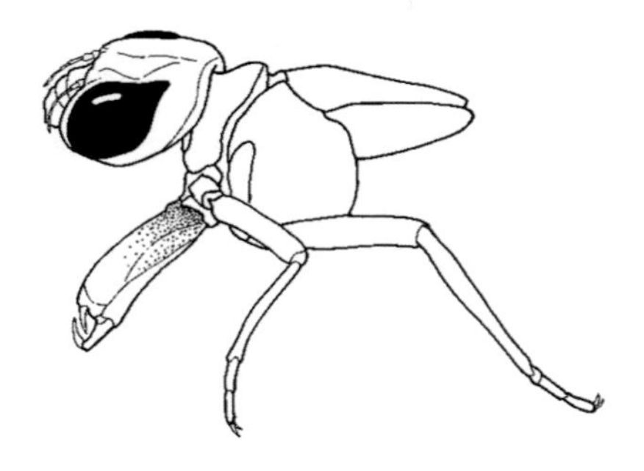

46b. Labium not so constructed....................47

47a. Abdomen with nonjointed false legs (pseudopods) arranged in pairs on several segments..........Few larval **LEPIDOPTERA**

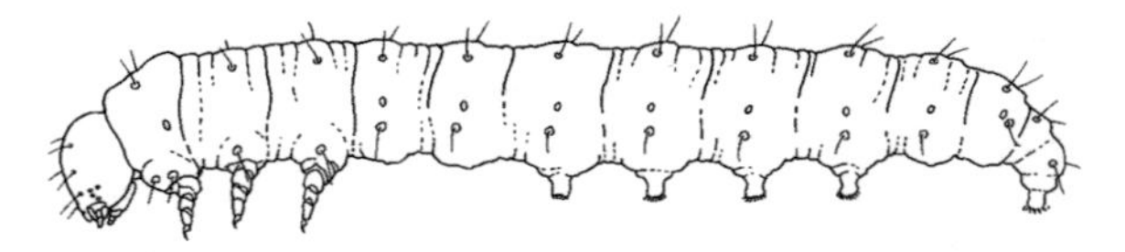

47b. Abdomen without pseudopods..............48

48a. Thorax in three loosely united divisions; antennae and tail filaments long and slender.................Larval **PLECOPTERA**

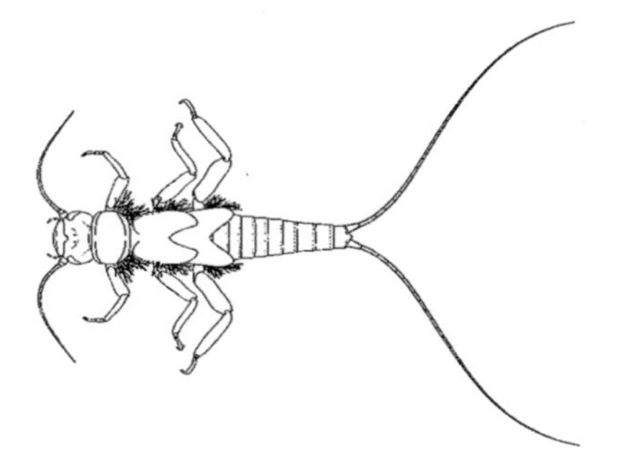

48b. Thoracic divisions without constrictions; antennae and tail filaments short (larvae of some aquatic Diptera and Trichoptera also key here)...................Larval **COLEOPTERA**

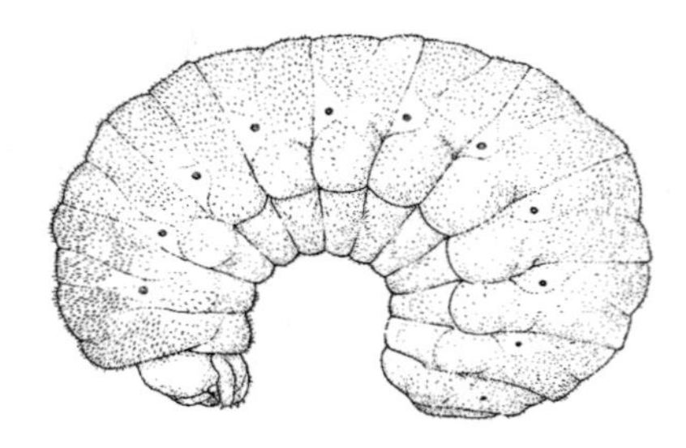

49a. Mouthparts retracted into head and difficult to see; antennae sometimes absent; venter of abdomen with appendages; very delicate, small to minute animals.....................................50

49b. Mouthparts external, conspicuous; antennae always present; venter of abdomen rarely with appendages; body typically larger than a few millimeters...52

50a. Head pear-shaped; antennae absent; abdomen without long cerci, pincers, jumping apparatus, or basal ventral "sucker" .... **PROTURA**

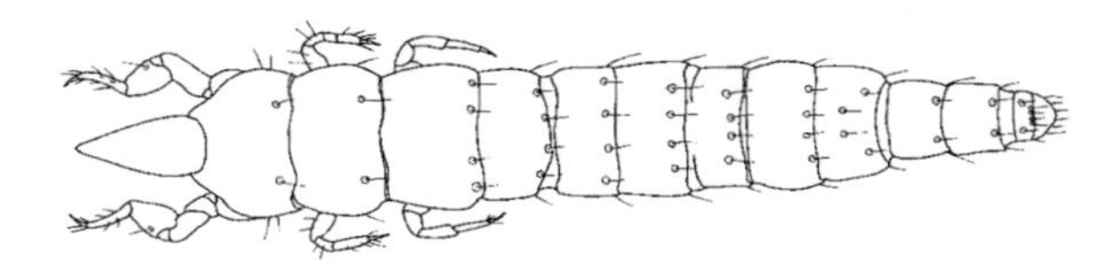

50b. Head usually not pear-shaped, antennae present; abdomen with long cerci, pincers, or basal ventral "sucker"..........................51

51a. Abdomen with six or fewer segments, with forked "sucker" at base below and usually with conspicuous jumping apparatus near apex; abdomen lacking conspicuous long cerci or pincers; eyes usually present though often reduced in size................**COLLEMBOLA**

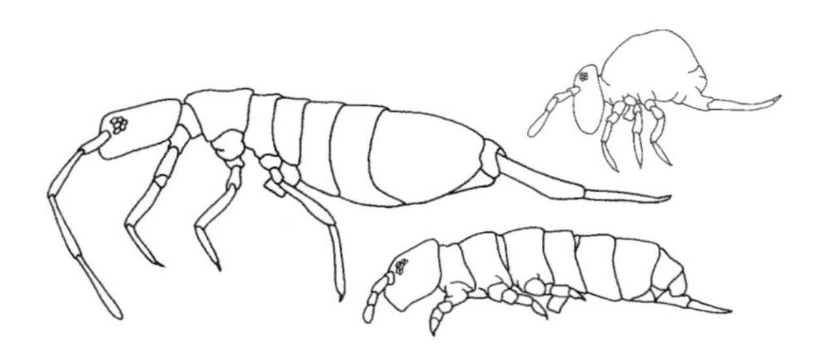

51b. Abdomen with more than eight evident segments but lacking a "sucker" at base; abdomen ending in long, many-segmented cerci or strong pincers; eyes and ocelli absent....................... ....**DIPLURA**

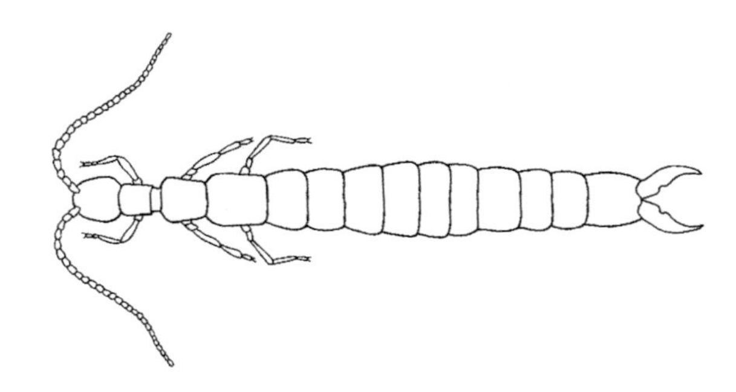

52a. Mouthparts with mandibles adapted for chewing...53

52b. Mouthparts in form of proboscis adapted for sucking...74

53a. Body usually covered with scales; abdomen with three prominent tail filaments and at least two pairs of ventral appendages (styli)......................**THYSANURA**

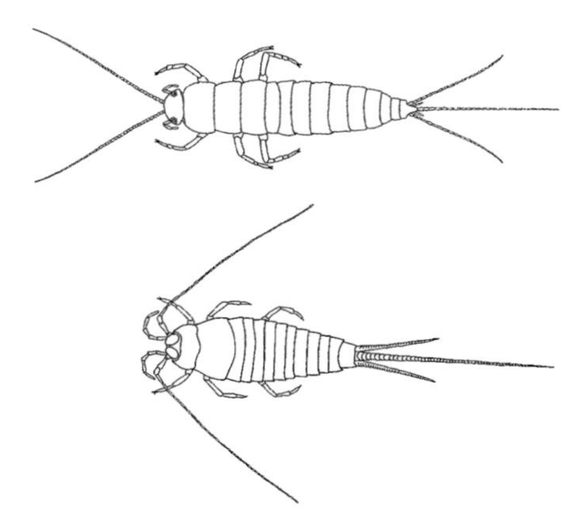

53b. Body not covered with scales; abdomen without three tail filaments or ventral styli 54

54a. Abdomen bearing ventral pairs of false, nonjointed legs (pseudopods) that differ from true legs on thorax; thorax and abdomen not distinctly separated; body caterpillar-like; larval forms...55

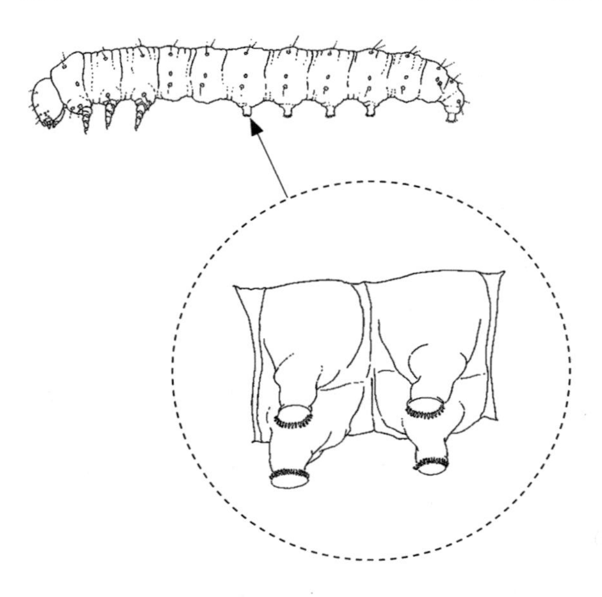

54b. Ventral surface of abdomen without legs or pseudopods; other characters different...57

55a. Abdomen with five or fewer pairs of pseudopods, none on first, second, or seventh segments; pseudopods tipped with many tiny hooklets and rarely present on second and seventh segments..................... Most larval **LEPIDOPTERA**

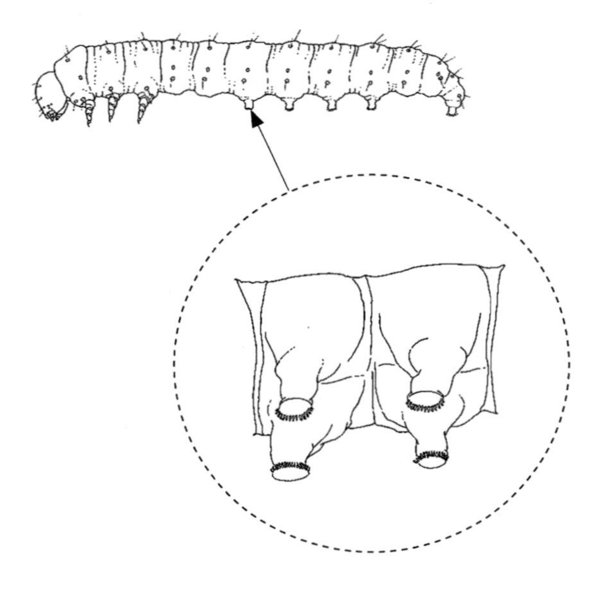

55b. Abdomen with 6–10 pairs of pseudopods, not tipped with tiny hooks; one pair of pseudopods on second segment.............56

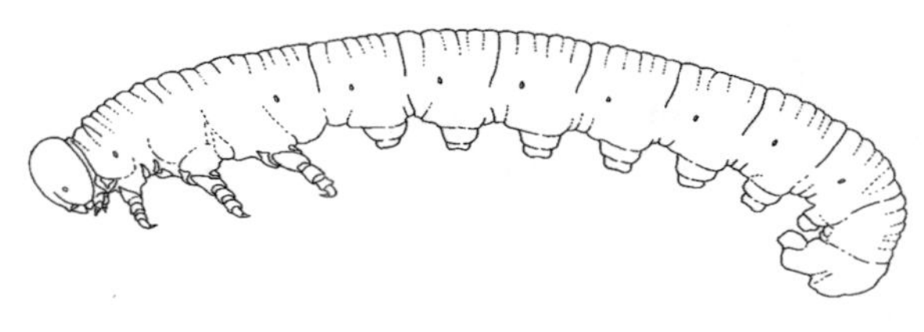

56a. Head with single ocellus (stemmatum) on each side............................Some larval **HYMENOPTERA**

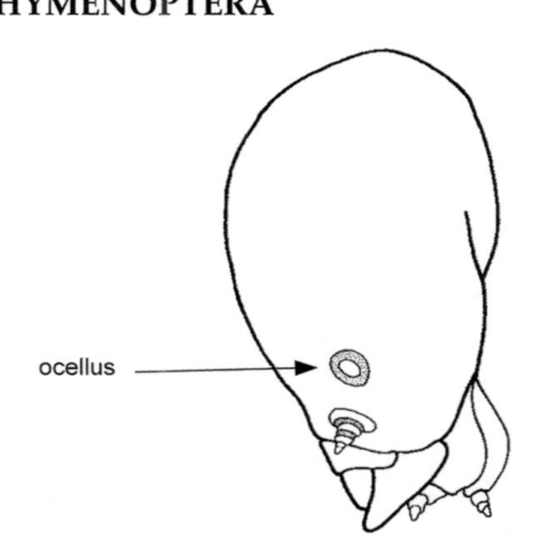

56b. Head with several ocelli (stemmata) on each side.........Larval **MECOPTERA**

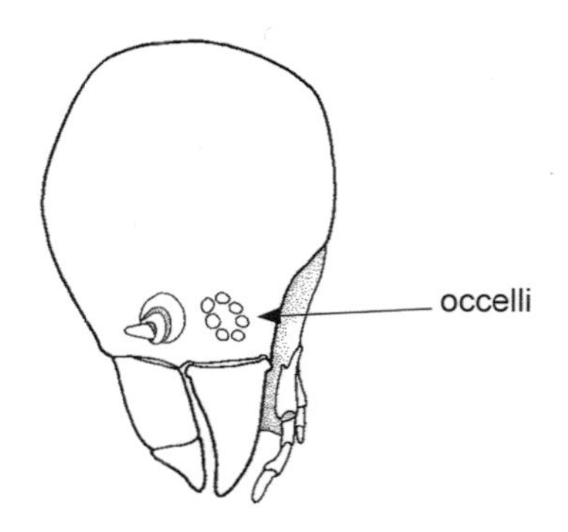

57a. Antennae long and distinct; adult or immature forms..58

57b. Antennae short; larval forms...................71

58a. Abdomen ending in strong pincerlike forceps; prothorax free.........**DERMAPTERA**

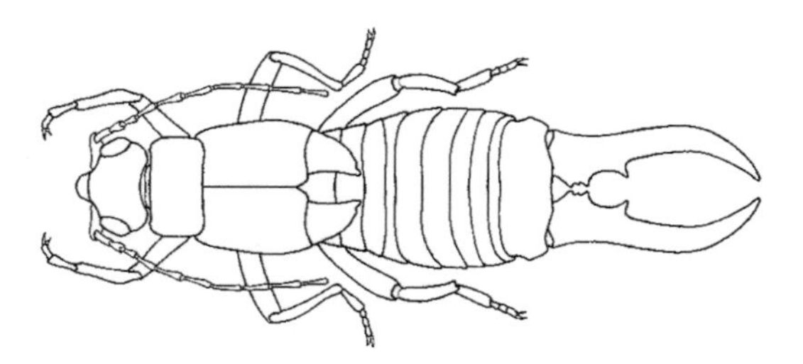

58b. Abdomen not ending in forceps; prothorax free or fused to mesothorax....59

59a. Adult abdomen strongly constricted at base, forming a petiole; prothorax fused with mesot horax...........................**HYMENOPTERA**

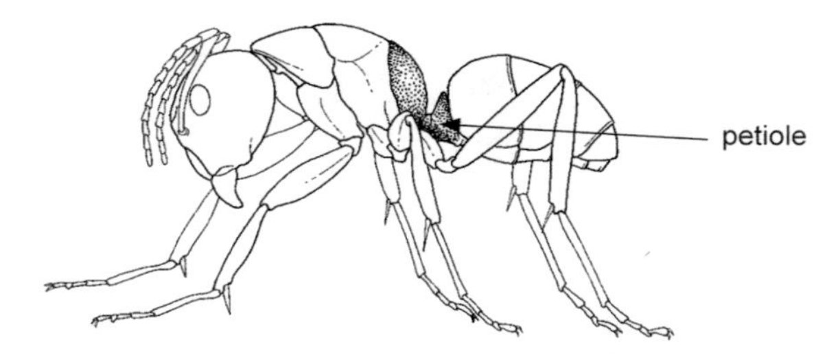

59b. Adult abdomen not strongly constricted at base, broadly joined to thorax.............60

60a. Head produced into beak with mandibles at apex.............**MECOPTERA**

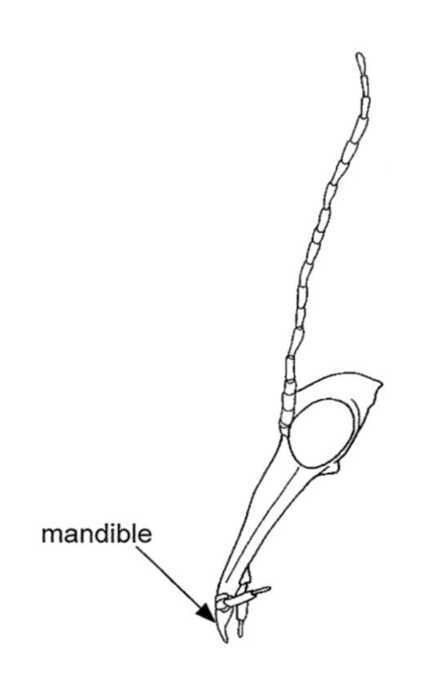

60b. Head not produced into beak..................61

61a. Very small insects with soft body; tarsi two- or three-segmented...........................62

61b. Usually very much larger insects; tarsi usually with more than three segments, or body hard and cerci abs ent.......................63

62a. Cerci absent.........................**PSOCOPTERA**

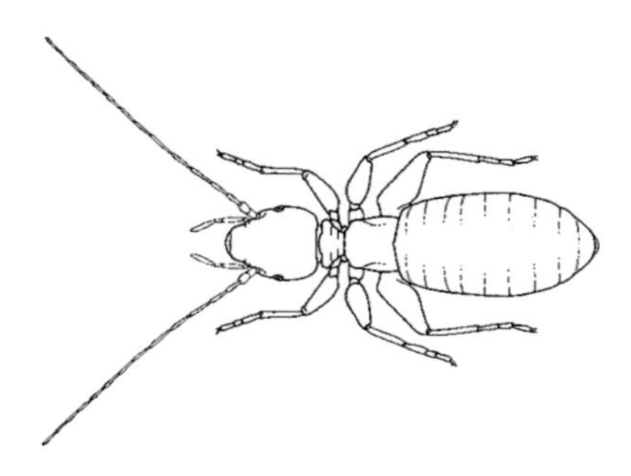

62b. Cerci of single segment, prominent ....**ZORAPTERA**

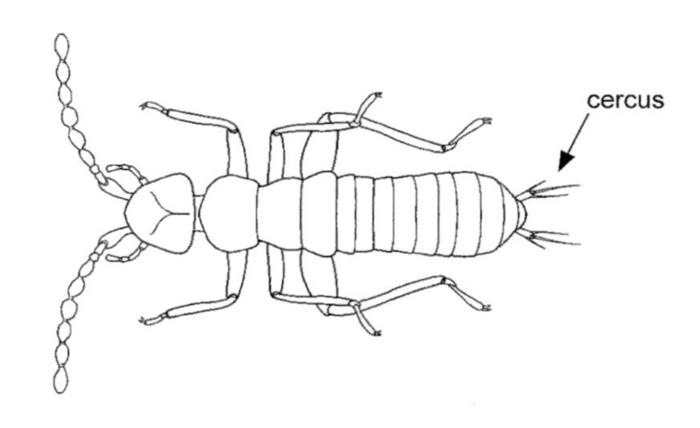

63a. Hind femora enlarged and adapted for jumping; wing pads of immatures inverted, hind wing pads overlapping forewing pads...................**ORTHOPTERA**

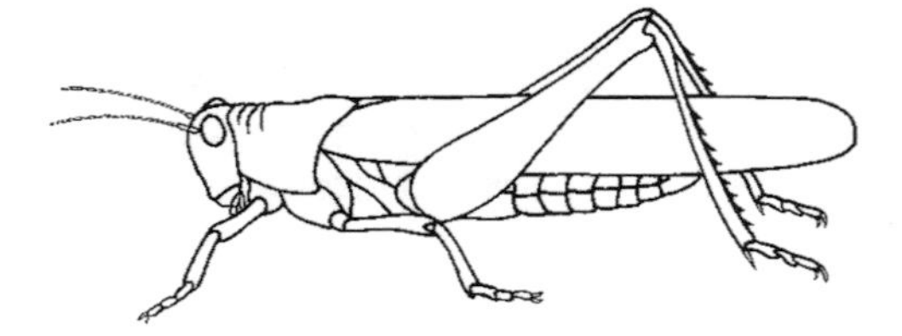

63b. Hind femora not enlarged or adapted for jumping; wing pads, if present, in normal position..64

64a. Prothorax much longer than mesothorax; opposable surfaces of front femora and tibiae with long spines adapted for grasping prey (raptorial insects) ....**MANTODEA**

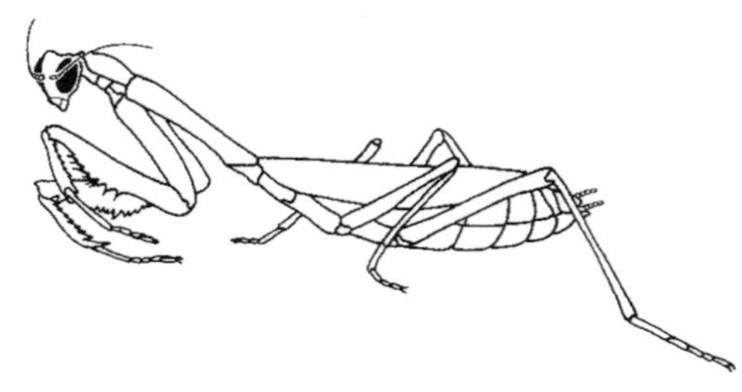

64b. Prothorax not greatly lengthened; front legs rarely raptorial.................................65

65a. Cerci absent; body often strongly sclerotized; antennae usually with 11 segments...........**OLEOPTERA**

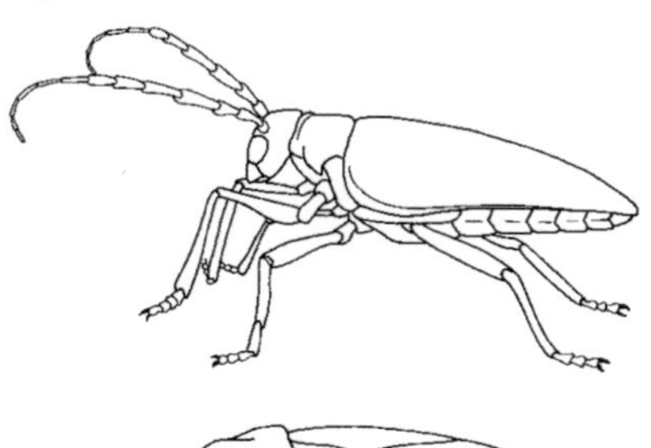

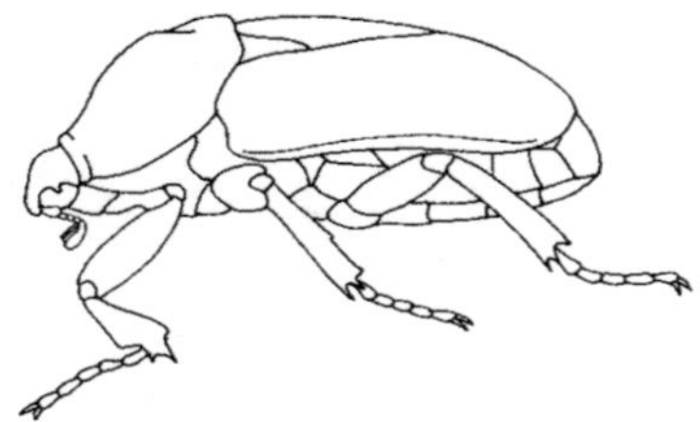

65b. Cerci present; body usually not strongly sclerotized; antennae usually with more than 15 segments....................................66

66a. Cerci with more than three segments..67

66b. Cerci with one to three segments........69

67a. Body flattened or dorsoventally compressed, oval in outline; head deflected downward, with mouthparts directed caudad.. **BLATTODEA**

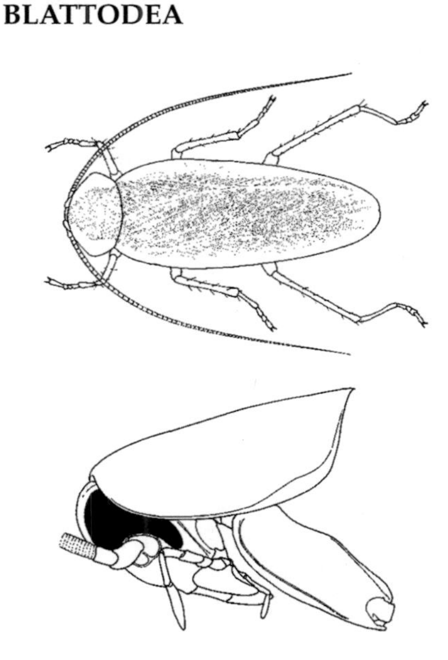

67b. Body not flattened, elongate in outline; head nearly horizontal, with mouthparts at extreme anterior margin of head........68

68a. Cerci long, five- to eight-segmented; ovipositor long, swordlike; tarsi five-segmented; found in cold habitats remote from human habitation..... ....**GRYLLOBLATTODEA**

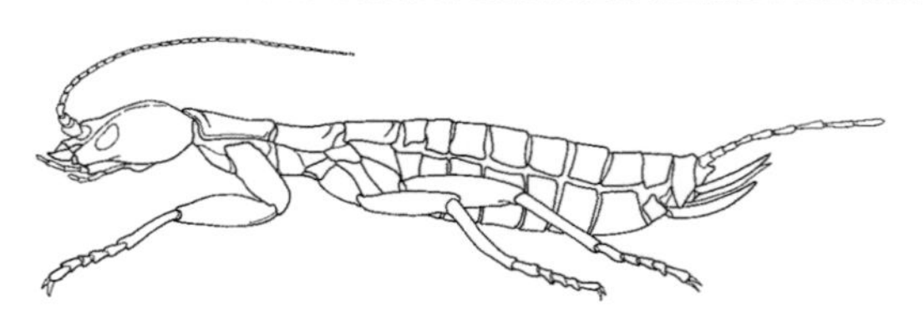

68b. Cerci short, one- to five-segmented; ovipositor absent; tarsi four-segmented; found in tropical and subtropical habitats or in association with humans.................... **ISOPTERA**

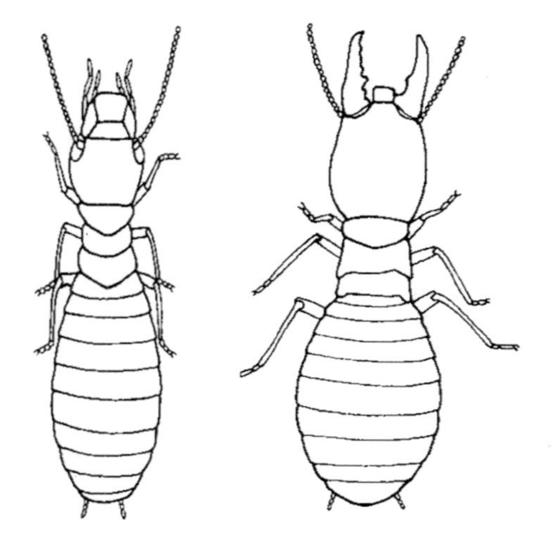

69a. Tarsi five-segmented (three-segmented in Timema [in Pacific coast states], most antennal segments several times longer than wide); body large and sticklike; not communal or social insects....................... **PHASMATODEA**

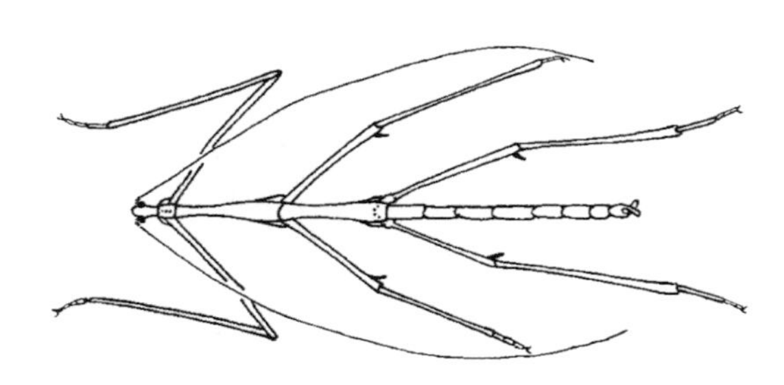

69b. Tarsi two- or three-segmented; antennal segments beadlike; body usually small, elongate but not sticklike; communal or social insects..70

70a. Front basitarsi swollen and containing silk-spinning gland for producing web in which insects live communally; cerci conspicuous; sexually dimorphic insects but without castes................... **EMBIIDINA**

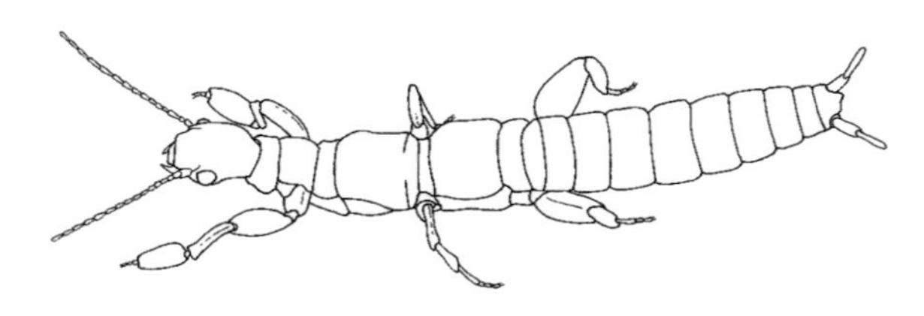

70b. Front basitarsi not swollen and not producing silk; cerci inconspicuous; polymorphic social insects with castes....................**ISOPTERA**

71a. Body cylindrical, caterpillarlike.............72

71b. Body more or less compressed, not caterpillar-like..73

72a. Head with six ocelli (stemmata) on each side; antennae inserted in membranous area at base of Mandibles.........Some larval **LEPIDOPTERA**

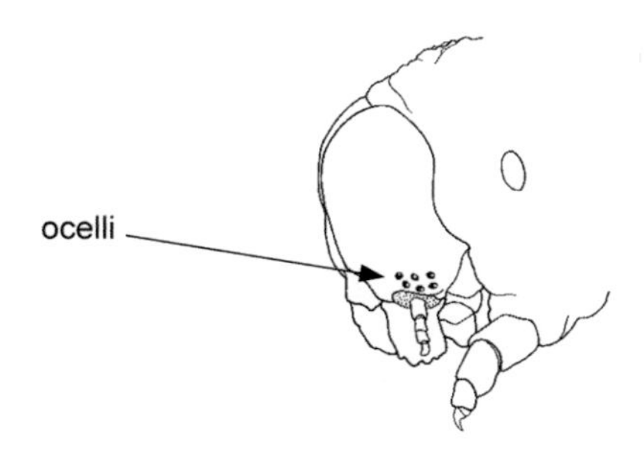

72b. Head with more than six ocelli on each side; third pair of legs distinctly larger than first pair....................Larval Boreidae **MECOPTERA**

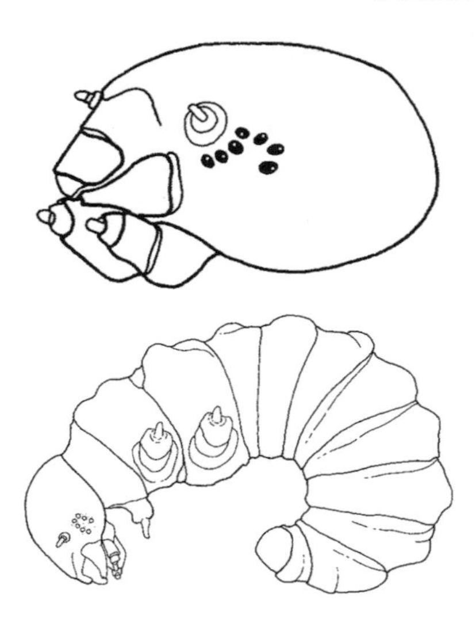

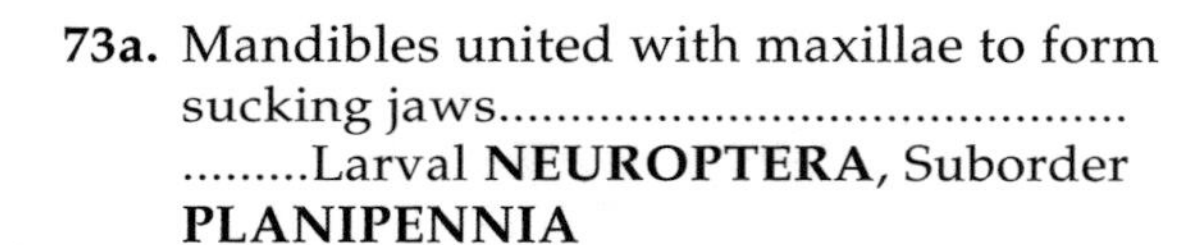

73a. Mandibles united with maxillae to form sucking jaws.. ....Larval **NEUROPTERA**, Suborder **PLANIPENNIA**

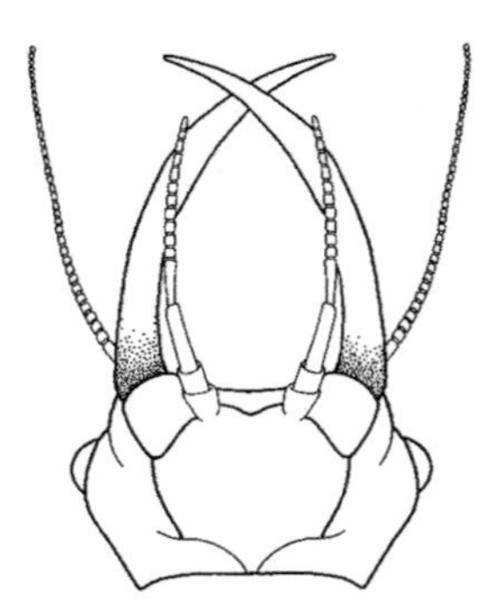

73b. Mandibles nearly always separate from maxillae..............................Larval **COLEOPTERA; NEUROPTERA,** suborder **RAPHIDIODEA;** suborder **RAPHIDIODEA; STREPSIPTERA; DIPTERA**

74a. Body densely covered with scales and hairlike setae; proboscis, if present, coiled under head.. **LEPIDOPTERA**

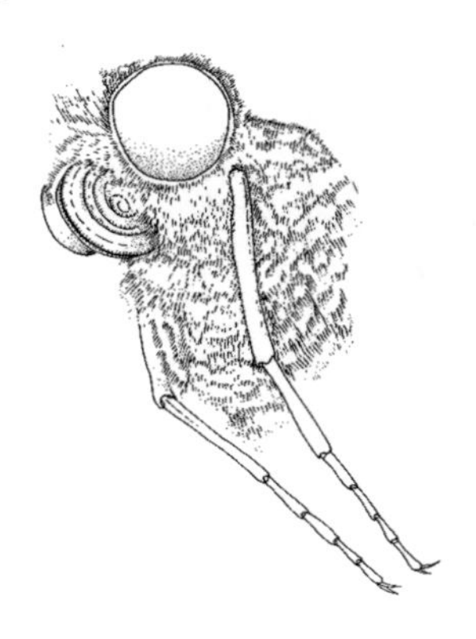

74b. Body bare, with scattered hairlike setae or waxy coating; mouthparts not coiled under head...75

75a. Last tarsal segment bladderlike, without claws; mouth forming a triangular, nonsegmented beak; very small insects .... **THYSANOPTERA**

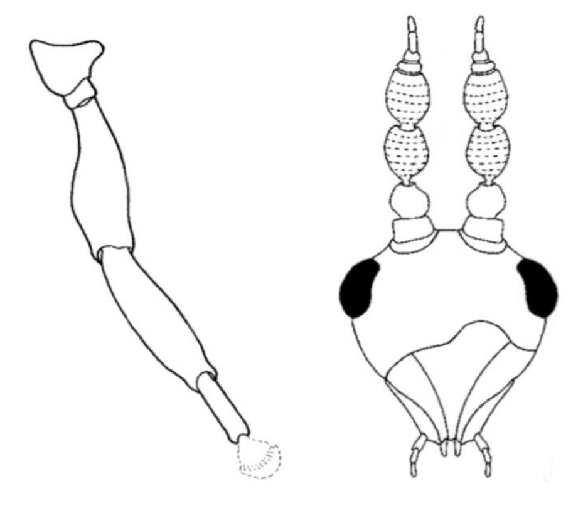

75b. Last tarsal segment not bladderlike, with distinct claws; other characters different..76

76a. Prothorax small, hidden when viewed from dorsal aspect.. **DIPTERA**

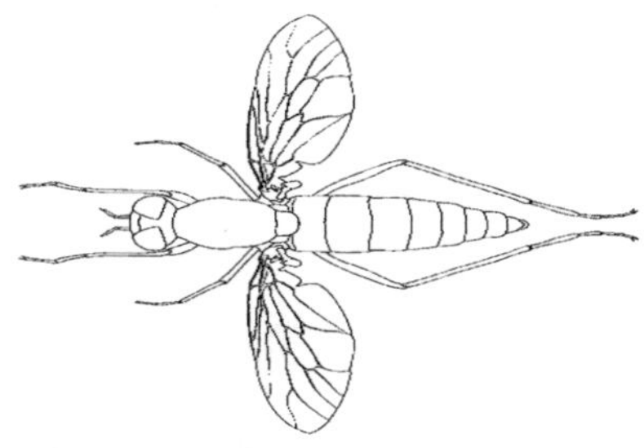

76b. Prothorax evident when viewed from dorsal aspect...77

77a. Beak arising from anterior part of head..................**HEMIPTERA**

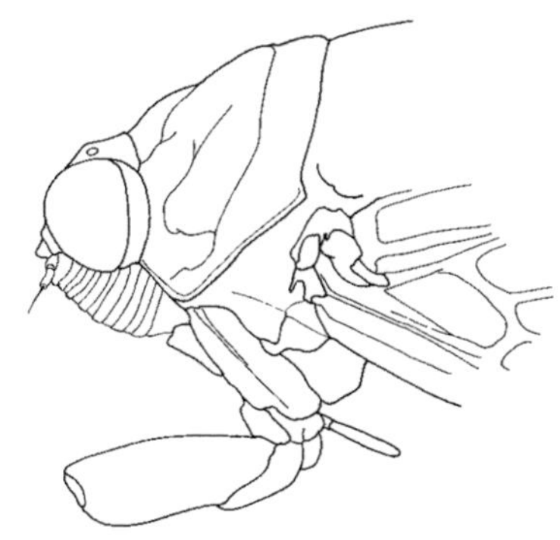

77b. Beak arising from lower posterior part of head..............**HOMOPTERA**

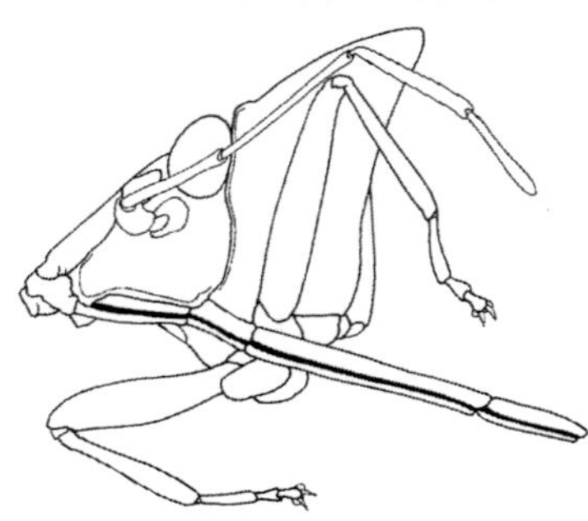

78a. Legless grubs or maggots; movement by wriggling.. Larval **DIPTERA** (If aquatic wrigglers, see larvae and pupae of mosquitoes); **HYMENOPTERA; LEPIDOPTERA; COLEOPTERA; SIPHONAPTERA; STREPSIPTERA** (in body of wasps or bees with flattened head exposed)

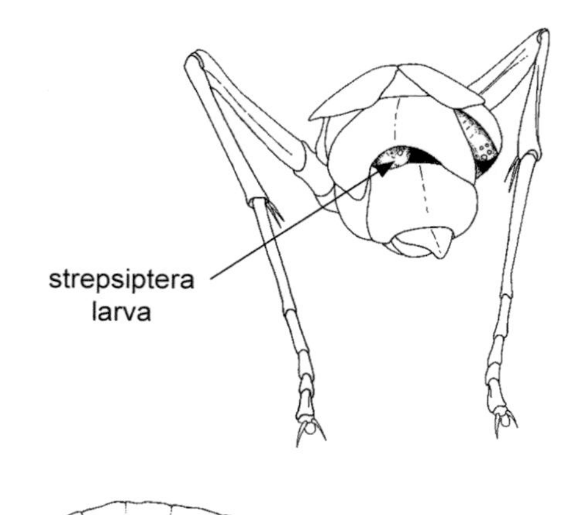

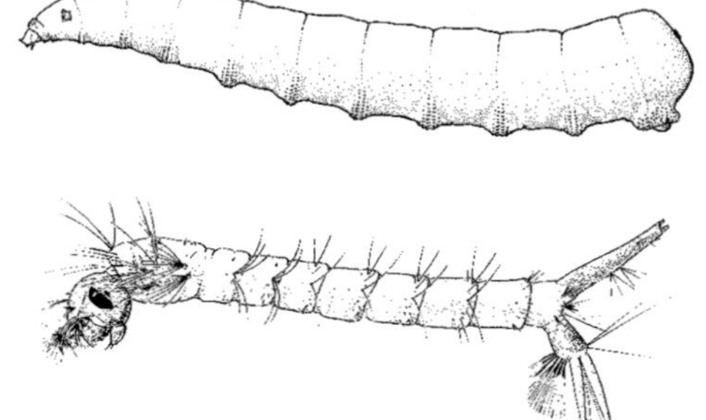

78b. Legless or if legged then each leg with one terminal claw..79

79a. Small animals with little resemblance to most insects; filamentlike mouthparts inserted in plant tissue; usually covered with waxy scale, powder, or cottony tufts..............**HOMOPTERA**

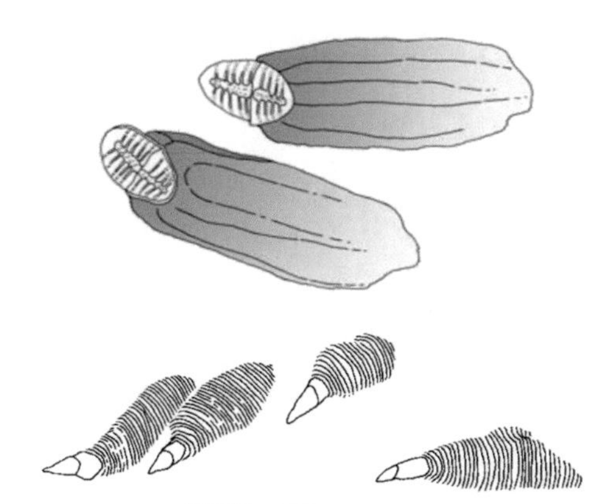

79b. Body unable to move or able to bend from side to side; mouthparts variable; body enclosed in tight integument, sometimes wholly covering body or sometimes with appendages free, but rarely movable; sometimes enclosed in cocoon (pupae)..80

80a. Legs, wings, etc., more or less free from body; biting mouthparts visible..............81

80b. Integument enclosing body holding appendages tightly against body; mouthparts evident as proboscis, without mandibles...83

81a. Prothorax small, fused with mesothorax; body sometimes enclosed in cocoonPupal **HYMENOPTERA**

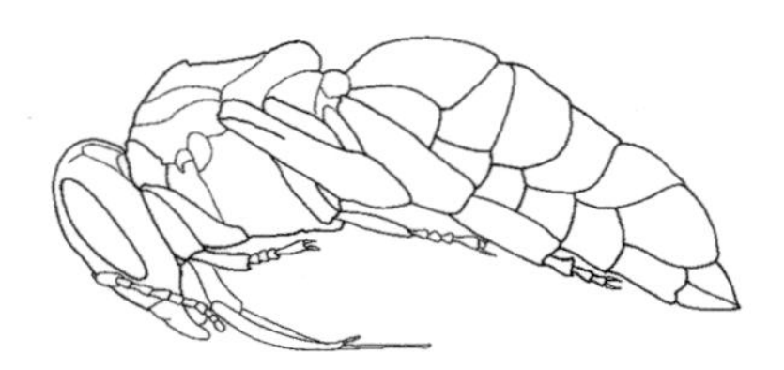

81b. Prothorax larger and not fused with mesothorax; cocoon development variable..82

82a. Wing cases with few or no veins; pupation rarely occurs in silken cocoon.. Pupal **COLEOPTERA**

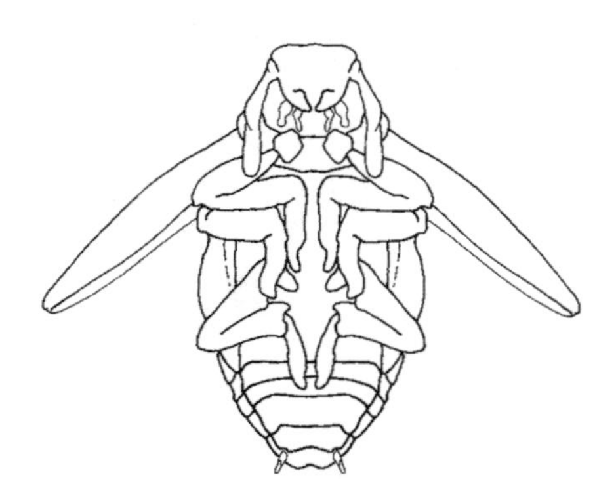

82b. Wing cases with several branched veins; pupation usually occurs in silken cocoon ....Pupal **NEUROPTERA**

83a. Proboscis usually long, rarely absent; four wing cases, one covering each wing; body often in cocoon.............................Pupal **LEPIDOPTERA**

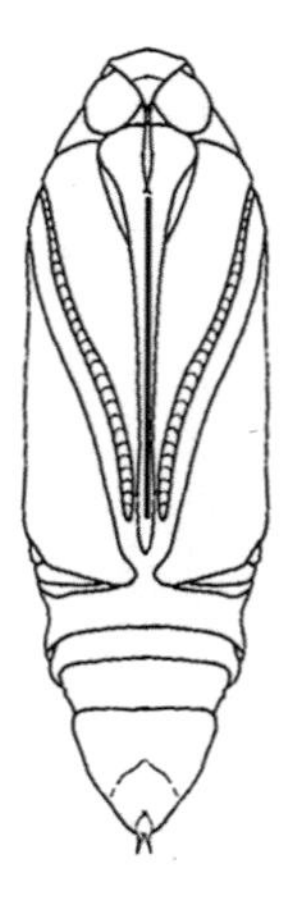

83b. Proboscis usually short; two wing cases, one covering each forewing; body not in silken cocoon, but often tightly enclosed in hardened last larval integument................... Pupal **DIPTERA**

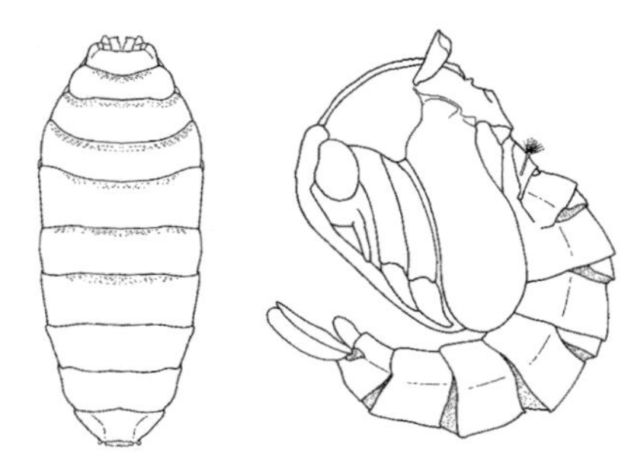

As demonstrated above, keys that are illustrated have much more utility for practicing diagnosticians than text-only keys. Illustrations that depict specific key characters are very valuable to diagnosticians.

SUMMARY

Limitations of Dichotomous Keys

Insects have traditionally been identified based upon differences and/or similarities of external morphology, and most dichotomous keys are constructed such that one morphological feature is studied at a time. Based on the outcome of the first couplet or pair of choices, a second, then third feature is examined in sequence. Eventually a dichotomous key leads to the identification of a specific insect.

Keys can be simple or complex. All insect taxonomy and systematics courses employ dichotomous keys and teach students how to use them. For pure taxonomy, these keys are essential (Pankhurst, 1991, Quicke, 1993). For diagnosticians, dichotomous keys are valuable, but strict and exclusive reliance upon them becomes an inefficient and tedious exercise, especially if not needed.

Dichotomous keys are constructed based on the assumption that a user obtains a near-perfect specimen and starts at the very beginning of the key (entry point). In most cases, diagnosticians and taxonomists really need only to separate one or two insects from each other. The necessity of a single entry point for a dichotomous key becomes overly restrictive in these cases. In addition, taxonomists, who construct keys, naturally have a very intimate familiarity with the insects they are attempting to describe and differentiate. While couplet descriptions such as 'proboscis medium length and slender, often clavate' may mean something for the expert in that group of insects, it has little meaning for a diagnostician who does not have a reference point as to what qualifies as 'medium length' or 'slender.' This results in a 'dead end' (no identification possible), at once both troublesome and frustrating. This dilemma results in the often-used statement *'keys are compiled by those who do not need them for those who cannot use them'* (Walters & Winterton, 2007).

Illustrated Keys

As can be immediately seen in the pictorial key to the orders of insects, illustrations and photographs inserted into or referenced by each couplet can dramatically enhance the ease and utility of a taxonomic key. Providing illustrations of the character in question not only helps the user locate and identify the structure, but also gives a reference point that non-illustrated keys do not. Definitions and references can be inserted in a similar way.

Use of computers and the ease with which a photo, illustration or definition can be incorporated in to a dichotomous key has revolutionized the ways in which illustrated keys can be used. The nature of the dichotomous key is based in interactivity, which is what computers do best. Even traditional dichotomous keys become value-added when simple hypertext (HTML) links are incorporated. Therefore, illustrated interactive keys have become a preferred method of insect identification for diagnosticians.

Matrix-Based Keys

When a small number of outcomes (a narrow group of potential insects to identify) are determined beforehand, these can be placed into a matrix and then, by working backwards, a series of 'either/or' or character statements can be composed that will separate them. We refer to these as matrix-based keys.

The 'either/or' character statements effectively dismiss or reject outcomes that do not fit the character described. Computerized sorting programs search the database for the selected characters and then prune those not exhibiting that character (Walter & Winterton, 2007).

In practice, matrix-based interactive keys allow a user to select a single character (physical or behavioral) at a time or multiple characters (using checkboxes) to eliminate all other possibilities except one. Obviously, the smaller the set of potential outcomes, the easier it is to construct and use such a key.

The use of computers to navigate such keys is ideal. Using hyperlinks to move between text and images and from image to image increases the impact of character illustrations and the user's understanding of technical language. Computer links to photographs, images, illustrations and glossary definitions add enormous power to matrix-based keys.

In summary, matrix-based interactive keys are composed of three basic parts: a database in the form of a matrix of distinguishing characters, a computer program that queries the database, and an interactive user interface that allows for information input and/or user choices (Dallwitz, 2005, 2006).

The major advantages of computer-interactive keys are that they are easily updated and corrected, and if published on the Web, changes are instantly distributed to potential users.

The matrix-based alternative as an identification tool affords multiple entry points and many paths to an identification. This is particularly valuable to a diagnostician who strongly suspects the identity of a specimen, but desires a confirmation. Jumping into the key at a later point or using a dedicated key only for those particular characters is efficient in terms of time and energy.

Finally, whereas dichotomous keys published in journals and books are fixed in time and place, interactive keys (not just those which are matrix-based) are updatable as taxonomy changes. As new taxa are described they can be incorporated into the key with minimal effort and may be instantaneously distributed via the Internet.

The advent of computer-assisted insect diagnostics is certain to revolutionize the way in which diagnosticians identify insects. A practical shift from traditional keys to matrix-based computer interactive keys is likely.

For diagnosticians, the major advantage of matrix-based keys is that they afford many paths to a correct identification and make extensive use of hypertext to link to images, glossaries, and other support material (Walter & Winterton, 2007).

Molecular Diagnostics

Insect identification based upon modern tools such as DNA analysis, special preparations and dissections, internal morphology or reproductive genitalia is beyond the expectations of most practicing insect diagnosticians. Even so, diagnosticians should be aware of these procedures and techniques. Methods of preservation and mounting such special samples are provided in Chapter 3.

In molecular diagnostics, DNA sequences are treated as distinguishing characters in the same way that morphological differences are used

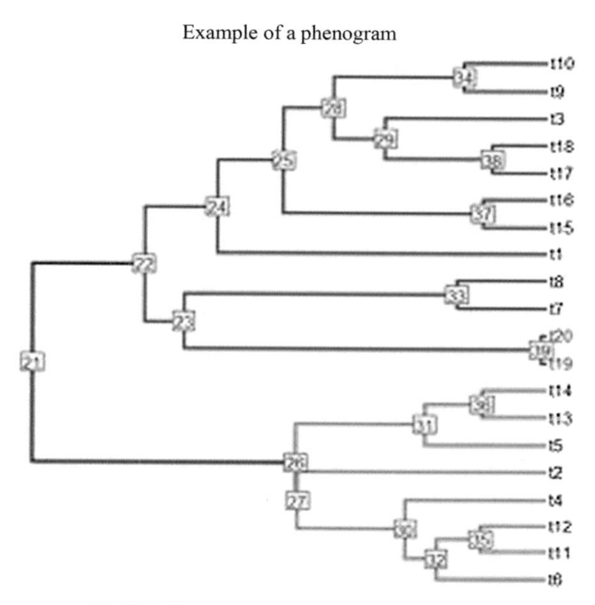

FIGURE 4.34 Example of a phenogram.

in traditional or matrix-based keys to separate related specimens. DNA's code is a series of four chemical bases, described with 'letters,' called A, C, T and G. The sequence in which these letters occur over and over again is unique enough to separate one organism from another; each species of insects possesses a unique DNA code. The codes or sequences can be used in a comparative way to determine relatedness of unknown specimens if compared to a reference library of sequences (Hebert et al., 2003). The result is an interactive key by which the input is sequence data and the output is an identification based on genetic distance. Obvious limitations of this protocol include incomplete sequence reference libraries, as well as the effort required to extract DNA and test the specimens. Signature sequences, such as those derived from rRNA, also serve to determine separation between groups.

In this way, genes and proteins can be used as molecular clocks, which determine the divergence from a common ancestor. Such comparisons are most often depicted using a phenogram. A phenogram is simply a diagram showing taxonomic relationships among organisms based on overall similarity of characteristics, without regard to evolutionary history or assumed significance of specific characters. Phenograms (see Fig. 4.34) are usually generated by computer and are based on genetic distance (relatedness) between organisms (Walter & Winterton, 2007).

Molecular techniques are currently limited by the relatively small library of genetic codes. As these increase, the power of this technique will also improve. Molecular techniques do hold promise, especially for uncovering cryptic species and for associating immature stages and sexes within a species. Advances in the use of DNA and molecular techniques may drastically change the way that insects and other arthropods are identified in the future (Walter & Winterton, 2007).

> Traditional dichotomous keys have long been the basis of insect identification; however, in practical usage, they all have the inherent problem of having a single entry point and only one set of choices (no deviations) which leads to the correct identification.

Sight Identification

As diagnosticians become intimately familiar with a particular insect, they may easily and accurately identify it at a glance, saving valuable time and resources. Some samples are expected every year at a certain time of season. Other insect pests are predictable depending on weather conditions. Pests that favor wet conditions, for example, can be anticipated if weather conditions have been wet and humid. Other insects favor dry and hot conditions. Occasionally, new insects either move into an area or extant insect populations, for one reason or another, seem to explode. After verifying the identity of such an insect once or twice using dichotomous keys, it becomes natural to identify it by sight for the succeeding samples. Over time and with experience, sight identification may be expected for a majority of samples that come into a diagnostic laboratory.

Reference Comparisons

In addition to dichotomous keys and sight identification, there are many resources available to entomologists that assist in diagnosing an arthropod specimen or damage sample. For example, comparison of a submitted sample to a specimen or series of specimens in an insect reference collection or museum is valuable. Many insect collections are housed in major universities, often land-grant colleges, and are curated by designated university specialists. If a diagnostician is fortunate enough to have access to such a collection, it can be a very valuable asset.

Any insect collection that is properly prepared and maintained can be a valuable reference for a diagnostician. Often a diagnostician will collect, preserve and label specimens which were previously submitted. Over time, such accumulated specimens can become a very valuable reference for a diagnostician. Many of these specimens may lack the traditional and proper label information (locality, date, name of collector, etc.) that is required of museum-bound specimens; however, when comparing a submitted sample to a series of previously identified specimens, a reference collection can be of great value.

Diagnosticians can add value to a personal reference collection by including facts about the specimen that might assist in diagnosis, including descriptions of damage, weather, behavior, specific locations, and controls recommended.

In many cases an independent damage reference collection is also of great worth. Many samples submitted to diagnosticians contain only damage or even descriptions of damage, along with the question, 'What caused this and should I be concerned?' A diagnostician who has collected, preserved and organized samples of damage can benefit greatly from a damage reference collection. Photographs and leads (references to printed materials) enhance such a collection.

A collection of insect text books, descriptions of insects in early scientific as well as current taxonomic literature, field guides such as White & Borer (1998) Arnett & Jacques (1981) and Marshall (2006) popular magazine articles and reputable internet sites is of great worth to a diagnostician. Innovations in digital photography, computer graphics, and electronic publishing are leading to the increased use of photographic images in electronic field guides as well. Together, these form a personal reference library for a diagnostician.

Diagnosticians often use visual comparisons as their routine method of sample identification. Visual comparison of a sample to photos or descriptions of existing, named specimens in a reference library, or to actual preserved specimens in a reference collection, is commonplace.

Identification techniques using visual comparisons can be enhanced by the use of supplementary photographs or illustrations that highlight key diagnostic characters when 'look-a-like' or 'confusing' specimens are considered. This is the approach that is often used in innovative field guides and similar texts, as was modeled in Chapter 2.

WEBSITES

Websites are also excellent sources of information on insect identification. They excel due to their inclusion of color photographs and their ability to target groups of pests, and offer identification aids that are difficult to deliver and access in other ways. The following list includes only a few of these sites currently avialable online.

- Bugwood Network: www.bugwood.org/
- Featured Creatures: www.idlab.ento.vt.edu/

- Plant Pest Identification Aid: vegipm.tamu.edu/imageindex.html
- Pest Identification Site: entweb.clemson.edu/pesticid/saftyed/pstident.htm
- Beneficial Insects Gallery: www.ipm.ucdavis.edu/PMG/NE/index.html
- Aphids: aphid.aphidnet.org/
- Grasshoppers: itp.lucidcentral.org/id/grasshopper/adult/Media/frmsetRLGH.htm
- USDA Invasive Pests: idtools.org/is
- Insect Identification: www.wvu.edu/~agexten/ipm/identify/insectid.htm
- Natural Enemies: www.nysaes.cornell.edu/ent/biocontrol/
- BugGuide: bugguide.net

APPLICATION SOFTWARE

Developing application software (apps) has become a very popular method of education. Apps that are devoted to pest identification are becoming more common. An example is the Purdue Plant Doctor App Series (www.purdueplantdoctor.com).

This App was developed by university experts to help users identify and manage plant problems caused by a variety of factors, including insects.

It guides users to:

- Identify plant problems by matching damaged plant parts with thousands of high-resolution color photos.
- Check diagnoses with detailed descriptions of damage and stages of problem development linked to each photo.
- Provide the latest unbiased recommendations from University experts on how to manage plant problems.

This and many other apps have become very powerful tools both for the diagnostician and for clients generally. As they continue to develop they will become even more engaging, automated, integrated and mobile. They are sure to improve how diagnosticians do their job well into the future.

DIAGNOSING NON-TYPICAL (MYSTERY) SAMPLES

Diagnosticians frequently do not have the luxury of seeing a perfect specimen or a series of specimens in the life stage preferred for diagnosis. In many cases only descriptions or samples of damage are submitted. These make diagnostics challenging and not always certain. On the other hand, an experienced diagnostician who can make a well-educated guess, based on seemingly little information, is invaluable for growers, consumers, and homeowners.

Phenotypic Variations

Individual differences in insect appearance may be due to genetics, food resources or other factors. Apparent differences within insect species are relatively rare, but certainly are worth noting because they can throw off an otherwise well made key. Sometimes size and color differences are noticeable in immature stages but disappear after pupation. In other cases, the differences are only apparent in the adult stage.

Consider the differences in the colors of the bean leaf beetles pictured below (Figs 4.35–4.37). All belong to the same species: *Cerotoma trifurcata*, and yet the example of color variation is wide.

Differences in color as well as markings is also apparent in many insects, tempting lay people to surmise that different species are present. The photograph in Figure 4.38 depicts many of the color, spotting and thoracic marking differences within the single species: *Harmonia axyridis*.

FIGURE 4.35–4.37 Color differences of the bean leaf beetle adult.

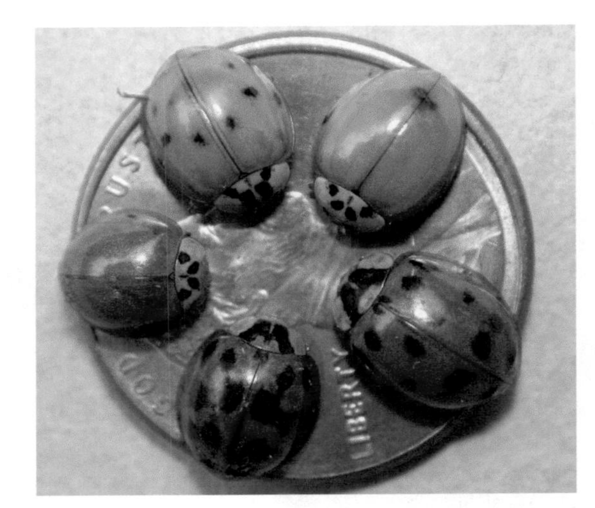

FIGURE 4.38 Color and marking variations of the Asian ladybeetle.

Therefore, making diagnoses of adult or immature specimens, based entirely on color or size, is tentative.

Sexual Dimorphism

Sexual dimorphism is the apparent difference between males and females of the same species. Phenotypic differences between sexes are very common in birds and to a lesser extent in other animals. Most insects do not display obvious sexual dimorphic differences, however, those that do can sometimes make species identification confusing.

Diagnosticians usually recognize that dimorphism is not uncommon in spiders and some butterflies. Other species such as beetles and dobsonflies can show differences as well. Often keys do not clearly depict both sexes and it is up to the diagnostician to remember that differences exist.

For example dobsonfly adults are very large brown insects with long, transparent wings covered with black and white dots. The wings extend past the abdomen, resulting in a total length of body and wings of about three inches. Female dobsonflies have large, strong jaws capable of delivering a painful bite. Male dobsonflies are even more remarkable by possessing very long, slender, curved mandibles that extend about one inch in front of the head. Although fearsome in appearance, males are unlikely to bite people (Figs 4.39 and 4.40).

Dobsonflies have aquatic immature stages but adult forms are strong fliers and are attracted

FIGURE 4.39 Dobson fly, Sexual dimorphic mandibles, female.

FIGURE 4.40 Dobson fly, Sexual dimorphic mandibles, male.

to lights at night. Although they are not numerous, when they are found they are a curiosity and thus are commonly submitted to diagnostic laboratories.

Diagnosing Insect Damage, Signs and Symptoms

Diagnosticians must be able to tentatively identify insects based on other evidence when insects themselves are not present. Often frass, holes or symptoms expressed by an infested plant are the only clues that a diagnostician is given.

Diagnosticians soon learn that some of their most important diagnoses are based on signs and symptoms rather than on specimens. By contrast, identifiers and taxonomists have the luxury of a perfect specimen from which to make their identification.

Consider the photos below submitted for identification. While there are no animals present in the photos, the damage depicts the problem very nicely. Even without seeing any, a diagnostician can be quite certain that the turf is infested with white grubs.

White grubs do some damage to grasses but what damage they cause is sometimes more than compounded by other animals coming in to forage for the grubs (Fig. 4.41).

It is clear in this photograph that either raccoons, skunks or opossums have been digging for grubs in this area. The turfgrass ripped up and strewn all around is evidence of foraging by these animals.

Peck holes are similar evidence of grub foraging but in this case caused by birds (Figs 4.42 and 4.43).

FIGURE 4.41 Turfgrass damage by racoons.

FIGURE 4.42 Bird damage in turfgrass.

FIGURE 4.43 Bird peck-holes in turfgrass.

Together, these photos show the result of what may have been somewhat tolerable grub injury to a lawn, – made intolerable by animal foraging activity.

Because the grass is largely dead at this point the only solution is to rake it up and replant or lay sod. Laying sod gives a quick fix to the problem but not all animal foraging damage is this severe. Keep in mind that any grass that is not torn up will survive and thus give a head start to an over-seeding strategy.

Diagnosticians may receive samples or photos of damaged wood or other products. Recognizing the signs of the insect or the symptoms expressed by a damaged plant may be all that a diagnostician has to go on. Submitted samples may only consist of photographs or descriptions of damage, signs and symptoms of infestations. Diagnosticians learn to recognize damage signs and symptoms by sight, just as they recognize and identify insects by sight.

Many insects transmit plant diseases. Recognizing the symptoms of the disease such as bacterial wilt in cucumber plants (Fig. 4.44) is a valuable diagnostic skill.

Experienced diagnosticians can identify various wood-boring insects based on size and shape of holes or appearance of frass that is associated with the damage. In many cases finding the insect is not possible because it is deep inside the wood where it cannot be extricated or else it has emerged and is now long gone. In other cases, diagnosticians must recognize different signs such as sap oozing from trunks (Fig. 4.45).

Bear in mind that not all holes in trees are the result of insect boring. Some holes, are not due to insects at all but are the result of wood peckers or sapsuckers (Fig. 4.46).

FIGURE 4.44 Bacterial wilt transmitted by spotted cucumber beetles.

FIGURE 4.45 Peachtree borer damage to tree trunk.

FIGURE 4.46 Sapsucker injury to tree trunk.

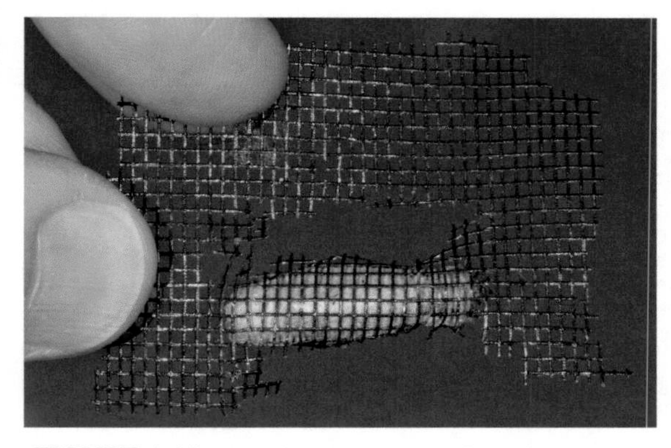

FIGURE 4.47 Pupal case in damaged window screen.

Diagnosticians must recognize that exceptions to every rule exist. For example, damage to screen windows is often blamed on insects when in most cases the insects that are blocked from entry do not have the chewing mouthparts necessary for cutting the screen. Occasionally, however, exceptions occur as in the case of this caterpillar that pupated inside the screen (Fig. 4.47).

To make a diagnosis of the two 'mystery' photographs (Figs 4.48 and 4.49) requires that a diagnostician be familiar with the biology and behavior of a carpenter bee as well as that of a woodpecker.

Carpenter bees construct their brood chambers by boring into seasoned lumber from which homes, decks, posts, and outbuildings are made. The tunnels enter into wood and then typically make a 90 degree turn in the tunnel to run with the grain of the wood for 6–8 inches. The bee then creates individual brood chambers along this tunnel, deposits an egg and provisions each

FIGURE 4.48 Carpenter bee brood chambers exposed by woodpecker feeding.

FIGURE 4.49 Individual cells and associated frass left by carpenter bees.

cell with enough food for the emerging larva to subsist. The result is a tunnel made up of 6–10 brood cells, housing individual developing larvae.

Woodpeckers belong to the family Picidae and are experts at locating insect larvae that are tunneling inside solid wood. Once located the birds utilize specialized bills and hammering techniques to excavate the wood and expose the larvae beneath.

Diagnosticians who are aware of the biology of both the woodpeckers and the carpenter bees will immediately diagnose the cause of the damage in the photographs as woodpecker/carpenter bee damage.

When insects feed on plants they often leave feeding signs that can be used for diagnosis. Holes in leaves, trunks and branches can be diagnostic.

Plant responses to insect feeding can also be used for diagnosis. Symptoms include, wilting, stunting, chlorosis as well as many other visual symptoms.

To diagnose causes of plant injury a diagnostician must recognize what a normal plant is. This is the basis by which abnormal growth symptoms are measured.

For example, galls are abnormal outgrowths of plant tissue that results from insect infestations. Insect galls are induced by chemicals injected by certain insects into plant tissues. The gall itself is actually plant tissue that forms itself around a small chamber in which the immature gall insect lives.

Size, shape and color of the gall is often diagnostic for the insect involved. Diagnosticians soon recognize or find illustrated keys for the most common galls of various tree species (Figs 4.50–4.52).

FIGURE 4.50 Horned oak galls.

FIGURE 4.51 Hackberry nipple galls.

FIGURE 4.53 Praying mantid egg mass.

FIGURE 4.52 Maple bladder galls.

FIGURE 4.54 Katydid egg cluster on twig.

Eggs and Pupae

Samples of eggs or pupae are usually quite difficult to diagnose. Few keys are available or complete enough to rely on for identification. Hatching eggs or rearing pupae until eclosion, is sometimes necessary for accurate diagnosis. Very common samples or those with which diagnosticians have had personal experience can be more easily identified (Figs 4.53–4.55).

FIGURE 4.55 Stinkbug egg cluster on leaf.

Unusual Insects

Many clients claim that they have found a rare, never-before-seen, insect. Often they will back their claims by stating that they have lived in the area for up to one hundred years and no one living has ever seen anything like it. It therefore must be a prehistoric specimen, new to science. In most every case these specimens turn out to be nothing out of the ordinary. They still get the interest of a diagnostician, however. Who would not want to be part of the discovery of a new insect?

Diagnosticians know intuitively that there really is no such thing as a 12-legged, flying insect. It is difficult to convince a telephone caller of this when they are actually looking at a 12-legged, flying insect (Fig. 4.56).

FIGURE 4.56 Mating pair of lovebugs.

Diagnosticians do encounter rare samples on occasion. That is what makes diagnostics fun. One never knows what may be submitted next.

Case in point, few diagnosticians will misidentify bagworms when a complete bag or photo of a bag is submitted. The bags are very characteristic and even the defoliation signs on a plant are diagnostic when accompanied by dates, locations and plant species.

On the other hand, most diagnosticians have never had an adult bagworm moth submitted for identification (Fig. 4.57). These are not commonly found and most often not associated with the typical 'bags' found on trees or the damage that they cause.

FIGURE 4.57 Adult bagworm moth.

Diagnosticians understand the biology of the insects that they deal with and use this knowledge to help identify what a problem is and what to expect the damage to be.

In the case of bagworms, simply understanding that the caterpillars molt inside the bags and the adults actually emerge there, helps to explain some of the bagworm's mystery. In fact, the female adult has no wings and never leaves the bag.

Behavioral Anomalies

Diagnosticians often diagnose insects based on their familiarity of insect biology and behavior. If an insect is behaving in an expected manner or is occurring in a particular place, it can be identified as a certain insect. Exceptions always occur.

For example insects such as this bagworm, can be found in very unusual places (Fig. 4.58). Bagworms are unexpected pests in soybeans.

Sometimes very peculiar behaviors are noted by clients. What a client may describe as a foot long, spotted, writhing slug moving across their lawn may actually be a mass of fungus gnat maggots migrating (Fig. 4.59).

Mysterious behaviors make problem-solving a challenge. The photos in Figures 4.60 and 4.61 were submitted with the claim that a

FIGURE 4.58 Bagworm feeding in soybeans.

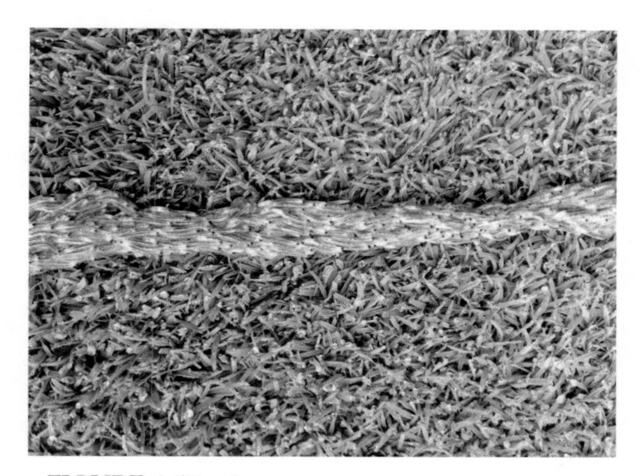

FIGURE 4.59 Fungus gnat maggot mass movement.

spider had decorated its web by meticulously weaving small plastic jewelry into the design.

It was not until several conversations later that it was determined that the spider web was actually located in the window of a grade school arts and crafts classroom. Apparently a child left a pile of the tiny jewels on the window ledge that were likely blown up into the web by a gust of wind when a door or window was opened. Such a theory is not nearly as intriguing as a Charlotte's web copy-cat, but infinitely more plausible.

FIGURE 4.60 Ornate spider web (1).

FIGURE 4.61 Ornate spider web (2).

In another case a wolf spider, family Lycosidae, was found alive. It was very unusual due to its distinctive blue rather than brown coloration. It was surmised by the client that it must be a rare genetic abnormality resulting in a blue spider (Fig. 4.62).

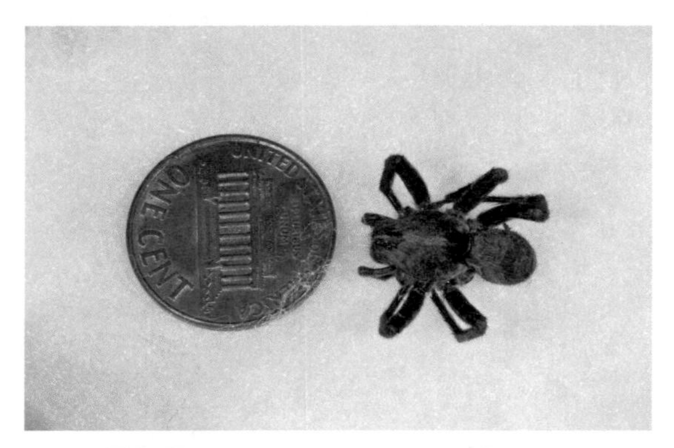

FIGURE 4.62 Blue-colored wolf spider.

The specimen was submitted and clearly identified as a wolf spider. It was only upon more intensive investigation, however, that the client revealed that the spider had been collected from the top of a pile of mulch. The spider was discovered when a blue tarp, covering the mulch was removed. It then became clear that the paint from the tarp had degraded over time and had rubbed off onto the spider, creating the blue coloration.

Color Irregularities

Occasionally we see photographs of insects that are so bizarre or odd that we really have to question their validity. Certainly the ease of posting and the distribution potential that the web offers a freakish photograph, makes the temptation of altering a photograph for increased sensation, more than an unusual occurrence.

The combination of high school level technical capabilities, an active imagination and a bit of extra time, makes the electronic manipulation of digital photographs relatively frequent. As a result, it is not uncommon to see photographs of outlandishly large or menacing looking insects, that rival even the grocery store tabloids. In most cases, equally fantastic stories accompany them.

The photograph in Figure 4.63 was submitted for confirmation. It was actually a photograph of a common bed bug initially. However, to make the bug look more startling, someone had added color enhancements; a green face, blue eyes and a red ominous looking mouth.

FIGURE 4.63 Digitally altered bed bug image.

Diagnosticans must be careful in responding to digital photographs with unknown origins. Certainly they do not want to validate or propagate inaccurate stories or digital images. If there is a question, common sense and a quick comparison to published text books or field guides is in order. Bear in mind that the discovery of a new insect is a laudable but extremely uncommon experience, and to find a large and scary one, chasing people out of their homes and down the streets, tromping down homes and devouring small children, well... that would be a plus.

Unusually colored insects do occur. These instances are uncommon but fascinating for a diagnostician. For example, the photos below depict a red wheelbug (*Arilus cristatus*) (Fig. 4.64) next to a photo of a normally colored adult (Fig. 4.65).

FIGURE 4.64 Color change of newly eclosed wheelbug: (*Arilus cristatus*): red wheelbug.

FIGURE 4.65 Wheelbug: normally colored adult.

Over time, the red color was replaced by the more typical gray and brown coloration, but for a time it truly was a red wheelbug.

Below are two photographs of equally bizarre colorations (Figs 4.66 and 4.67). A pink-colored katydid and a wolf spider carrying a blue egg sac. Such color aberrations do occur from time to time in nature and are the result of a genetic anomalies, NOT new species.

FIGURE 4.67 Wolf spider carrying a blue egg sac.

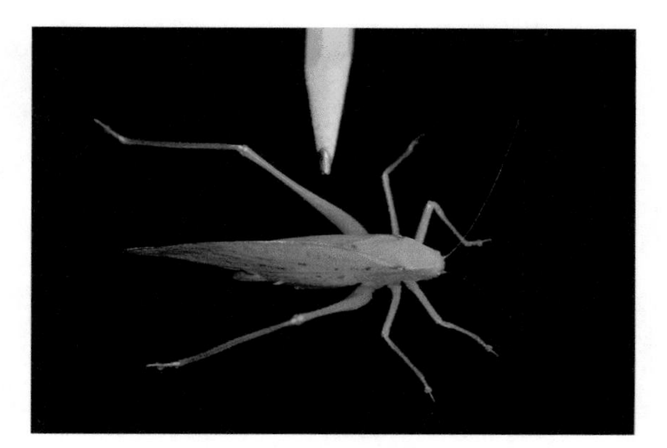

FIGURE 4.66 Pink katydid.

CHAPTER

5

Pest Insects

Key Pests

Many insects are considered key pests. These are pests that occur regularly and in most cases require management. Key pests are predictable because they occur every season. Each commercially grown agricultural crop typically has two or three key pests. For example, corn rootworms, corn borers, and cutworms are a threat to field corn every year. Without management they will cause unacceptable yield loss. Therefore, these are all considered key pests of field corn.

Key pests of fruits and vegetables are similar whether the crops are grown commercially or in the garden. Key pests of ornamental plants include pests found on trees, flowers, shrubs and turfgrass. Similarly, key pests of home-stored foods (pantry pests) are predictable regardless of where a person lives. Key pantry pests include ants, cockroaches, grain beetles and Indian meal moths. Key structural pests in a building are termites, carpenter ants and cockroaches. Key home-invading pests (occasional invaders) include pests that seek shelter inside a home, become a complete nuisance but do not breed there. Key human health pests are insects that bite such as mosquitoes, biting flies and ticks.

Grouping key pests is helpful for diagnosticians but is a somewhat arbitrary process. Some insects may be key pests in multiple areas, such as the brown marmorated stink bug. It can cause serious injury to fruits, vegetables and field crops and if that were not enough, it invades homes during the fall and winter.

Key pests may vary from one geographical region to the next. For example, chinch bugs may be a key pest in the mid-western states but not so in other parts of the country.

A key pest may not always be considered a key pest because management thresholds are not always consistent. For example, cutworms may be a key pest in crops or on highly visible athletic turfgrass fields but not on minimally valuable plants or less manicured backyard lawns.

Other insects such as Asian lady beetles may be nuisance pests during the fall and winter because they congregate inside homes and buildings but during the growing season they are considered very valuable predators because they feed on small aphids and other potential crop pests.

Even though we must recognize that key pest designations may change slightly from one local to another and from one commodity to another, it is instructive for diagnosticians to gain a general appreciation for all key pests. In fact, recognizing key pests is the baseline upon which diagnosticians can measure other pests. For instance, comparing insects submitted for identification against key pests can aid in making management recommendations.

Contemporary Insect Diagnostics
http://dx.doi.org/10.1016/B978-0-12-404623-8.00005-3

Diagnosticians should become intimately acquainted with key insect pests in all of their life stages. They should also become familiar with the damage that these pests cause so that they can recognize them immediately. The following key insect pests include some of those most commonly submitted to diagnostic laboratories. Each is represented by a photograph and description of the adult and immature stages, a photograph of damage symptoms and general management recommendations.

The key insects presented below include those that compete with humans for food (field crops, livestock, commercial and home horticulture and stored food products), insects that damage or destroy property including ornamental plants and pests that affect human health and comfort.

For ease of reference, these are combined and are presented in alphabetical order by common names.

INSECTS THAT COMPETE WITH HUMANS FOR FOOD

Commercial Agriculture Crops

Common Name: alfalfa weevil
Scientific Name: Curculionidae: *Hypera postica*
Status: key insect pest of alfalfa

Common Name: Aphid, plant louse
Scientific Name: Aphididae: several species
Status: can be very serious pests of many crops

Common Name: armyworm moth
Scientific Name: Noctuidae: *Pseudaletia unipuncta*
Status: pest of grasses, small grain crops, and corn

Common Name: black cutworm moth
Scientific Name: Noctuidae: *Agrotis ipsilon*
Status: pest of field and garden crops

Common Name: cabbage butterfly
Scientific Name: Pieridae: *Pieris rapae*
Status: pest of vegetable crops and some agriculture crops

Common Name: corn earworm, cotton bollworm, tomato fruitworm
Scientific Name: Noctuidae: *Heliothis zea*
Status: a major agricultural pest, particularly of sweet corn

Common Name: European corn borer
Scientific Name: *Pyradilae: Ostrinia nubilalis*
Status: pest of field corn

Common Name: Japanese beetle
Scientific Name: Scarabaeidae: *Popillia japonica*
Status: pest of corn and soybeans

Common Name: leafhopper
Scientific Name: Cicadellidae: several species
Status: pests of crops
Common Name: spittlebug
Scientific Name: Cercopidae: several species
Status: minor pest of agricultural forage crops

Common Name: western corn rootworm
Scientific Name: Chrysomelidea: *Diabrotica virgifera* LeConte
Status: serious pest of corn

Commercial and Home Horticulture Crops

Common Name: aphid, plant louse
Scientific Name: Aphididae: several species
Status: can be very serious pests of many horticulture plants

Common Name: armyworm moth – adult
Scientific Name: Noctuidae: *Pseudaletia unipuncta*
Status: pest of some garden crops

Common Name: black cutworm moth
Scientific Name: Noctuidae: *Agrotis ipsilon*
Status: pest of field and garden crops

Common Name: cabbage butterfly
Scientific Name: Pieridae: *Pieris rapae*
Status: pest of vegetable crops

Common Name: cicada, locust
Scientific Name: Cicadidae: several species
Status: can damage tender tree limbs on fruit trees

Common Name: codling moth
Scientific Name: Tortricidae: *Cydia pomonella*
Status: pest of fruits and nuts

Common Name: Colorado potato beetle
Scientific Name: Chrysomelidae: *Leptinotarsa decemlineata*
Status: pest of some vegetable crops, particularly potato

Common Name: corn earworm, cottom bollworm, tomato fruitworm
Scientific Name: Noctuidae: *Heliothis zea*
Status: a major pest of sweet corn as well as other vegetable crops

Common Name: hornworm, tomato hornworm; tobacco hornworm
Scientific Name: Sphingidae: *Manduca* sp.
Status: pest of various vegetable plants

Common Name: Japanese beetle
Scientific Name: Scarabaeidae: *Popillia japonica*
Status: pest of many vegetables and fruits

Common Name: leafhopper
Scientific Name: Cicadellidae: several species
Status: pests of vegetables

Common Name: Mexican bean beetle
Scientific Name: Coccinellidae: *Epilachna varivestis*
Status: pest of beans and peas

Common Name: spittlebug
Scientific Name: Cercopidae: several species
Status: minor pest horticultural crops

Common Name: squash bug
Scientific Name: Hemiptera: *Anasa tristis*
Status: pest of vegetable crops

Common Name: stink bug
Scientific Name: Pentatomidae: several species
Status: pest of vegetable crops

Common Name: thrips
Scientific Name: Thysanoptera: several species
Status: pest of many plants, particularly in greenhouses

Common Name: vinegar fly, fruit fly
Scientific Name: Drosophilidae: *Drosophila* sp.
Status: pest of ripe and fermenting fruits, new species (Spotted wing drosophila) also attacks non-ripened fruits

Stored Food Products

Common Name: dermestid beetle
Scientific Name: Dermestidae: several species
Status: pest of stored products

Common Name: Indian meal moth
Scientific Name: Pyralidae: *Plodia interpunctella*
Status: pest of stored food products, grains

Common Name: rice weevil
Scientific Name: Curculionidae: *Sitophilus oryzae*
Status: pest of stored cereal products

Common Name: vinegar fly, fruit fly
Scientific Name: Drosophilidae: *Drosophila* spp.
Status: pest ripe fruits and vegetable

INSECTS THAT DESTROY PROPERTY

Structures

Common Name: carpenter bee
Scientific Name: Apidae: *Xylocopa virginica*
Status: can be a pest of wooden structures

Common Name: American cockroach
Scientific Name: Dictyoptera: *Periplaneta americana*
Status: common pest in homes and buildings

Common Name: bed bug
Scientific Name: Cimicidae: *Cimex lectularius*
Status: infests bedrooms and furniture

Common Name: carpenter ant
Scientific Name: Formicidae: several species
Status: occasionally a pest of homes and buildings

Common Name: dermestid beetle
Scientific Name: Dermestidae: several species
Status: infests homes

Common Name: flea, cat flea
Scientific Name: Pulicidae: *Ctenocephalides felis*
Status: infests homes where pets live

Common Name: lady beetle
Scientific Name: Coccinellidae: several species
Status: annoying household invader

Common Name: stink bug
Scientific Name: Pentatomidae: several species
Status: a nuisance pest in homes

Common Name: termite – white ant
Scientific Name: Blattodea: several families
Status: pest of homes and buildings

Insects that Damage Ornamental Plants

Common Name: aphid, plant louse
Scientific Name: Aphididae: several species
Status: can be very serious pests of many ornamental plants

Common Name: armyworm moth
Scientific Name: Noctuidae: *Pseudaletia unipuncta*
Status: pest of turfgrasses and sweet corn

Common Name: bagworm
Scientific Name: Psychidae: *Thyridopteryx ephemeraformis*
Status: common pest of evergreens and shrubs

Common Name: black cutworm moth
Scientific Name: Noctuidae: *Agrotis ipsilon*
Status: pest of garden crops and turfgrass

Common Name: cicada, locust
Scientific Name: Cicadidae: several species
Status: can damage tender tree limbs, especially damaging on newly transplanted trees

Common Name: emerald ash borer
Scientific Name: Buprestidae: *Agrilus planipennis*
Status: devastating introduced pest of ash trees

Common Name: gypsy moth
Scientific Name: Erebidae: *Lymantria dispar*
Status: introduced pest of hardwood trees

Common Name: Japanese beetle
Scientific Name: Scarabaeidae: *Popillia japonica*
Status: pest of many plants including fruits turfgrasses and ornamentals

Common Name: leafhopper
Scientific Name: Cicadellidae: several species
Status: pests of flowers, grasses, vegetables, and trees

Common Name: spittlebug
Scientific Name: Cercopidae: several species
Status: minor pest of horticultural crops

Common Name: thrips
Scientific Name: Thysanoptera: several species
Status: pest of many plants particularly in greenhouses

INSECTS THAT THREATEN HUMAN HEALTH AND COMFORT

Health Threats

Common Name: American cockroach – adult
Scientific Name: Dictyoptera: *Periplaneta americana*
Status: spreads germs to people

Common Name: bed bug – adult
Scientific Name: Cimicidae: *Cimex lectularius*
Status: blood-feeding pest of humans

Common Name: blow fly – adult
Scientific Name: Calliphoridae: several species
Status: an annoyance around homes, spreads disease

Common Name: dermestid beetle – adult
Scientific Name: Dermestidae: several species
Status: infestations can trigger asthma

Common Name: flea, cat flea – adult
Scientific Name: Pulicidae: *Ctenocephalides felis*
Status: nuisance pest of humans and pets

Common Name: gypsy moth – adult
Scientific Name: Erebidae: *Lymantria dispar*
Status: can be allergic to some people

Common Name: house fly – adult
Scientific Name: Muscidae: *Musca domestica*
Status: nuisance pest of people, can spread disease

Common Name: yellowjacket – adult
Scientific Name: Vespidae: several species
Status: painful and potentially lethal sting

Annoying/Nuisance Pests

Common Name: carpenter bee – adult
Scientific Name: Apidae: *Xylocopa virginica*
Status: nuisance pest of people

Common Name: blow fly – adult
Scientific Name: Calliphoridae: several species
Status: an annoyance around homes, spreads disease

Common Name: cicada, locust – adult
Scientific Name: Cicadidae: several species
Status: noise can be very annoying

Common Name: house fly – adult
Scientific Name: Muscidae: *Musca domestica*
Status: nuisance pest of people, can spread disease

Common Name: lady beetle – adult
Scientific Name: Coccinellidae: several species
Status: an annoying household invader

Common Name: stink bug – adult
Scientific Name: Pentatomidae: several species
Status: nuisance pest in homes

Common Name: thrips – adult
Scientific Name: Thysanoptera: several species
Status: can nip people, annoyance

Common Name: vinegar fly, fruit fly – adult
Scientific Name: Drosophilidae: *Drosophila* spp.
Status: nuisance pest in homes and restaurants

FIGURE 5.1 Alfalfa weevil adult.

Scientific Name: Curculionidae: *Hypera postica*
Status: key insect pest of alfalfa
Damaging Stage: mostly larval; occasionally adult
Description: The alfalfa weevil is a major pest of alfalfa and often requires chemical treatments to manage. The adult beetle is a small, dark gray or brown beetle approximately 6 mm long with a prominent brown snout and a distinct dark band that extends down the back.

FIGURE 5.2 Alfalfa weevil larva.

Description: Immature alfalfa weevils are always legless and have black head capsules. In late spring, the eggs hatch and the larvae begin to feed. For the first few days, the larvae feed within the stem but then move to the leaf buds at the tips of the stems. The first two instars are generally brown in color, but the third and fourth are always green and have a characteristic white line down the middle of the back. Immature alfalfa weevils grow to approximately 9.5 mm in length.

FIGURE 5.3 Alfalfa weevil damage: defoliation.

Injury: Alfalfa weevil larvae are important pests of alfalfa because they stunt the growth of the plants and reduce harvest potential. As defoliation progresses their feeding gives the leaves a skeletonized appearance.
Management: Alfalfa weevil populations must be monitored in relation to the development of the plant. When economically significant numbers of weevils occur, pesticide use is justified.

FIGURE 5.4 American cockroach adult – sometimes called water bugs.

Scientific Name: Dictyoptera: *Periplaneta americana*
Status: common pest in homes and buildings
Damaging Stage: nymph and adult
Description: The American cockroach is a distinctive insect with an elliptical-shaped body and thick spines on the tibia. Adults are usually between 25 and 37 mm long and have long, thread-like antennae. They have a characteristic red-brown coloration with a lighter yellowish border around the thorax.

FIGURE 5.5 American cockroach life cycle.

Description: Female cockroaches carry their eggs in cases (oothecae). White-brown nymphs hatch from the cases but develop the more typical red-brown color over time. An immature cockroach can molt as many as thirteen times in one year. Wing pads start to develop in the third or fourth instar. American cockroaches usually hide during the day and feed on decaying organic matter at night.

FIGURE 5.6 American cockroach damage: fecal contamination.

Injury: American cockroaches contaminate foods and food storage areas. They often leave behind fecal pellets that resemble mouse droppings. They tend to favor areas with damp conditions and are the most common roach found in sewers. This allows them to transmit a number of disease-producing organisms, including food poisoning, dysentery and diarrhea.
Management: American roaches must be controlled in structures where they cause health hazards. Exclusion, baits and residual pesticides are used in their control.

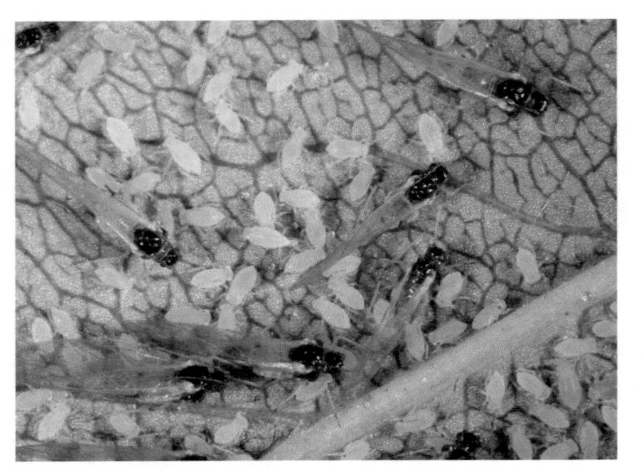

FIGURE 5.7 Aphid winged adult, sometimes called plant louse.

Scientific Name: Aphididae: several species
Status: can be very serious pests of many plants
Damaging Stage: nymph and adult
Description: Aphids are small, soft-bodied insects that can be green, yellow, brown, red, or black, depending upon the species. The body is usually pear-shaped with long legs and antennae and may or may not have wings. Most species have a characteristic pair of points called cornicles protruding from the posterior ends of the abdomen in an exhaust pipe-like fashion.

FIGURE 5.8 Aphid nymphs.

Description: Aphids have many generations per year, reproduce asexually and give birth to live young. Immature aphids appear very much like the adult except that they are smaller and never have wings. Adults and nymphs feed together in large colonies and so may become serious plant pests very quickly.

FIGURE 5.9 Aphid damage: wilting, chlorosis, sooty mold.

Injury: Aphid nymphs and adults may cause plant damage in three important ways: (1) as they suck out plant juices they cause leaf wilting, curling, and chlorosis (yellowing): (2) their piercing sucking feeding methods facilitates the transmission of important plant diseases: (3) their partially digested liquid excrement called 'honeydew' serves as a base upon which sooty mold can grow. This interferes with photosynthesis.
Management: Conservation of natural enemies is one of the most effective methods of aphid management. Maintaining plant health and vigor is also important. Alternative and chemical controls are a last resort.

FIGURE 5.10 Armyworm adult moth.

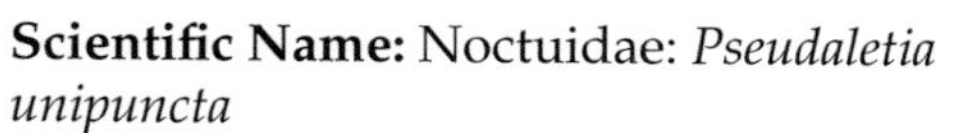

Scientific Name: Noctuidae: *Pseudaletia unipuncta*
Status: pest of grasses, small grain crops, and corn
Damaging Stage: caterpillar
Description: Armyworm adults are light brown-gray moths with a conspicuous white spot about the size of a pinhead on each front wing. The wingspan is approximately 60 mm across.

FIGURE 5.11 Armyworm caterpillar.

Description: Armyworm eggs are greenish white in color and are laid in masses. Often, many hatch at the same time which produces 'armies' of these caterpillars. They appear to march in masses, devouring all of the plants in their path. Full-grown caterpillars are gray and are approximately 37 mm long. White, orange, and dark brown stripes run the length of the abdomen on each side. The head capsule is light orange and can sometimes be mottled.

FIGURE 5.12 Armyworm damage: skeletonization, defoliation.

Injury: Armyworms can be serious pests on a number of grasses, small grain crops and corn. However, they also feed on and sometimes damage alfalfa, beans, clover, flax, millet and sugar beets. Young caterpillars skeletonize leaf blades, while older caterpillars can consume the entire leaf.
Management: Armyworm management depends upon the stage of the crop, number of armyworms and their size. These factors together determine their potential to cause economic or aesthetic harm and should be the basis for management decisions.

FIGURE 5.13 Bagworm adult moth.

Scientific Name: Psychidae: *Thyridopteryx ephemeraformis*
Status: common pest of evergreens and shrubs
Damaging Stage: caterpillar
Description: Adult bagworm moths are seldom encountered. The small football-shaped bags are the most noticeable form of the insect and are commonly found hanging from leaves and twigs. Bagworm eggs hatch in midsummer and the larvae crawl out of the bottom of the bag. There may be as many as 300 eggs per bag.

FIGURE 5.14 Bagworm caterpillar.

Description: Larvae are light brown or tan, although some may have a mottled appearance. The small caterpillars spin silken strands that are either caught by the wind and dispersed or are wrapped around tree branches. From there they begin creating small silk shelters, woven together with bits of foliage (bags) from their environment. Bagworms live within these bags for protection and enlarge them as they grow. They may grow to 50 mm or more in length.

FIGURE 5.15 Bagworm damage: defoliation.

Injury: Bagworms prefer juniper, arborvitae, spruce, pine, and cedar but also occasionally attack deciduous trees. They can cause severe damage to trees and shrubs as they defoliate the branches and are particularly damaging on evergreen plants that do not replace their needles.
Management: Picking off and destroying the bags is often best if populations are light and plants are small. Timing is critical for bagworm chemical management. Using pesticides or alternative controls at egg hatch and while the larvae are exposed, is the most effective strategy.

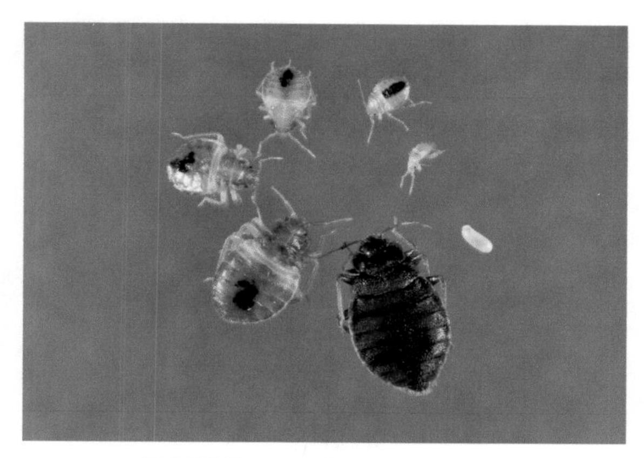

FIGURE 5.16 Bed bug life cycle.

Scientific Name: Cimicidae: *Cimex lectularius*
Status: blood-feeding pest of people
Damaging Stage: nymph and adult
Description: Bed bug adults are reddish-brown, oval, wingless insects that measure about 12 mm in length. The flattened abdomen has a banded appearance. Bed bugs have five nymphal instars, each requiring a blood meal and approximately 1 week to complete.
Bed bugs have a long history with humans and are well known. In the recent past they were largely eradicated from the United States. However, very recently they have made a dramatic comeback and are now considered serious urban pests in the United States and throughout the world.

FIGURE 5.17 Bed bug nymph immediately after feeding.

Description: Females bed bugs lay one to seven eggs each day and may potentially lay more than 100 eggs in a lifetime. Immature bed bugs are similar in shape to adults but are much smaller. They are usually flattened and colorless, except immediately after feeding when they turn a purple-red color and swell up.

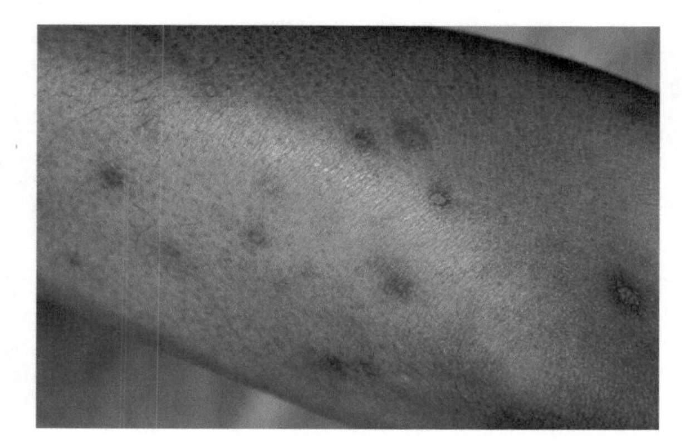

FIGURE 5.18 Bed bug bites on human arm.

Injury: Although they do not transmit diseases they are responsible for considerable physical irritation and emotional stress due to their blood-feeding activities. Bed bugs usually bite people at night while they are sleeping. Although bite symptoms vary, most people develop an itchy red welt.
Management: Eradicating bed bugs is very difficult and often requires a combination of laundry, heat treatments, encasements, monitoring, as well as pesticide tactics.

FIGURE 5.19 Black cutworm moth.

Scientific Name: Noctuidae: *Agrotis ipsilon*
Status: pest of field and garden crops, and turfgrass
Damaging Stage: caterpillar
Description: The drab-colored moths are similar in size and shape to other species of the cutworm family. The distinguishable marking is a small, black slash or dagger near the outer edge of the front wings. Females lay their eggs in cracks in the soil. The eggs hatch within a week then the larvae feed on host plants for about a month. Adults and larvae are nocturnal.

FIGURE 5.20 Black cutworm caterpillar.

Description: Black cutworms caterpillars are gray to nearly black in color with a distinct pale stripe extending down the center of the back. Newly hatched larvae are approximately 6 mm long but reach 50 mm the time they are ready to pupate.
Pupae and adults overwinter in the soil but cannot survive the winter in northern United States. The moths return each year from the South via strong air currents and storms that blow them northward.

FIGURE 5.21 Black cutworm defoliation damage.

Injury: Severe damage usually does not occur until black cutworms reach the fourth instar. Early instar cutworm feeding involves cutting small irregular holes in plant leaves. Larger black cutworms completely sever the plants at the soil line. Grass plants, including corn and turfgrasses are most susceptible to cutworm damage. Multiple generations of cutworms may occur beginning very early in the spring.
Management: Cutworm management depends upon the value of the crop, stage of development, number of cutworms and their size. These factors determine their potential to cause economic or aesthetic harm and should be the basis of management decisions.

FIGURE 5.22 Blow fly adult.

Scientific Name: Calliphoridae: several species
Status: annoyance around homes and can spread disease
Damaging Stage: adult;
Beneficial Stage: maggot
Description: Adult blow flies are metallic blue, green, copper, or black in color and otherwise may resemble house flies in appearance. The hair on the last antennal segment is feathery and can be used as a diagnostic character.
Blow fly adults deposit eggs in and around dead animals and animal refuse. When the eggs hatch, the larvae feed on decaying flesh, rotting vegetation or matted hair.

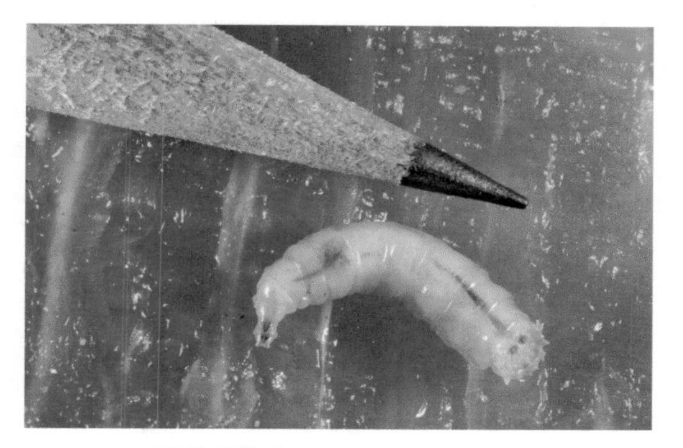

FIGURE 5.23 Blow fly maggot.

Description: Female flies lay eggs on or near suitable habitats. Maggots hatch from the eggs within two days and develop through three instars before pupating in the soil. Blow fly maggots are cream-colored, have a pointed head end and are approximately 12 mm long. Adult flies emerge 10 to 17 days after the formation of the pupal cell. They may complete many generations per year.

FIGURE 5.24 Blow fly benefit: recycle organic wastes.

Value: Blow flies are valuable because they help break down and recycle animal tissue and manure. They are also valuable to forensic investigations in determining time of death.
Injury: Blow flies can be responsible for transmitting pathogens such as *Salmonella, Shigella, Enterococcus*, and *Chlamydia*.
Management: When required, blow fly populations may be controlled using sanitation and exclusion practices. Specific pesticides must be used very cautiously around humans or human foods.

FIGURE 5.25 Cabbage butterfly adult.

Scientific Name: Pieridae: *Pieris rapae*
Status: pest of vegetable crops
Damaging Stage: caterpillar
Description: The wings of cabbage butterflies are white-green and forewings have black tips. There are two submarginal black spots in females and one in males. The butterflies overwinter as pupae and emerge in early spring. There is one generation per year.

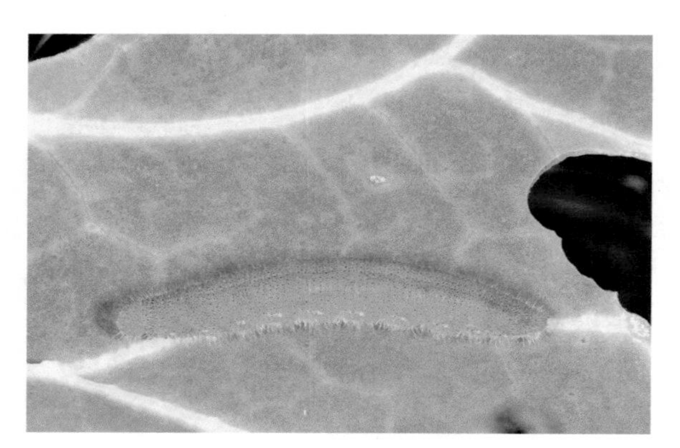

FIGURE 5.26 Cabbageworm caterpillar.

Description: Caterpillars are called cabbageworms. They are green with a yellow stripe down the middle of the back and grow to approximately 31 mm in length. They have four pairs of prolegs in addition to three pairs of true legs near the head.

FIGURE 5.27 Cabbageworm damage: defoliation and skeletonization.

Injury: Cabbageworms feed on plants of the mustard family (crucifers), particularly cabbage and broccoli. They also are attracted to dandelions and other flowers.
Adults lay eggs on the undersides of leaves. Caterpillars tend to feed on leaf tissue in between the veins. They often hide next to or underneath the leaf veins.
Management: Frequent sampling for cabbageworm when plants begin to head may allow for timely interventions with pesticides to protect yields.

FIGURE 5.28 Carpenter ant adult.

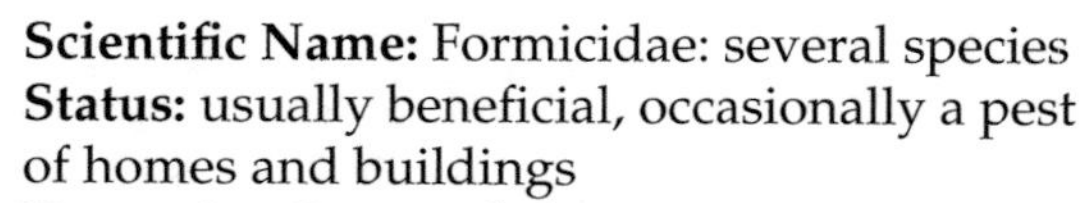

Scientific Name: Formicidae: several species
Status: usually beneficial, occasionally a pest of homes and buildings
Damaging Stage: adult
Description: Carpenter ants range in size from 6 to 20 mm in length, depending on whether the insect is a queen or a worker. Queens are large and black with some red, brown, or yellow spots occurring on parts of the body and legs. The smaller workers are brown and have a large head and a small thorax.

FIGURE 5.29 Carpenter ant winged reproductive.

Description: Carpenter ants live in social colonies with castes including one queen and many workers. When colonies mature, winged reproductives may be produced. After mating, fertilized queens form nests and lay eggs.
Ant larvae are legless and grub-like. They are approximately 7 mm long and are cream-colored. Larvae remain in the nest and are cared for by the workers.

FIGURE 5.30 Carpenter ant wood damage.

Biology: Most carpenter ants are beneficial. Outdoors, they may offer plant protection by consuming potential pest insects and they also facilitate the breakdown of dead trees on a forest floor.
Injury: Carpenter ants do not eat wood but they often nest in it. They may cause damage to structures when they nest in or hollow out wood boards for shelter.
Management: Chemical controls are usually the best option for a carpenter ant infestation inside a building when the nest can be located. Ant baits are effective when properly used to control nuisance foraging ants.

FIGURE 5.31 Carpenter bee adult.

Scientific Name: Apidae: *Xylocopa virginica*
Status: beneficial as pollinators but can be a pest of wooden structures
Damaging and Beneficial Stage: adult
Description: A carpenter bee closely resembles a bumble bee except that the upper surface of its abdomen is bare and shiny black, lacking the yellow and white hair that is so distinctive in bumble bees. They may cause aesthetic damage to exposed bare wood surfaces in buildings and decks.

FIGURE 5.32 Carpenter bee boring damage in wooden deck board.

Description: Carpenter bees are solitary and do not form colonies as do bumble and honey bees. They deposit eggs into tunnels bored into wood along with provisions of pollen from trees and flowers. Larvae are grublike and remain in the tunnels until mature thus are seldom encountered.

FIGURE 5.33 Wood frass from carpenter bee boring.

Benefit: Carpenter bees are valuable pollinators of many plants.
Injury: Carpenter bees do not feed on wood. Damage to wooden structures occurs when the adults drill holes into and excavate tunnels for shelter and nesting. They may bore into a wide range of woods used by people including wooden fences, utility poles, siding, decks, firewood, trees, and lawn furniture.
Management: Chemical controls are somewhat effective if applied repeatedly every few weeks to the entrance of holes. Physical removal of boring adults is also effective.

FIGURE 5.34 Cicada adult (annual).

Scientific Name: Cicadidae: several species
Status: annoying; can damage tender tree limbs
Damaging Stage: adult
Description: Cicadas have prominent wide-set eyes, short antennae, and clear wings held roof-like over the abdomen.
Two major groups of cicadas are common. Annual cicadas can be more than 50 mm long, have very unique green or brown bodies and large eyes. Annual cicadas emerge each year during the summer months.
Periodical cicadas are 25–50 mm in length, have black bodies and orange wing veins. They spend the majority of their lives underground but emerge in great numbers after 13 or 17 years, depending on the brood.

FIGURE 5.35 Cicada emerging from nymphal cast skin.

Description: After mating, female cicadas cut slits into the tender bark of young twigs in which to lay their eggs. When the eggs hatch, the nymphs drop to the ground and bury themselves down into the soil until they find a tree root where they can attach and begin to suck out juices. They reappear above ground to complete their last molt before emerging as winged adults.

FIGURE 5.36 Cicada damage to tree twigs and branches.

Injury: When cicadas emerge in large numbers, they may leave significant numbers of unsightly cast skins on tree trunks, fence posts or the sides of structures. When there is a large emergence, the loud calls of the cicadas can be annoying and deafening. Mostly, however, real cicada damage occurs as a result of these insects killing small tree branches and twigs when they lay their eggs.
Management: Wrapping or caging susceptible trees (young or recently transplanted) is the only effective exclusion strategy.

FIGURE 5.37 Codling moth adult.

Scientific Name: Tortricidae: *Cydia pomonella*
Status: pest of fruits and nuts
Damaging Stage: caterpillar
Description: Codling moth adults are approximately 12 mm long with mottled gray wings that are held tent-like over their bodies. They can be distinguished from other moths by a dark, copper-brown band at the tips of their wings.

FIGURE 5.38 Codling moths are also called apple worms.

Description: Codling moth larvae are pink or white with brown head capsules and are approximately 20 mm long.
Adults emerge in late spring and lay eggs on fruit, nuts, leaves, and spurs. The eggs hatch within two weeks and the larvae bore into and feed inside the fruit of their host tree. After they complete development the larvae drop from the trees to find pupation sites in the soil. There may be two to three generations per year.

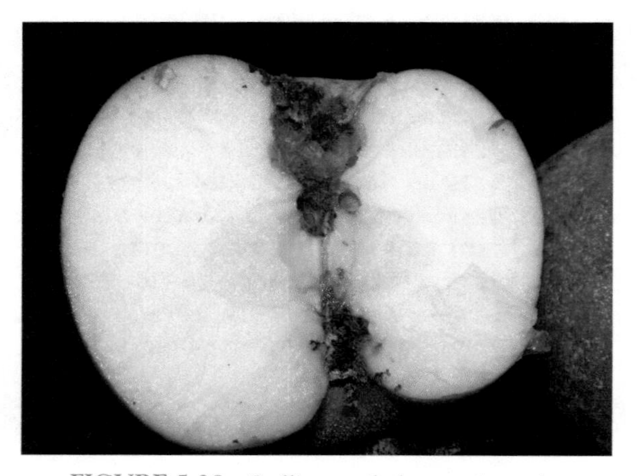

FIGURE 5.39 Codling moth damage in apple.

Injury: Larvae typically tunnel to the core of the fruit. The tunnel entrances are characteristically surrounded with red-brown, crumbly frass deposits. Infested fruit is considered 'wormy' and unfit for human consumption.
Management: Insecticides, mating disruption, and cultural controls are used to keep codling moth populations in check.

FIGURE 5.40 Colorado potato beetle adult.

Scientific Name: Chrysomelidae: *Leptinotarsa decemlineata*
Status: pest of some vegetable crops, particularly potato
Damaging Stage: larval
Description: Colorado potato beetles are major pest of potatoes throughout the U.S. They became a pest in Colorado in 1859 when they moved from their native host to feed on cultivated potatoes. They then began to spread eastward and have since become a major potato pest throughout the world.
Adults are approximately 13 mm long and have characteristic convex elytra marked with black and yellow-white stripes. The head is tan-orange with black markings.

FIGURE 5.41 Colorado potato beetle larvae are often called potato bugs.

Description: Colorado potato beetle larvae are soft-bodied, slug-like, and appear hump-backed. Early stages are red with two rows of black spots on each side. As they mature they turn a salmon-pink color.
Female beetles lay eggs in clusters of ten to thirty on the undersides of leaves. After the eggs hatch, the larvae remain in feeding groups while young but disperse through the plant leaves as they age. There are one or two generations per year.

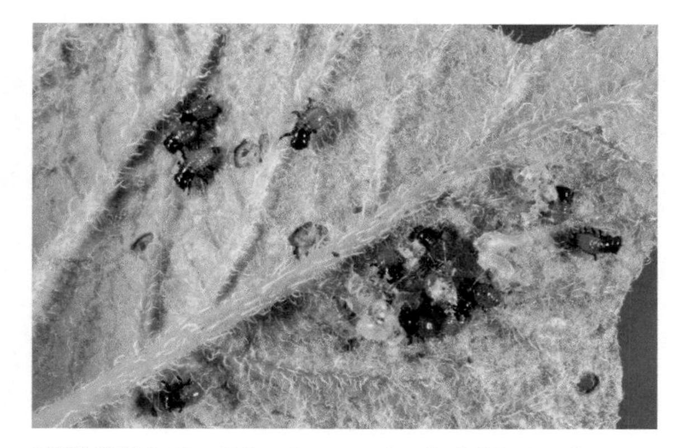

FIGURE 5.42 Colorado potato beetle defoliation damage.

Injury: Colorado potato beetle larvae feed on potato foliage until it is completely devoured. In addition to potato, they also feed on eggplant, tomato, pepper and other garden plants.
Management: Crop rotation, residue sanitation, and mulching, as well as pesticides and alternative controls are required for Colorado potato beetle management.

FIGURE 5.43 Corn earworm moths are also called cotton bollworms or tomato fruitworms.

Scientific Name: Noctuidae: *Heliothis zea*
Status: a major agricultural pest, particularly of sweet corn, cotton and tomatoes
Damaging Stage: caterpillar
Description: Adult moths are light yellow/ brown in color and with a wing span of 35 mm. Their hind wings are creamy white with dark edges. They live for 2–4 weeks and may lay as many as 3000 eggs during that time. Usually two generations occur in the northern U.S. but many more in the South.

FIGURE 5.44 Corn earworm caterpillar.

Description: Larvae can be light green, pink, dark brown, or nearly black. They have alternating light and dark stripes that run longitudinally down the body. Usually, there are double dark stripes toward the center of the back. When mature, larvae reach a length of approximately 35 mm.

FIGURE 5.45 Corn earworm damage.

Injury: After eggs hatch, the larvae feed on leaves, tassels, whorls and within fruits. In corn, young larvae may feed on the silks but the greatest damage is inflicted on the ears when kernels are filling. Extensive damage is characterized by large amounts of frass around the feeding sites.
Management: Corn earworm management practices should be timed to coincide with egg hatch and silk development.

FIGURE 5.46 Dermestid beetles are also called carpet beetles, skin beetles and hide beetles.

Scientific Name: Dermestidae: several species
Status: pest of stored products
Damaging Stage: larval and adult
Description: Adult beetles range from 2–15 mm in length and can vary in shape from elongated to oval. They are usually dark brown but may be patterned with lighter scales. There are many species of dermestid beetles, several of which may become pests.

FIGURE 5.47 Dermestid beetle larvae.

Description: Dermestid larvae are brown in color and covered in long red-brown setae. They have two distinctive spines on the end of the abdomen that curve backward. Eggs usually hatch within two weeks. Larvae develop through five or six instars before pupating. The pupal stage usually lasts less than two weeks. There may be many generations per year.

FIGURE 5.48 Dermestid beetle damage to stored products.

Value: In outdoor environments, dermestid beetles fill a valuable role as decomposers of organic materials. Some species are commonly found in decaying animal carcasses.
Injury: Most damage is caused by the larval stage. Dermestid larvae feed on a number of stored foods as well as valuable animal products such as wool, silk and leather.
Management: Dermestid beetles are most effectively managed using preventative and sanitation techniques. When populations are out of control, fumigations and conventional pesticides may be used.

FIGURE 5.49 Emerald ash borer adult.

Scientific Name: Buprestidae: *Agrilus planipennis*
Status: devastating exotic pest of ash trees
Damaging Stage: larval
Description: Emerald ash borers are conspicuous because of their flat-heads, large black eyes and their bullet-shaped, dark metallic-green bodies. They measure 12 mm in length and are 3 mm wide. The adult beetles emerge in early to midsummer.

FIGURE 5.50 Emerald ash borer galleries.

Description: Females begin laying eggs about two weeks after emergence. Larvae hatch in one to two weeks, are cream-colored and have flat, broad, segmented bodies. The larvae bore through the bark of ash trees and feed for several weeks beneath the bark, leaving characteristic S-shaped tunnels. Pupation occurs in the springtime. Emerald ash borers also leave a characteristic D-shaped exit hole in the bark when they emerge as adults.

FIGURE 5.51 Emerald ash borer caused dieback – leading to tree death.

Injury: Tunneling by the larvae eventually kills the tree. Trees typically begin to die back from the top of the canopy and symptoms progress downward leading to the death of the tree in 2-4 years.
Management: Preventing the spread of infested ash trees is the best control strategy. Once infested, ash borers are difficult to control. Systemic insecticides show some promise but are a last resort.

FIGURE 5.52 European corn borer adult.

Scientific Name: Pyradilae: *Ostrinia nubilalis*
Status: pest of field corn
Damaging Stage: caterpillar
Description: Adults are approximately 12 mm in length. They hold their tan-colored wings in a triangular shape at rest. European corn borers overwinter as larvae in corn stalks left from the previous growing season. As springtime temperatures increase, the larvae enter the pupal stages for two weeks before developing into adults. The adults generally emerge in late summer. There are two generations per year.

FIGURE 5.53 European corn borer larva.

Description: European corn borer larvae are either a light tan or pink color. They have distinctive small, round brown spots on each segment. Mature larvae can reach a length of 25 mm. European corn borers feed on all above-ground tissues of the corn plant but are known for tunneling within the stalk and ear thereby weakening the plant and reducing potential yield.

FIGURE 5.54 European corn borer damage on corn.

Injury: European corn borers attack the tassels, ears, and stalks of corn plants. Once they enter into the stalks, corn borers form characteristic cavities that interfere with water and nutrient movement in the plant and weaken it structurally. Damage progresses from shot holes to tunneling to plant lodging to death.
Management: Chemical controls are effective if they are used preventively early in the growing season. Recently transgenic corn (*Bt* corn) has been developed that includes a protein from a biological control agent and offers significant improvements in corn borer control.

FIGURE 5.55 Flea adult.

Scientific Name: Pulicidae: *Ctenocephalides felis*
Status: nuisance pest of humans and pets
Damaging Stage: adult
Description: Fleas are 1.5 mm in length, very agile, dark-colored, wingless insects with tube-like mouthparts that are adapted to feed on the blood of mammals. Their bodies are laterally compressed, and they have long legs that are well adapted for jumping. After a blood meal, female fleas lay their eggs in the hair of their host. Inside homes, eggs fall from the hair onto bedding, floor surfaces, rugs, and furniture wherever the animal spends a majority of its time. Eggs hatch in a few days.

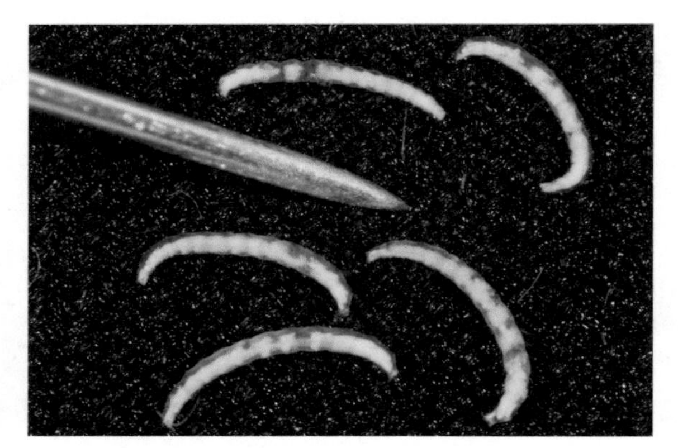

FIGURE 5.56 Cat flea larva.

Description: Flea larvae resemble tiny white worms having no legs and a distinct brown head with no eyes. Even at maximum length, flea larvae are very small, measuring approximately 5 mm.
Flea larvae, unlike adults, do not move around much and do not feed on blood. They tend to avoid light by hiding in cracks and crevices. The larvae primarily feed on animal dander in their habitat and develop over the course of several weeks. The pupae mature to adulthood in a silken cocoon in about two weeks.

FIGURE 5.57 Flea bites cause significant irritation and health concerns in animals and humans.

Injury: Adult fleas are nuisance pests to both humans and their pets. Flea saliva is irritating to the skin and small, red welts can form where bites have occurred. Fleas can also transmit tapeworms, cause anemia and create other minor medical issues such as allergic reactions.
Management: Although fleas can bite humans, they prefer to live on domesticated animals. When adult fleas are found, both the pet and the premises must be treated simultaneously. Pets should be properly treated as soon as fleas are discovered. Infested areas (where the larvae reside) should be properly cleaned and chemically treated concurrently.

FIGURE 5.58 Gypsy moth male and female.

Scientific Name: Erebidae: *Lymantria dispar*
Status: introduced pest of hardwood trees
Damaging Stage: caterpillar
Description: Adult gypsy moths have wings with variable patterns of black spots and bands. Males (right) have brown wings and feathery antennae whereas females (left) are cream-colored, and have thread-like antennae. Gypsy moth adults do not feed but larvae may feed on several hundred different species of trees. Gypsy moth taxonomy has been very fluid but as of 2012, they have been classified in the family Erebidae.

FIGURE 5.59 Gypsy moth caterpillar.

Description: Gypsy moths survive the winter in the egg stage and hatch in the spring when temperatures are above 15°C. Eggs are laid in mid- to late summer. They complete one generation per year. Female moths lay egg masses indiscriminately on trees, houses, and other structures in late summer. Often campers unknowingly spread this pest when egg masses are attached to recreational vehicles. Caterpillars characteristically have five pairs of raised blue spots along the back followed by six pairs of red spots.

FIGURE 5.60 Gypsy moth defoliation of trees.

Injury: Heavy infestations can be responsible for the complete defoliation of host trees. This damage, however, does not directly cause death but leaves trees susceptible to secondary organisms such as borers and root rots that are attracted to and finally kill the weakened trees. Caterpillars may also cause allergic reactions in humans. The hairs cause skin rashes, particularly during the month of May, when larvae are small.
Management: Biological and chemical controls can suppress populations if applied early. Homeowners can also plant gypsy moth resistant trees to replace damaged ones.

FIGURE 5.61 Hornworms are called either tomato hornworms and tobacco hornworms.

Scientific Name: Sphingidae: *Manduca* sp.
Status: pest of certain vegetable plants
Damaging Stage: caterpillar
Description: Hornworms are large, robust moths that have long, narrow front wings. Their bodies are spindle-shaped and pointed at both ends. They have gray-and-white mottled wings and abdomens lined along each side with five (tomato hornworm) or six (tobacco hornworm) conspicuous, orange-yellow spots. Females lay their eggs singly on leaves. Larvae feed for three or four weeks before burrowing into the soil to pupate. The pupae overwinter, and the adults emerge in the spring. There are two generations per year.

FIGURE 5.62 Hornworm caterpillars.

Description: The distinctive green caterpillars can grow to more than 75 mm in length. A unique large, thick, pointed structure, or 'horn,' protrudes from the upper abdominal end and is responsible for the name 'hornworm.' Tobacco hornworms have seven straight, oblique white lines edged with black on the upper borders that run laterally along their abdomen. Tomato hornworms have eight V-shaped marks. Due to their size, tobacco hornworms can consume large quantities of plant material. However, hornworms are not considered to be serious pests of commercial crops, probably due to the activities of natural enemies.

FIGURE 5.63 Hornworm defoliation and fruit destruction.

Injury: Hornworms damage foliage, blossoms, and green fruits of tomatoes, tobacco, potatoes, eggplants, and peppers. Larvae develop through five instars. The fifth instar is responsible for the majority of the defoliation that occurs.
Management: Conserving naturally occurring, parasitic wasps provides adequate biological control in most cases. This, together with picking off and destroying caterpillars in tomato and other garden crops, is usually sufficient. Rototilling to destroy the pupae is also helpful. *Bt* is effective against early instars. Chemical pesticides are also very effective when justified.

FIGURE 5.64 House fly adult.

Scientific Name: Muscidae: *Musca domestica*
Status: nuisance pest of homes and farms
Damaging Stage: adult;
Beneficial Stage: maggot
Description: Adults are 12 mm long and have a gray thorax with four, dark, longitudinal lines on the back. The undersides of their abdomens are yellow, and their bodies are covered with hair. They have red compound eyes. Females lay their eggs singly in moist environments.

FIGURE 5.65 House fly maggot.

Injury: Female house flies lay large numbers of eggs near suitable larval food sources (moist garbage, animal excrement, or decomposing plant material). House fly maggots are cream- or white-colored and cone-shaped. The head contains one pair of dark hook-like mouthparts. Maggots develop during a two-week period, and the pupal stage lasts less than one week. Adults may live for two months. There can be as many as twelve generations per year.

FIGURE 5.66 House flies are attracted to refuse and can be a nuisance and potential disease vector.

Value: In natural systems house flies are helpful in breaking down and recycling organic wastes.
Injury: Human health problems can occur with the movement of flies from animal or human feces to food made for human consumption. House flies can be responsible for transmitting pathogens such as *Salmonella*, *Shigella*, *Enterococcus*, and *Chlamydia*.
Management: The most effective way to control house fly populations is to implement sanitation and exclusion practices. Specific pesticides must be used very cautiously around humans or human foods.

FIGURE 5.67 Indian meal moth adult.

Scientific Name: Pyralidae: *Plodia interpunctella*
Status: pest of stored food products especially grains
Damaging Stage: caterpillar
Description: Indian meal moth adults are approximately 10 mm in length. The overall body color is generally brown-gray, but the outer half of the wing is abruptly bronze-colored. Female moths lay their eggs singly or in clusters on suitable larval food. There are four to six generations per year.

FIGURE 5.68 Indian meal moth larva.

Description: Larvae have brown head capsules and are a dirty white or cream color. Sometimes, they may have a slight pink, green, or yellow tint and may reach approximately 15 mm in length.
The larvae produce silken blankets for protection while feeding. Larval development time varies with temperature and type of food material. Before pupating, the larvae leave the food source.

FIGURE 5.69 Indian meal moth in stored grains.

Injury: Indian meal moth larvae feed on many kinds of stored food products. As the larvae mature, they leave behind silken threads that bind to food particles. This webbing is often what attracts attention, either in stored grains or stored pantry foods.
Management: Improved sanitation and prevention practices are the best way to control Indian meal moth populations. This involves removing old infested products and thoroughly cleaning all containers before new product is added. This is true for farm-stored grains or for infestations in homes or pantries. Chemical controls including fumigation should only be used in cases of extremely high infestations.

FIGURE 5.70 Japanese beetle adult.

Scientific Name: Scarabaeidae: *Popillia japonica*
Status: pest of many plants, grasses, and ornamentals
Damaging Stage: grub and adult
Description: The Japanese beetle is about 13 mm long with shiny copper-colored wing covers and an iridescent green thorax and head. The abdomen has a row of white hair tufts on each side.
Adults emerge from the ground in midsummer, and may feed on more than 400 different trees, shrubs, flowers, vegetables, and crops. Females deposit their eggs in the soil. The eggs hatch about two weeks after deposition, normally between July and August.

FIGURE 5.71 Japanese beetle annual white grubs.

Description: Japanese beetle larvae are C-shaped white grubs that live in the soil. They have a brown head capsule and three pairs of legs. The larvae overwinter in cells beneath the soil surface. In the spring, the grubs resume feeding and prepare to pupate. There is one generation per year.

FIGURE 5.72 Japanese beetle defoliation by adults.

Injury: Adult Japanese beetles skeletonize leaves and render them of little value to the plant. Japanese beetle grubs feed on the roots of grasses.
Management: When plant leaves are decimated by adult beetle feeding control options should be considered.
Six to eight Japanese beetle larvae recovered per square foot of soil is considered the average action threshold for turfgrass.
Japanese beetle adults and larvae can be managed by a number of biological and chemical controls.

FIGURE 5.73 Lady beetle adult.

Scientific Name: Coccinellidae: several species
Status: very beneficial insect but an annoying household invader
Beneficial Stage: larval and adult
Description: Lady beetles are small, round, and dome-shaped. The most well known lady beetles have black markings on red, orange, or yellow forewings, but some are black. Lady beetles are extremely beneficial insects but can be an annoyance when they appear in large numbers in the home. One species of lady beetle (multicolored Asian lady beetle) has a peculiar behavior of congregating and passing the winter in man-made structures.

FIGURE 5.74 Lady beetle larva.

Description: Lady beetle larvae can grow to approximately 10 mm in length. They are usually black with orange spots and are covered with spines. To some they resemble tiny alligators with three pairs of legs.
In the spring, overwintering adults find food, and the females lay their eggs, often near aphid colonies. The eggs hatch in three to five days, and the larvae feed on aphids or other small insects. After two to three weeks, they pupate. Adults emerge within a week. There may be five to six generations per year.

FIGURE 5.75 Lady beetle congregations in homes can be a serious nuisance.

Value: Larvae feed on aphids, soft-scale insects, mealybugs, spider mites, and other pests. One mature lady beetle larvae can eat its weight in aphids (approximately 400 aphids) in one day.
Injury: Asian lady beetles invade homes during the fall time. This causes significant annoyance to homeowners and may become a health (allergy) problem.
Management: Cultural controls such as sealing entry points in the fall and vacuuming emerging beetles are effective against low populations. Pesticides, applied to the outside of a home, also may be needed.

FIGURE 5.76 Leafhopper adult.

Scientific Name: Cicadellidae: several species
Status: pests of crops, flowers, grasses, vegetables, and trees
Damaging Stage: nymph and adult
Description: Potato leafhoppers are serious pests of agricultural crops. Adults are generally yellow or green and some are marked with various color patterns. They range from 3–8 mm in length and are wedge-shaped.
Many different species of leafhoppers may damage trees and crops such as potatoes, beans, apples, grapes, and clover.

FIGURE 5.77 Leafhopper nymph.

Description: Leafhoppers overwinter as eggs or adults. Adults insert eggs into leaf veins or stems and eggs hatch after about 10 days. Nymphs feed in the same manner as the adults. Nymphs are similar to adults but do not have fully developed wings. Generally, leafhoppers complete one generation per year but some species can complete up to six.

FIGURE 5.78 Leafhopper damage includes wilting, chlorosis and disease transmission.

Injury: There are many species of leafhoppers. All can be pests if populations build up to high levels. Leafhoppers feed by inserting their mouthparts into the plant and sucking the juices. This can cause injury in the form of wilting and yellowing-often called hopperburn. Leafhoppers also can transmit deadly plant diseases.
Management: Outside of agriculture, control methods are rarely required for leafhoppers due to the large number of natural enemies that are usually present to suppress populations.

FIGURE 5.79 Mexican bean beetle adult.

Scientific Name: Coccinellidae: *Epilachna varivestis*
Status: pest of beans and peas
Damaging Stage: larval and adult
Description: The copper-colored adults resemble large lady beetles. They are 6 mm long with eight black spots on each wing. Adults overwinter and emerge in midsummer. After feeding, the females lay their eggs on the underside of foliage. They hatch within a week during warm weather. There are two to three generations per year.

FIGURE 5.80 Mexican bean beetle life stages.

Description: Mexican bean beetle larvae are quite unique in appearance. The entire body is covered with rows of stout-branched spines. When the larva is newly hatched, the entire body (including spines) is bright yellow, but as they mature, the spines become darker at the tips. Larvae also have a sucker-like apparatus at the hind end for attachment to feeding surfaces. The larvae feed for two to five weeks before pupating. When pupating, a larva fastens the tip of its abdomen to a part of the plant and sheds its larval skin. The pupal stage lasts for a week before the adult emerges.

FIGURE 5.81 Mexican bean beetle leaf damage.

Injury: The Mexican bean beetle is one of the only harmful members of the lady beetle family. Adults and larvae feed on the leaves of all kinds of beans (snap, lima, pole, kidney, pinto, navy, and bush). Mexican bean beetle leaf damage gives the leaves a skeletonized appearance. Pods and stems can be attacked also.
Management: Cultural, biological, and chemical control options are all successful in suppressing beetle populations.

FIGURE 5.82 Mosquito adult.

Scientific Name: Culicidae: several species
Status: serious blood-sucking pests, transmits diseases
Damaging Stage: adult (female)
Description: Mosquitoes are one of the most important pests that afflict mankind. They can transmit a number of deadly diseases, including malaria, dengue fever, yellow fever, West Nile virus, and encephalitis. Several species of mosquitoes exist. Most are 6–12 mm long and have only two wings, both covered in scales. Females have an elongated proboscis.

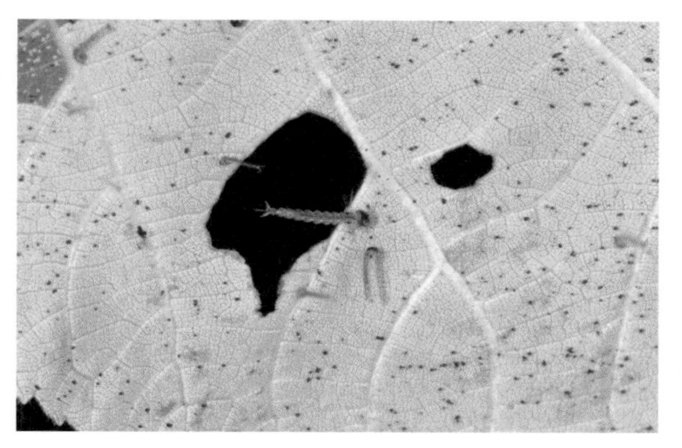

FIGURE 5.83 Mosquito larvae are called wigglers.

Description: Most mosquitoes lay their eggs together in rafts on the surface of water. They hatch into larvae in about two days. Mosquito larvae are usually black or brown and are called 'wigglers' because of their distinctive swimming style when disturbed. Larvae breathe through a long siphon at the surface of the water. They are slightly C-shaped with an enlarged front end. The larvae live in the water until they pupate. Mosquito larvae primarily feed on organic material, bacteria, and microscopic plants in the water. Most mosquito species overwinter as eggs.

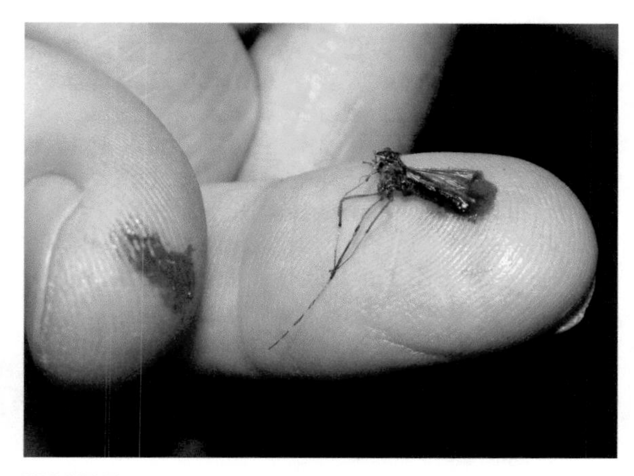

FIGURE 5.84 Mosquito vectored disease transmission is a serious threat.

Injury: Disease transmission and irritation due to mosquito bites.
Management: Avoid mosquito-infested areas at times when they are most active. If contact is likely, wear long-sleeved shirts and pants. The best way to reduce mosquito populations long-term is to eliminate their aquatic habitat. Non-chemical control methods include mosquito-eating fish, Bt, and oil applications. Pesticides applied to the water are effective but must be used very carefully. Fogging for adult mosquitoes only provides short-term control and should only be used as a last resort.

FIGURE 5.85 Rice weevil adult.

Scientific Name: Curculionidae: *Sitophilus oryzae*
Status: pest of stored cereal products
Damaging Stage: larval and adult
Description: The rice weevil is small (2.5 mm) but has a long, curved snout almost one-third of the total length of the insect. The body is red-brown to black in color with four light yellow or red spots on the corners of the wing covers. Adults chew into the grain kernels from the outside and lay their eggs inside the grain. Adults often live for seven to eight months. There are usually four generations per year.

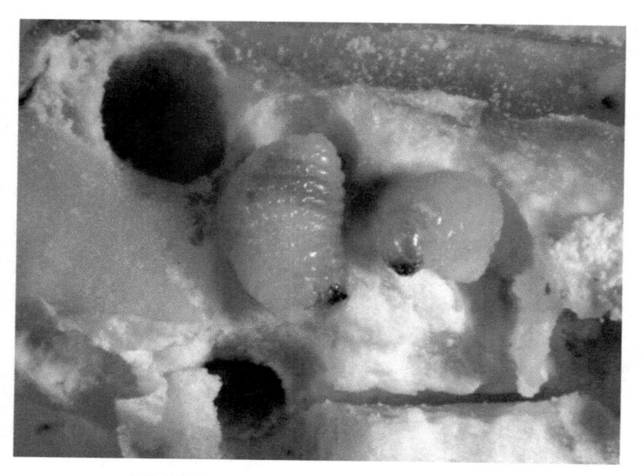

FIGURE 5.86 Rice weevil larvae.

Description: Rice weevil larvae are cream-colored with small, tan-colored, head capsules. They are legless, humpbacked, and rarely seen because they stay inside hollowed grain kernels. Larvae develop through several instars and also pupate inside the grain kernels. They may complete a generation in a month in warm conditions.

FIGURE 5.87 Rice weevil grain contamination and destruction.

Injury: Rice weevils are generally pests of wheat, oats, rye, barley, rice, and corn. Adult females drill a hole into the grain kernel and lay their eggs in the cavity. The hole is plugged with a sap-like secretion. Once the eggs hatch, the larvae bore toward the center of the kernel, where they feed and pupate.
Management: The best way to control an infestation is to locate the source and eliminate it. Insecticides should not be used in some instances because food for human consumption is involved.

FIGURE 5.88 Spittlebug adult.

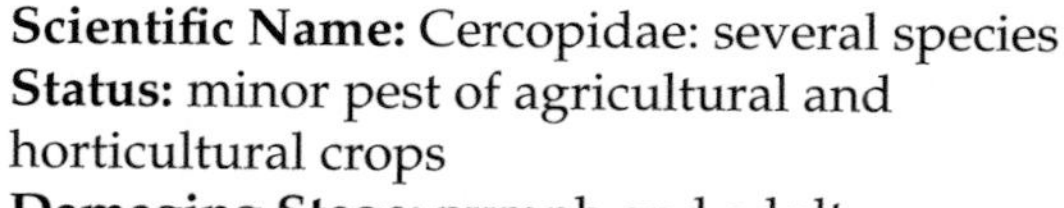

Scientific Name: Cercopidae: several species
Status: minor pest of agricultural and horticultural crops
Damaging Stage: nymph and adult
Description: Spittlebugs derive their name from the white, frothy spittle the nymphs produce. Adults resemble leafhoppers but can be quite large, measuring approximately 10 mm in length. The body color varies from brown to orange. Common species have dark wings with two red stripes that cross the back. They complete two or three generations per year.

FIGURE 5.89 Spittlebug nymph.

Description: Spittlebug nymphs resemble small wingless adults. They produce a characteristic white foamy substance that surrounds them as they feed on the sap of a host plant. Spittlebugs can be found on both herbaceous and woody plants.
In late summer, adults lay their eggs, which then overwinter. The eggs hatch in early spring, and the nymphs go through five instars before emerging as adults.

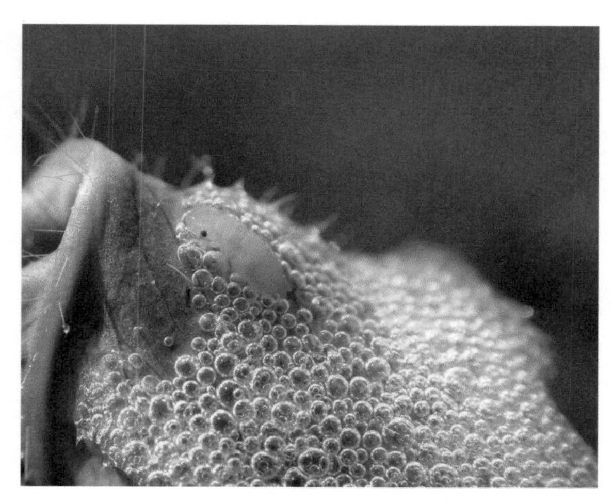

FIGURE 5.90 Spittlebug feeding results in plant dieback.

Injury: In large populations, twigs and branches may be damaged enough to cause some dieback. Herbaceous plants suffer from the sap feeding as well as the injection of phytotoxic salivary substances during feeding. Weakened pines can suffer in weather conditions that encourage disease.
Management: Spittlebugs rarely occur in large enough numbers to cause enough injury to justify pesticide applications. More often, the spittle masses are considered an aesthetic nuisance because they are unsightly rather than damaging. Chemical controls are the best option for a severe spittlebug infestation.

FIGURE 5.91 Squash bug adult.

Scientific Name: Hemiptera: *Anasa tristis*
Status: pest of vegetable crops
Damaging Stage: nymph and adult
Description: Squash bug adults and nymphs attack all cucurbit vine crops, particularly squash, pumpkin, cucumber, and melon. Adults are rather large (14 mm long), winged, brown-black, flat-backed insects that give off a disagreeable odor when crushed. Adults overwinter in the shelter of dead leaves, vines, boards, or buildings. They emerge in the spring to lay masses of eggs on the undersides of leaves. Only one generation develops each year, and new adults do not mate until the following spring.

FIGURE 5.92 Squash bug – nymph.

Description: Squash bug eggs hatch within ten days, and the nymphs pass through five instars in one month. Newly emerged nymphs have light-green-colored abdomens with black heads and legs. As they mature, the abdomen turns a brownish-gray. The final two instars grow from approximately 3 to 12 mm long and have noticeable wing pads.

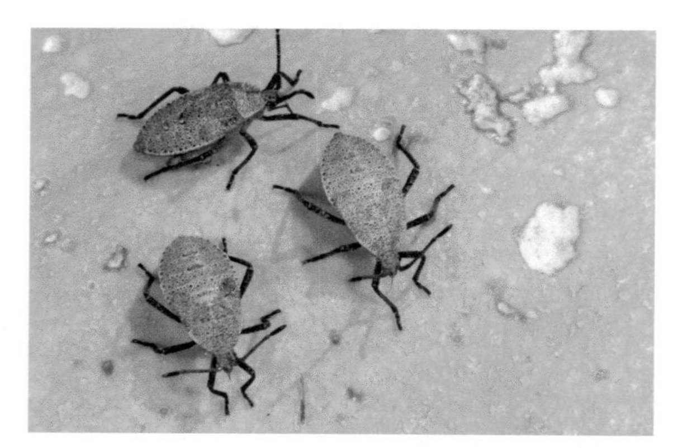

FIGURE 5.93 Squash bug damage is both plant wilting and disease transmission.

Injury: Squash bugs pierce vines with needle-like mouthparts to inject a toxic substance into the plant. Resulting diseases can cause the plant to yellow, wilt, and die.
Management: Diseases spread by squash bugs can result in severe yield loss. If plants show wilting early in the growing season, chemical control of squash bugs may be required. In most cases however, squash bug populations rarely require chemical control methods. Adults and nymphs can be knocked off plants into buckets of soapy water during a light infestation.

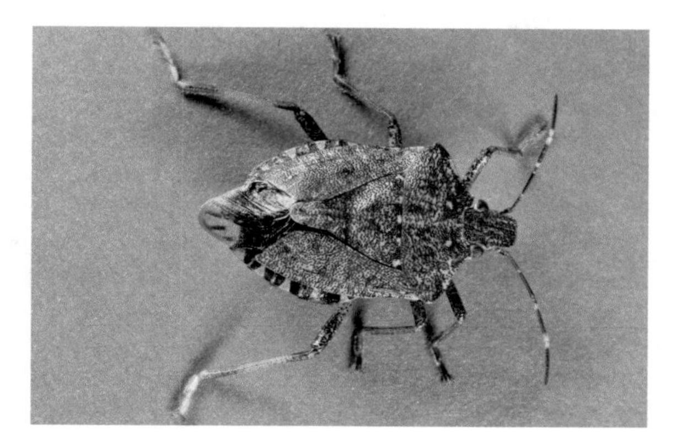

FIGURE 5.94 Stink bug adult.

Scientific Name: Pentatomidae: several species
Status: minor insect pest of crops and a nuisance pest in homes. Can also be a beneficial predatory insect.
Damaging Stage: nymph and adult
Description: Most stink bugs are green or mottled brown/gray in color and grow to about 12 mm in length. Stink bugs have broad shield-shaped bodies, with five points (accounting for their scientific name; Pentatomidae). They have a unique behavioral tendency to emit a strong, pungent odor when disturbed, accounting for their common name. Some species of stink bugs are predatory on other insects – thus beneficial. Others feed on plants.

FIGURE 5.95 Stink bug adult.

Description: Female stink bugs lay their eggs on the undersides of plant leaves or on stems. Eggs hatch and nymphs begin feeding in clusters by inserting their needle-like, mouth parts into plants. Feeding and appearance of nymphs is similar to that of adults. Nymphs are smaller than adults and lack fully developed wings, otherwise they may cause the same injury to a plant. Several generations of stink bugs may occur each year.

FIGURE 5.96 Stink bug eggs.

Value: Some stink bugs are predators of other insects.
Injury: Stink bugs are occasionally damaging pests of cotton, corn, soybeans, trees, shrubs, fruits, and sometimes vegetables. Damage occurs as leaf stippling, seed and fruit abortion, and possible diseases transmission. The recently introduced, brown marmorated stink bug may also be a home-invading nuisance pest.
Management: Stink bugs may be managed with pesticides and exclusion practices.

FIGURE 5.97 Termites are sometimes called white ants.

Scientific Name: Blattodea: several families
Status: pest of homes and buildings
Damaging Caste: worker
Description: Termites are small insects that are white, tan, or black. They are less than 12 mm long. Termites feed in sound, dry wood and create tunnels that run along the grain of the wood. They are very serious structural pests because they destroy the wood that is used for buildings. Workers are able to eat cellulose-containing materials such as wood because they have microorganisms in their intestines that assist in digesting the cellulose.

FIGURE 5.98 Termite soldier and worker castes.

Description: Termites are social insects with three castes (reproductives, workers, and soldiers), each with separate functions in the colony. The reproductives are responsible for producing large numbers of offspring. The workers and soldiers are sterile, wingless, and blind. The workers build the nest, forage for food, and care for the young. The soldiers have enlarged heads and mandibles and defend the colony from intruders.

FIGURE 5.99 Termite damage to a home.

Injury: Termite control costs in the U.S. exceed $2 billion per year and structural damage falls into the hundreds of billions of dollars. Often mud tubes or piles of discarded wings are the first evidence of termite infestation. Weakened and discolored boards and actual presence of the workers themselves are direct evidence of termite infestations.
Management: Termite infestations should be treated by a professional with proper training and equipment. Great advances have been made in both chemical barrier treatments as well as baits.

FIGURE 5.100 Thrips adult.

Scientific Name: Thysanoptera: several species
Status: pest of many plants, particularly in greenhouses
Damaging Stage: nymph and adult
Description: Thrips are slender and minute (1 mm long) and have long fringes on the margins of both pairs of their long, narrow wings. Magnification is required to make out important morphological features. Thrips range in color from translucent white to dark brown.
Depending on the species, thrips can produce from one to eight generations per year.

FIGURE 5.101 Thrips nymphs.

Description: Female thrips deposit eggs in slits made in the leaf tissue. The eggs hatch within a week into active nymphs. The developmental period from egg to adult ranges from eleven days to three weeks. Parthenogenesis (asexual reproduction) occurs in many species.
Nymphs resemble adults in appearance and feeding behavior but lack the coloration and the wings that the adults have.

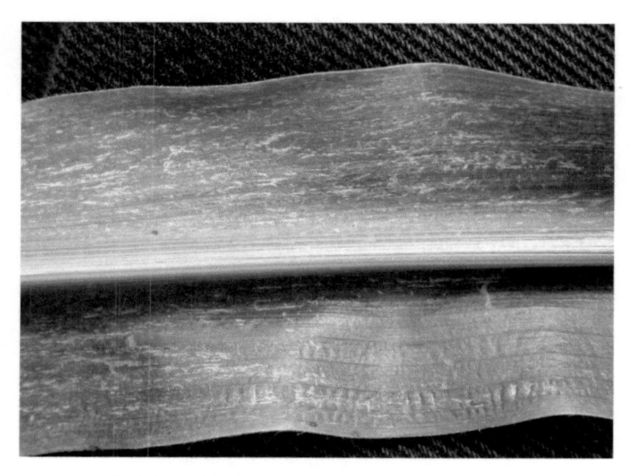

FIGURE 5.102 Thrips damage on leaf.

Injury: Thrips prefer to feed on the youngest leaves of a plant. The plant cells are punctured by the rasping-sucking mouthparts, and plant cell sap is withdrawn. Damaged foliage eventually wilts and dies.
Management: Sticky traps can be used to detect thrips population levels. If thrips numbers are large, control options should be considered. Integrated Pest Management strategies using a combination of cultural, biological, and less-toxic chemical controls generally work best to manage thrips infestations.

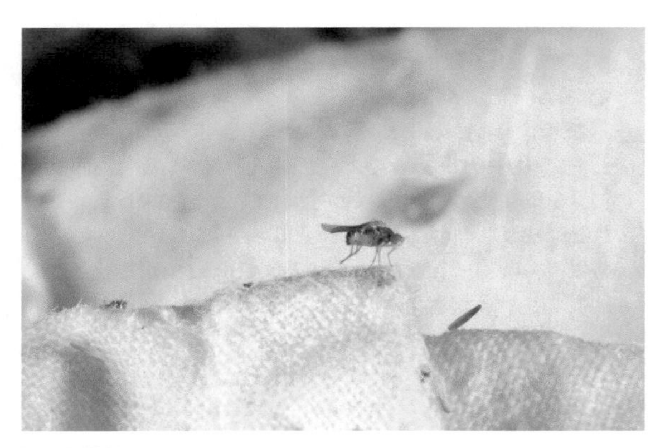

FIGURE 5.103 Vinegar flies are commonly called fruit flies.

Scientific Name: Drosophilidae: *Drosophila* spp.
Status: nuisance pest of homes and restaurants
Damaging Stage: larval and adult
Description: Vinegar flies are small (2.5 mm), delicate flies. They have red eyes, are yellow-brown in color, and have transverse black rings across their abdomens. Vinegar flies are most often found flying about ripening fruit, clogged drains, or dirty garbage cans where they mate and lay eggs. Populations build up very quickly. The time required to complete one life cycle ranges from eight to twenty days. They may complete from 10 to 13 generations per year.

FIGURE 5.104 Vinegar fly larvae.

Description: Adult females lay eggs near moist, fermenting food material, such as overripe fruit, rotten vegetables, or residues left in garbage containers, and drains. When the eggs hatch, the larvae feed near the surface of the fermenting food masses. The seldom-seen larvae are cream-colored, lack a sclerotized (hard) head capsule, and have a tapered head and extended, fleshy tubes on the last body segment.

FIGURE 5.105 Vinegar flies are a nuisance and cause food contamination.

Injury: Vinegar fly populations can be an annoyance to homeowners even when in low numbers. Food handling establishments must proactively deal with them, due to potential food contamination.
Management: Removal of breeding sites (rotting fruit, full garbage bins, food leftovers) is paramount to getting rid of vinegar flies. Fruits and vegetables should be stored properly in a refrigerator and thrown away if they become overripe.

FIGURE 5.106 Western corn rootworm beetle adults.

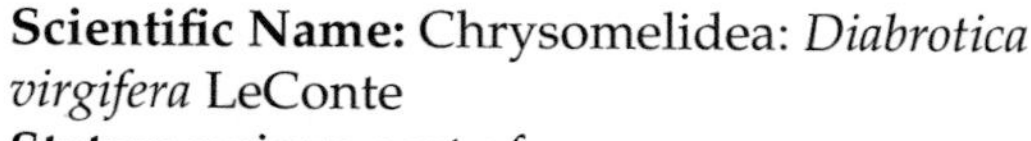

Scientific Name: Chrysomelidea: *Diabrotica virgifera* LeConte
Status: serious pest of corn
Damaging Stage: larval and adult
Description: Adults are approximately 6 mm long and have relatively long antennae. The coloration varies some but most are yellow with three black stripes running down the lengths of the wing covers. The wing covers of males may be entirely black except for narrow yellow margins and yellow tips.

FIGURE 5.107 Western corn rootworm larval damage on roots.

Description: Western corn rootworm larvae are cream-colored and approximately 10 mm long. They have a brown head capsule and are generally found in corn or soybean fields. Adults emerge and mate in midsummer, and females begin laying eggs about two weeks after emergence. The eggs hatch the following spring. The newly hatched larvae find their way to corn roots, bore in, and begin feeding. The larval stage lasts about three weeks, then they move into the soil to pupate.

FIGURE 5.108 Western corn rootworm root pruning damage.

Injury: Larvae feed on the roots and root hairs of corn and can reduce crop yield. Adults also feed on emerging silks and may interfere with pollination when their populations are high.
Management: Rootworms are managed by using pesticides applied to the soil at planting. Crop rotation practices also are a popular cultural method of regulating corn rootworm populations. However, new rootworm variants that can feed in soybean and corn fields are forcing growers to consider new control options such as planting transgenic corn varieties.

FIGURE 5.109 Yellowjacket adult wasp.

Scientific Name: Vespidae: several species
Status: painful and potentially lethal sting; otherwise beneficial
Damaging Stage: adult
Description: Yellowjackets are smaller than their close relatives – hornets and paper wasps – but they occur in larger colonies, sometimes with several hundred workers. They are usually bright yellow with black dots and stripes across their abdomens. While yellow jackets can sting and harm humans, they are invaluable predators of crop-damaging pests such as flies and caterpillars. When defending themselves or their nests, however, yellowjackets can become very aggressive and may sting repeatedly. The sting is always painful but even can be life-threatening.

FIGURE 5.110 Yellowjacket.

Description: A new queen leaves the nest during the fall, mates, and passes the winter under leaf litter or the bark of trees. In the spring, the queen starts a colony by building a gray paper nest, usually underground or in a wall void or other cavity, in which she lays eggs and cares for the developing larvae. Immature yellowjackets are white, grub-like larvae. They are rarely ever seen unless the nest is torn open. The queen remains inside the nest and focuses primarily on reproduction. By late in the season the nest can become very large and be home to many yellowjackets.

FIGURE 5.111 Yellowjacket nest.

Value: Predators of other possible pest insects.
Injury: A nest constructed on or near a home can be a potential threat to people. Nests well away from human activity should be left alone because of the beneficial nature of yellowjackets.
Management: Yellowjackets often forage in and around trash receptacles or ripened fruits. Simply removing or temporarily relocating these will be enough to reduce human/yellowjacket encounters. If yellowjacket nests are found near people, insecticides can be used to control them.

New and Emerging Pests

From time to time new insect pests appear. These come as either exotic pest species that are introduced from foreign countries or native insects that suddenly become pests.

Introduced Exotic Pests

Newly introduced pests are referred to as invasive or exotic pests.

Many non-native insects are introduced into the United States and other countries each year through international trade, commerce and tourism.

> Exotic pests are non-indigenous or non-native plants or animals that adversely affect the ecosystem, environment or economy of the location in which they invade.

Regulating the introduction of invasive arthropods is high priority. Millions of dollars are spent each year for inspections, surveys and quarantines. Notwithstanding all of the effort and resources applied to prevent the introduction of exotic insects, some invariably make it through our borders every year. Of those that get past the borders only some can be considered pests. Further, not all potential pests survive. Of the few that can survive in a new environment, however, many become very serious annual pests and cause billions of dollars in crop losses, and forest damage. The list of insect pests that have been detected and are now considered invasive species is staggering.

New pests have the potential of severely impacting our agricultural economy, our native plants and animals and our way of life. Pests introduced into a new area are usually free of the natural controls (predators, diseases, parasites) and competitors that keep them in check in their native countries. Without checks and balances their populations typically explode out of control resulting in severe damage to commercial agriculture and untold damage to the environment. Scientists scramble to find ways to eradicate or quarantine the pests. When that fails new methods of dealing with the pests are investigated including cultural management techniques, new biological controls and even genetically modifying plants. (Regulatory, cultural and biological controls are part of integrated pest management and are discussed in Chapter 7.) Over time, pest numbers typically become regulated and natural controls once again take over.

Successful invasive species have certain traits that enable them to outcompete native species. Their competitive edge is usually due to a superior rate of population growth and reproduction but there are other factors to consider.

Traits that invasive species possess include:

- Fast generation time.
- High reproductive ability.
- Ability to disperse quickly and effectively.
- Genetic resiliency – ability to adapt to new conditions.
- Tolerance of a wide range of environmental conditions.
- Wide range of possible foods (generalists).

It is important for diagnosticians to understand how and why new pests flourish upon introduction as well as to recognize them.

Diagnosticians must be first detectors and train their clientele to assist in this very important responsibility. Detecting new pests while they are still in low numbers affords opportunities to eradicate or quarantine the pests that are not available once the pest becomes entrenched. Limiting the spread of a new pest through early detection, quarantine and eradication can save untold revenues. After all, preventing is much easier than curing a problem.

Diagnosticians should be aware of newly introduced pests as well as pests that threaten our borders. Working closely with federal USDA and Plant Protection and Quarantine (PPQ) programs is paramount.

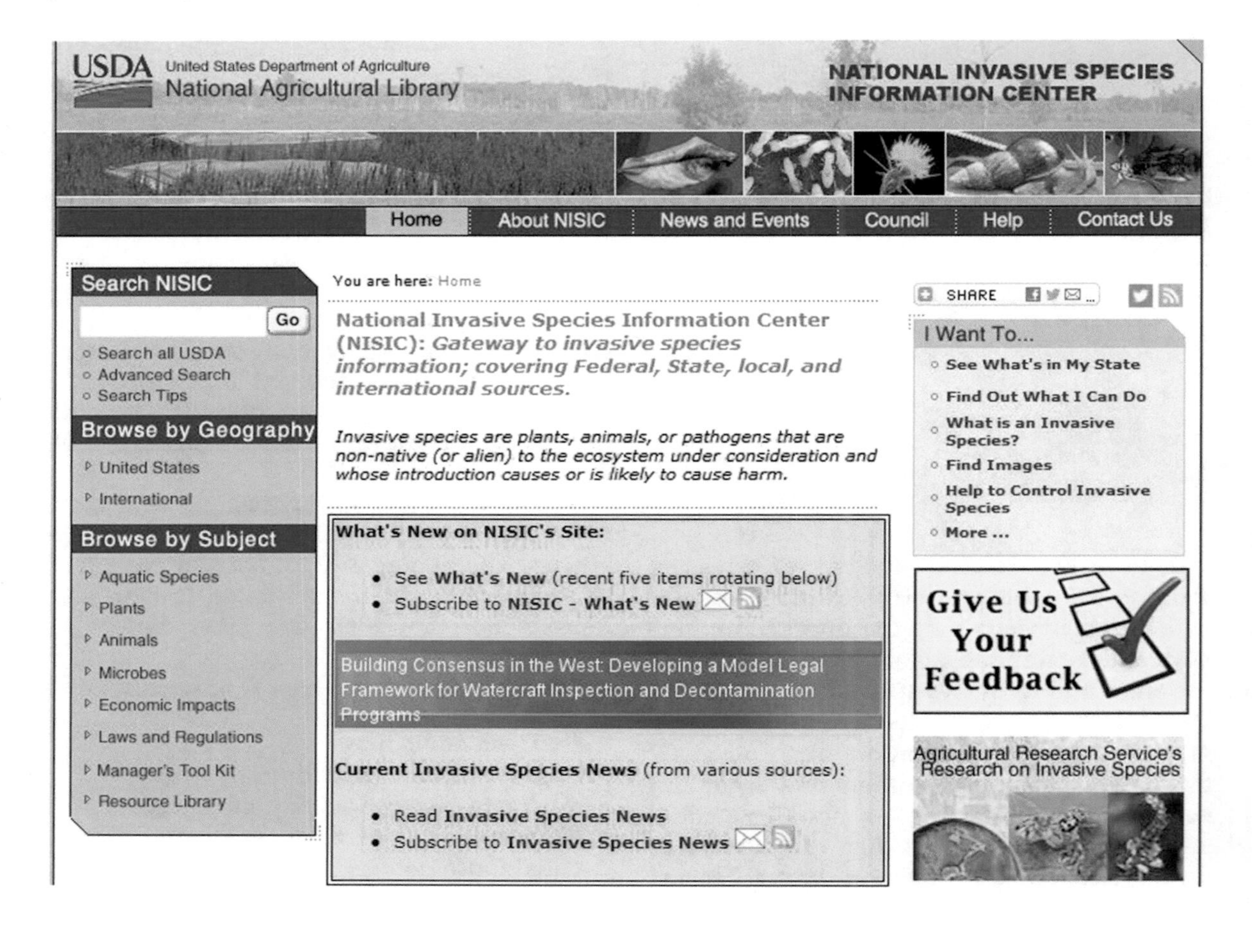

Individual states also track the invasion of pests across their borders and spend millions of dollars in an attempt to exclude them, eradicate them or quarantine them. State government programs are often executed with the assistance of insect diagnosticians. Diagnosticians must remain up to date on new pest alerts issued by regulatory agencies. They may be the first to encounter new pests, thus they must be aware of what to look for and what procedures should be followed upon the discovery of a suspected pest.

Any list attempting to include all invasive insects would not only be very large but also would be outdated very quickly because the threat of new insect introductions is constant. Nevertheless, for illustration purposes six diagnostic exotic pest summaries are provided below. Each has become or has the potential to become serious pests in the United States.

Asian Lady Beetle

The Asian Lady beetle (*Harmonia axyridis*) was intentionally introduced into the U.S. in an attempt to provide biological control of aphid pests. It was first reported as an established nuisance in 1988 in Louisiana and during the following years it quickly spread throughout the U.S.

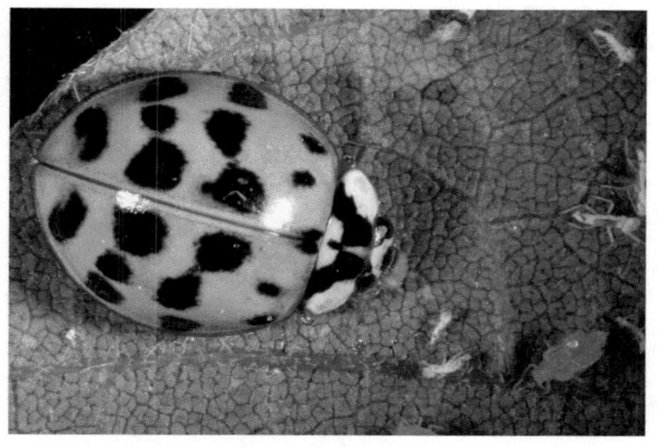

FIGURE 5.112 Asian lady beetle (*Harmonia axiridis*). Oval, convex, and approximately 15 mm in length.

It is considered one of the most invasive and obnoxious home invading pests because of its tendency to seek shelter in buildings, emit a foul odor when disturbed, and congregate during the late fall. It does not directly damage contents of or furnishings in the home and does not infest stored foods, but its presence is seriously annoying. (Recent research indicates that the Asian lady beetle may become a human health threat because large accumulations of dead beetles in wall voids and attics can trigger asthma attacks in some people.)

FIGURE 5.113 Asian lady beetles congregate and pass the winter in buildings.

Asian lady beetles are attracted en masse to buildings during the late fall where they squeeze through cracks and other openings that ultimately lead into wall voids or to the interior of the home. Even though they do not feed or breed in the building, they are a nuisance when they become active on warm winter days and begin flying about inside the house.

Japanese Beetle

Japanese beetles (*Popillia japonica*) were first introduced from Japan into New Jersey probably around 1916. They have since spread throughout the eastern and midwestern United States and Canada. They are now monitored very closely as they threaten to invade the western states. Intensive eradication of newly found populations, as well as restricting the movement of certain commodities from infested states is slowing the spread of this destructive pest. Japanese beetles feed on more than 400 different tree, shrub, flower, fruit, vegetable, and crop plants. What makes this pest particularly problematic is that, in addition to its appetite for such a wide range of foods, it causes damage both in its immature (grub) stage as well as its adult stage.

FIGURE 5.114 The Japanese beetle (*Popillia japonica*) is about 12 mm long with shiny copper-colored wing covers and an iridescent green thorax and head. The abdomen has a row of white hair tufts on each side.

FIGURE 5.115 Japanese beetle grubs have a white, C-shaped body, a brown head capsule and three pairs of legs.

FIGURE 5.116 Adult beetles feed on all parts of a plant including leaves, flowers and other reproductive parts. Grubs feed on grass roots including corn and turfgrass.

Gypsy Moth

Gypsy moths (*Lymantria dispar*) have become one of North America's most devastating invasive forest pests. They were introduced into the U.S. from Europe in the late 1860s near Boston, MA. as part of a silkworm experiment. They have since spread into southeastern Canada, northeastern United States and into parts of the upper Midwest and Great Lakes states.

FIGURE 5.117 Adult gypsy moths (*Lymantria dispar*) have wings with variable patterns of black spots and bands. Males (right) have brown wings and feathery antennae whereas females (left) are cream-colored, and have thread-like antennae.

FIGURE 5.118 Gypsy moth caterpillars characteristically have five pairs of raised blue spots along the back followed by six pairs of red spots.

Gypsy moth caterpillars feed on the foliage of hundreds of species of trees and shrubs in North America but prefer oak trees. When gypsy moth populations reach high levels, trees may become completely defoliated by feeding caterpillars.

FIGURE 5.119 Gypsy moth damage to forests.

Heavy infestations of gypsy moths over successive years can render them susceptible to secondary organisms such as borers and root rots that kill stressed trees. Defoliation of thousands of acres or forest and urban trees not only impacts tourism but also negatively affects wildlife populations and the ecosystem.

Gypsy moths spread naturally quite slowly but their spread is augmented by human activity. Female moths lay egg masses indiscriminately on trees, houses, and other structures in late summer. Often campers unknowingly spread this pest when egg masses are attached to recreational vehicles.

Emerald Ash Borer

Emerald ash borer (*Agrilus planipennis*) was introduced into the United States in Michigan in 2002 from Asia and eastern Russia. Emerald ash borer has spread into 22 states and Canada, largely through the movement of nursery stock and firewood. Larvae bore into the trunk of the tree, disrupt the flow of nutrients through the trees vascular system and ultimately kill the tree. Hundreds of millions of ash trees have already died as a result of this pest. Once Emerald ash borer is established, control efforts are difficult and unreliable. Preventing the spread of infested ash trees is the best management strategy to date.

FIGURE 5.120 Emerald ash borer (*Agrilus planipennis*) adult.

Emerald ash borer adult beetles have flatheads, large black eyes and bullet-shaped, dark metallic-green bodies. They measure 12 mm in length and are 3 mm wide.

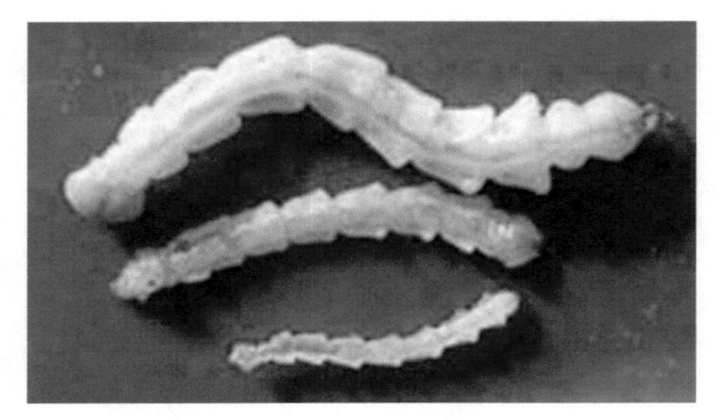

FIGURE 5.121 Emerald ash borer larvae.

Emerald ash borer larvae are cream-colored and have flat, broad, segmented bodies. The larvae bore through the bark of ash trees and feed for several weeks beneath the bark leaving characteristic S-shaped tunnels. Pupation occurs in the springtime.

FIGURE 5.122A,B Emerald ash borer S-shaped galleries (A), and D-shaped exit holes (B).

Emerald ash borers leave a characteristic D-shaped exit hole in the bark when they emerge as adults.

FIGURE 5.123 Tree dieback due to emerald ash borer infestation.

Trees typically begin to die back from the top of the canopy and symptoms progress downward leading to the death of the tree, usually within 2–4 years.

Spotted Wing Drosophila

One of the most recent invasive fruit pests in the U.S. is the spotted wing drosophila (*Drosophila suzukii*). This pest originated in Japan and was first discovered in California in 2008. What is alarming is the rate at which this pest has since spread throughout the country. It has now been detected in most states.

The spotted wing drosophila is projected to have a major impact on the American fruit industry, particularly soft fruits such as blackberries,

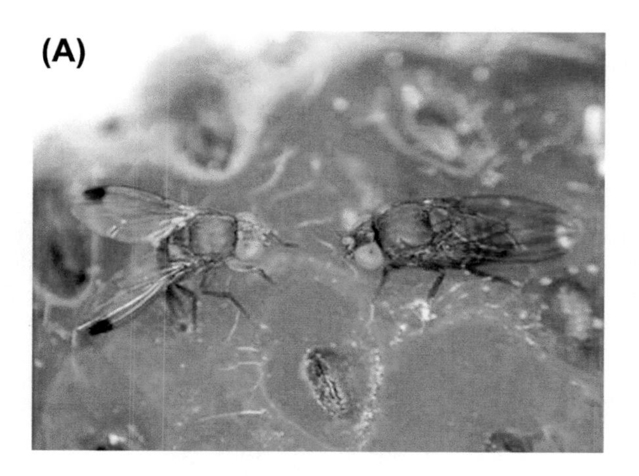

FIGURE 5.124A,B The spotted wing drosophila (A) is similar in appearance to the common vinegar fly (B).

plums, nectarines, raspberries, blueberries, grapes, strawberries, cherries, and peaches. However, it can also infest apples, apricots, and tomatoes that are either ripe or damaged.

Many native species of drosophila are attracted to ripe, over ripe, or decaying fruit. Homeowners are very familiar with them infesting kitchens and pantries when ripened home orchard fruits and garden produce is brought into the home during the fall. The spotted wing drosophila, however, is a much more serious pest because of its ability to infest healthy, non-ripened fruit. Unlike the common vinegar fly, it can infest fruits in the field. Fruits become most attractive as to egg laying females when they begin to turn color during ripening and sugar level increase.

The overall size, shape and appearance of the spotted wing drosophila is very close to common vinegar flies. Wings, legs and the ovipositor are diagnostic characters used to separate them. The spotted wing drosophila gets its name from a single spot (sometimes faint) on each wing in the male. It can be further identified by two black bands on each front leg.

The most diagnostic feature of the female is the obvious serrated or saw-like ovipositor that allows it to cut through the skin of fruits to lay eggs inside. Several recent web sites feature photographs depicting these diagnostic structures, including:

pest.ceris.purdue.edu/pest/php?code =IOAPAUA
pubs.cas.psu.edu/FreePubs/PDFs/x j0045.pdf
spottedwing.org
www.ipm.ucdavis.edu/PMG/PESTNOTES/ pn74158.html

After the eggs hatch larvae tunnel inside the fruit, causing it to rot and ruining its commercial value. Mature larvae are white in color, approximately the diameter of mechanical pencil lead (0.5 mm) and about 2 mm long. There may be several in and on each fruit.

To confirm suspected infestations look for small puncture wounds on the fruit that may or may not be leaking juice. Cut open, peel, or smash the fruit and look for larvae.

Brown Marmorated Stink Bug

The brown marmorated stink bug (*Halyomorphora halys*) is an invasive pest from China, Japan, Korea and Taiwan that was first introduced into eastern Pennsylvania sometime prior to 1998. The brown marmorated stink bug is known to be an agricultural pest in its native lands but has become a serious pest of fruits, vegetables and farm crops as well as a nuisance pest inside homes as it has spread across the United States.

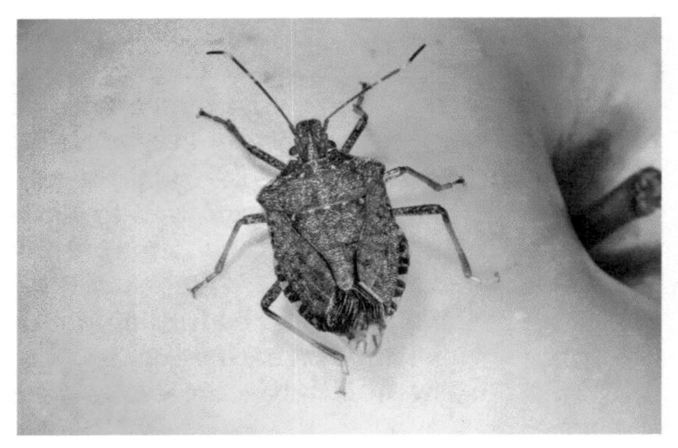

FIGURE 5.125 Brown marmorated stink bug (*Halyomorphora halys*) on apple.

FIGURE 5.127 Native brown stink bug.

Adult bugs measure about 17 mm in length, are almost as wide and are 'shield-shaped,' typical of the Pentatomidae family. Brown marmorated stink bugs have shades of mottled, brown on both the upper and lower body surfaces. These descriptions also fit other stink bugs in the U.S. including the brown stink bug *Euschistus servus* but can be distinguished by the alternating dark and light bands on the last two antennal segments, and exposed lateral margins of the abdomen.

FIGURE 5.126 Diagnostic characters of the Brown marmorated stink bug.

The name 'stink bug' refers to an obnoxious and pungent odor emitted by the scent glands if this insect is disturbed. This odor is characteristic of the family but is especially strong in the brown marmorated stink bug.

Adults emerge during the spring (late April to mid-May), mate and deposit eggs from May through August. The eggs hatch into small black and red nymphs that develop through five molts. Adults begin to search for overwintering sites starting in September through the first half of October.

Brown marmorated stink bug adults and nymphs feed on a wide variety of host plants, including fruits, ornamental plants, trees, weeds, soybeans and vegetables. They damage plants by inserting their long piercing-sucking mouth parts into the plant and withdrawing juices.

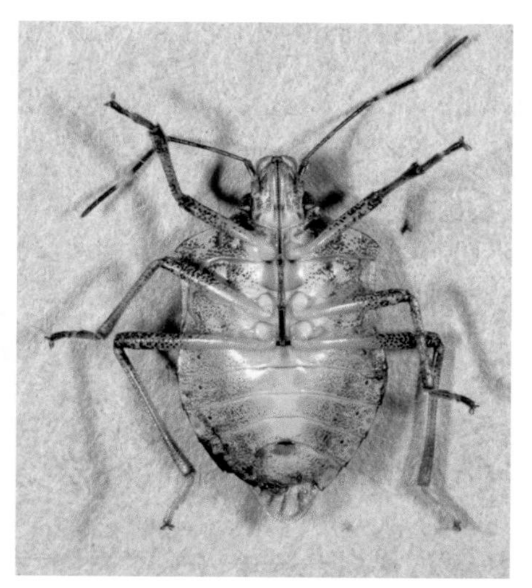

FIGURE 5.128 Underside of brown marmorated stink bug depicting piercing-sucking mouthparts.

Damage symptoms include shrunken and distorted fruits, necrotic spotting, and discoloration including 'cat facing' that render the fruits unmarketable as fresh products.

Brown marmorated stink bugs also become a nuisance pest when they are attracted to and congregate on homes on warm fall days in search of protected, overwintering sites.

FIGURE 5.129 Brown marmorated stink bug on window ledge.

Brown marmorated stink bugs seek cracks and crevices around windows, doors, siding, utility pipes, chimneys, and underneath fascia in which to squeeze themselves. This behavior results in brown marmorated stink bugs finding their way into wall voids and eventually into the inside of homes. Although they do not feed or breed inside homes, they become a pest because of their presence and the odor they create, particularly during warmer sunny periods throughout the winter and spring.

Mechanical exclusion is the best method to keep stink bugs from entering homes and buildings. Exterior applications of insecticides may offer some minor relief. Both live and dead stink bugs can be removed from interior areas manually or with the aid of a vacuum cleaner.

FACULTATIVE PESTS

Not all new pests are exotic. Insects that are native or those that have been present for long periods of time may suddenly become pests in places where they were never considered pests before. This is because environmental conditions change or pests change and adapt.

Arthropod populations are never static. They rise and fall in response to changes in their environment. Environmental conditions are also in a constant state of flux. Changes may benefit or hamper animals that live there. Population growth of individual species is a direct reflection of this change. When environmental conditions favor the growth of a pest population, an outbreak occurs. Changes in weather conditions are most often credited for insect population fluctuations, however there are several other factors that also play a role as well. Due in part to the unpredictability of weather patterns, it is not always possible to predict with confidence which insect species will benefit from a weather event and which will be reduced.

In other instances pests adapt. When pests adapt, pest outbreaks can follow. For example, development of resistance to pesticides has been documented among many species of arthropod pests. When resistance develops pests populations are free to grow as if the pesticides were not present. Genetic and behavioral changes, in addition to resistance, also may develop that allow pest populations to dramatically increase.

Insects and other arthropods can become new pests as environmental conditions change or as pests adapt.

Arthropods can become facultative pests if conditions are just right. That is why some are pests one year and not another. In a reverse strategy, humans may change the environment or the host plant such that they become an impediment to pests.

Hessian Fly

Early spring temperatures and precipitation directly affect planting dates of agricultural crops. The planting date, in turn, directly affects some pest populations. For example, the Hessian fly (*Mayetiola destructor*), a tiny but potentially serious wheat pest in the United States lays its eggs in young, emerging wheat plants. American wheat growers have learned that Hessian flies can largely be controlled by adjusting the wheat planning date to a time when the flies are no longer present (called the 'fly-free' date). Fly-free dates vary by latitude and also by yearly weather conditions. The further north and the colder the year, the earlier the fly-free date.

Planting wheat after the fly-free date avoids exposing large numbers of potential host plants during the times when adult flies are actively laying eggs. If the flies die before the new wheat plants emerge, egg laying success is halted and the crop is not infested.

In a similar but less programmed way, variations in time of plant development, reproduction and harvest may directly affect the reproductive success of many other pests and may thus influence population growth in any given year. Therefore, what was not a pest one year may become a pest in another year if host plant growth conditions permit.

Aquatic Insects

Other changes in the environment such as pollution or silt in aquatic systems can affect insect numbers. Aquatic insects that are sensitive to changes in water quality can either build up in numbers or can become reduced. For example, recent improvements in river, stream and lake water quality in the U.S. have resulted in an increase of mayfly, caddisfly, stonefly and midge numbers.

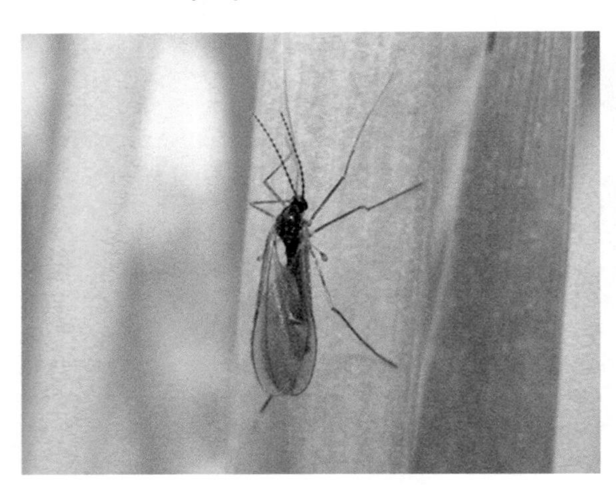

FIGURE 5.130 Hessian fly (*Mayetiola destructor*).

FIGURE 5.131 Aquatic, immature stage of mayfly (Ephemeroptera).

When and where these insects occur in very high numbers they can become pests in the adult stage. For example, mayflies and caddisflies are famous for emerging from the water as adults (flight periods) all at once, and in very high numbers.

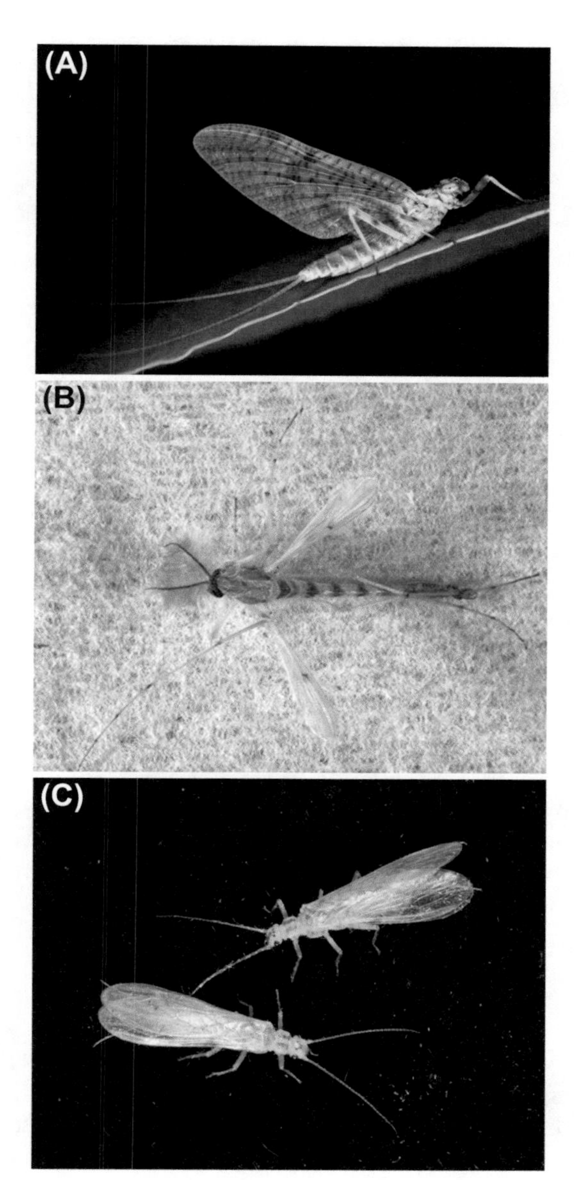

FIGURE 5.132A,B,C Adult stage of (A) mayflies, (B) midges and (C) stone flies.

Street or porch lights near large bodies of water can attract the flying adults by the millions resulting in a significant nuisance to people and properties nearby.

FIGURE 5.133 Aquatic insects found under porch lights in urban areas.

Abundance of water, especially stagnant water, has long been known to have a direct influence on mosquito populations. Without standing water, many mosquitoes do not propagate. Human vigilance plays a large role in reducing the availability of standing stagnant pools of water to mosquito populations.

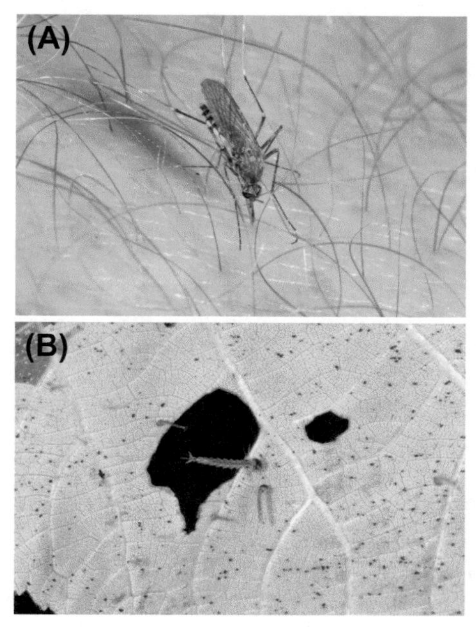

FIGURE 5.134A,B Mosquito (A) adult, (B) mosquito larva.

An interesting twist to the exotic pest introduction pattern is illustrated with the soybean aphid story.

Soybean Aphid

The soybean aphid *(Aphis glycines)*, an exotic pest from China that was introduced into the United States at about the turn of the century, has since become so numerous in parts of the country that migrating aphids have become nuisance as well as agriculture pests.

FIGURE 5.135 Soybean aphid.

In 2004, insects that normally are innocuous to people but are predators of the soybean aphid, suddenly became nuisance pests themselves. Syrphid flies, Asian lady beetles, lacebugs, and minute flower bugs are all known to be predators on soybean aphids. Their populations are tied very tightly with populations of aphids so much so that when aphid populations became high, populations of these predators also rose to the point that people began to complain of them as nuisance pests in and around homes.

Typical of predator prey relationships, these populations rise and fall in direct relation to the changes in soybean aphid populations.

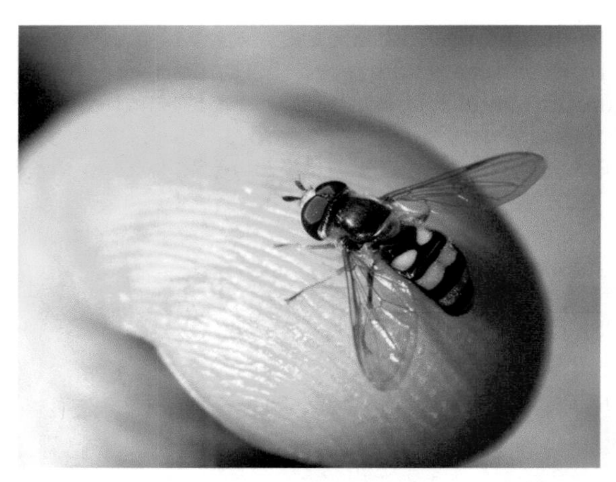

FIGURE 5.136 Immature syrphid flies (Syrphidae) often feed on aphids.

Syrphid flies are small, usually 6 mm or shorter, have two wings, and yellow and black or brown bands on the abdomen that resemble sweat bees. However, they are not bees at all and therefore cannot sting. Syrphid flies are known to hover around people, especially when perspiring, and may even land to lap up the sweat. This hovering behavior also gives them the name 'hover flies.'

In its immature stage, this species of syrphid is a predator of soybean aphids. Following the mass population build-up of the aphid, this fly also underwent a population explosion and became very common around homes. Although it is harmless to people, it was a concern because of its high numbers and its resemblance to bees. Many calls to the diagnostic laboratory were fielded regarding this population explosion.

PESTS THAT ADAPT

As discussed above, changing environmental factors, including availability of food resources, directly impact the success of a pest from year to year. Pests, because they are able to mutate and evolve, can also adapt from one environment to another. Below are some notable examples of pests that were once minor but now have become major pests.

Corn Earworm

Corn earworms (*Helicoverpa zea*) have been in the United States for as long as we have records. They naturally fed on a great variety of plants, including weeds.

FIGURE 5.137 Corn earworm (*Helicoverpa zea*) caterpillar feeding on corn.

Years ago, when farmers began planting huge monocultures of crops such as corn, cotton or tomatoes, these pests were suddenly granted a seemingly inexhaustible supply of food. The natural result was a population explosion of this pest. The new food rich environment allowed it to flourish and become a pest in several crops as is evidenced by its common names, 'corn earworm,' 'cotton bollworm' and 'tomato fruitworm.'

Colorado Potato Beetle

In another example, Colorado potato beetle (*Leptinotarsa decemlineata*) an insect native to the Americas, historically fed on a variety of plants before adapting the ability to feed on cultivated potatoes.

FIGURE 5.138 Colorado potato beetle (*Leptinotarsa decemlineata*).

Currently it is a key pest of potatoes in the United States and has become a serious potato pest in many other countries in the world. This all transpired because it adapted to a new food source. Note: Its adaptability is further evidenced by its becoming resistant to many pesticides once used to control it.

Corn Rootworms

Corn rootworms (*Diabrotica* sp.) have nearly always been a devastating pest of cultivated corn in the United States. The most effective cultural control for this pest has been crop rotation. Because corn rootworm larvae must feed on corn roots to survive, switching crops from one year to the next effectively eliminates their chances of survival. Without corn roots to feed on, they starve to death.

FIGURE 5.139 Corn rootworm (*Diabrotica* sp.) feeding on corn roots.

The crop most commonly rotated with corn is soybeans.

Corn/soybean rotations became the standard way of preventing damage due to rootworms and was much more effective than insecticides for many years.

However, in the late 1980s growers in the mid-western United States found that two different variants (subgroups) of corn rootworms had adapted to the practice of crop rotation.

The soybean variants of the western corn rootworm adapted behaviorally such that instead of laying eggs into the corn field where they had fed during the summer, they were found to preferentially fly into a nearby soybean field to lay eggs in the fall. This allowed their larvae to find themselves in a field that was likely to be planted in corn the following year.

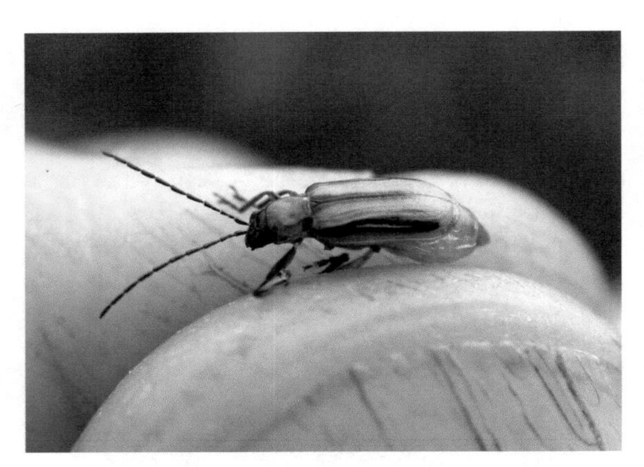

FIGURE 5.140 Corn rootworm (*Diabrotica virgifera*) behavioral variant.

At nearly the same time another corn rootworm variant, this time of the northern corn rootworm (*Diabrotica barberi*) began to circumvent the established management practice of crop rotation by using a different mechanism called extended diapause.

FIGURE 5.141 Northern corn rootworm (*Diabrotica barberi*) behavioral variant.

Eggs of this rootworm variant remained unhatched for a period of 1.5 years. This behavior effectively allowed them to skip the soybean year and hatch again when corn was planted in the traditional corn-soybean rotation.

Home Invading Pests

Changes in how people construct or manage buildings have also directly affected pests. Relatively recent changes in lumber drying requirements have had an impact on home pest invasions. Constructing buildings from green, non-kiln dried woods has led to a substantial increase in home infesting, foreign grain beetles (now also called new home beetles).

FIGURE 5.142 Foreign grain beetle (*Ahasverus advena*).

These beetles feed directly on the molds that develop on wood. If building lumber is not properly kiln dried or gets wet during construction, it can support mold especially in enclosed areas such as wall voids and attics. Foreign beetles will persist as nuisance pests until the lumber has adequately dried down, usually 12 years.

Filth Flies

Diagnosticians must be constantly aware of changes to the environment and how they might affect native pest populations. Introduction of livestock facilities or, more commonly, building residences in close proximity to such facilities, leads to increased complaints of nuisance and filth flies in homes. Changing regulations and rules regarding the disposal of manure on agriculture fields also directly impacts the number of complaints.

FIGURE 5.143 House fly (*Musca domesticus*).

Midge larvae live in the mud and detritus at the bottom of ponds and lakes but also in sewage treatment facilities. They feed as scavengers on decaying material and are an important component of the decomposition process. Midge species vary in their ability to survive in low oxygen and polluted environments. Some species are very common in sewage lagoons and can suddenly become very numerous and a pest to surrounding homes.

FIGURE 5.144 Midges.

Adult midges are slender insects ranging from 1/8 to 1/2 in length and have transparent wings. They closely resemble mosquitoes, but don't bite. Midges are strongly attracted to lights at night and can cover the walls and roofs of buildings. They may accumulate to several inches thick on nearby homes, posts, and fences.

Drastic increases in local populations of both flies and midges can be a direct result of changing livestock raising practices or sewage treatment facilities. When diagnosticians are confronted by dramatic population swings, they can usually find correlations with environmental conditions. Until these are understood, control recommendations can only be temporary.

Bed Bug

An obvious example of an insect pest that has taken advantage of changes to its environment is the common bed bug (*Cimex lectularius*). Bed bugs have lived in close association with man for thousands of years.

FIGURE 5.145 Bed bug (*Cimex lectularius*).

Bed bugs (*Cimex lectularius*) are small nocturnal parasitic pests of humans. Changes to their environment began in the 1940s when broad spectrum, long lasting pesticides such as DDT as well as improved laundry and housekeeping technologies are thought to have all but eliminated the bed bug from most developed countries in the world. However, later in that century drastic changes to the bed bug environment occurred again.

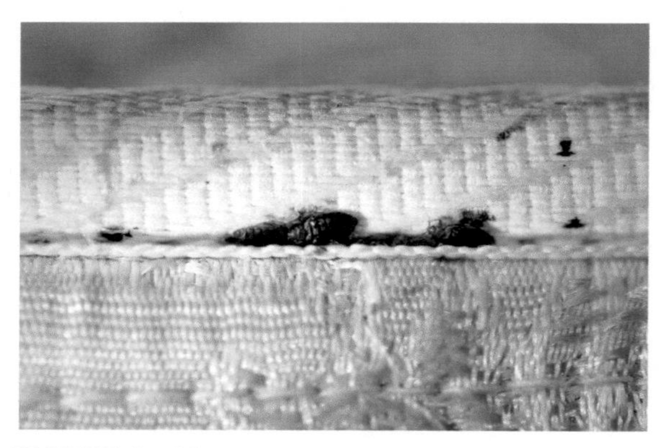

FIGURE 5.146 Bed bugs squeeze into bed crevices.

DDT was banned, travel between countries became more frequent, changing residences and trading of furniture became more commonplace. These changes directly impacted the bed bug yet again, only this time in a very positive way for the bug. Beginning with the 21st century, bed bug populations have rebounded in an epidemic way. The resurgence of what was once a nearly exterminated pest to become the fastest growing urban pest in the world within just a few years is remarkable.

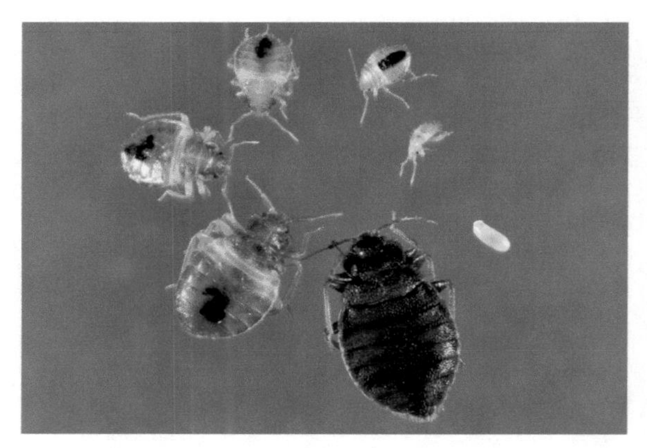

FIGURE 5.147 Bed bug egg, nymphs and adult.

Bed bugs are thigmotactic, meaning that they prefer to squeeze into cracks and crevices making them successful hitchhikers on furniture and suitcases. They have infested homes, schools, hotels, public transportation, retail businesses, theaters and libraries. To think that they were nearly eradicated from this country just a few years ago gives one an appreciation of the ability of an insect population to explode when the conditions in which it lives become favorable.

Look Alike Pests

DIAGNOSTIC RISKS

Many insects are very similar in appearance. These can be easily mistaken for one another. Diagnosticians should be aware of the risks of making an incorrect diagnosis, and err on the side of caution.

Evaluating the risk of an incorrect identification of two very similar appearing insects is part of the art of diagnostics. In some cases it is not so important.

For example, if a butterfly photo is submitted and from the distance that the photograph was taken, a diagnostician cannot be certain whether it is a monarch butterfly or a viceroy butterfly (a monarch butterfly mimic) the consequences of being wrong are minimal - neither of the butterflies are harmful, therefore neither need control.

On the other hand, if a bat bug is misidentified as a bed bug, the costs of being wrong are much greater. Control strategies are vastly different and significant financial as well as emotional expenses hang on a positive identification.

In cases where costs of making an incorrect diagnosis are potentially high, a diagnostician should request samples to be submitted that allow for positive identification. In many cases this means submission of the insect in a certain life stage or of a specimen in a condition that allows magnification of select characters.

Working diagnosticians are familiar with most key pests because they have been seen many, many times. Experienced diagnosticians soon begin to recognize these insects even if the specimen is badly damaged, broken or incomplete. Poor, blurry or even non-magnified photos can be identified by an experienced diagnostician based on general size, form or shape or sometimes other clues such as the habitat in which it is found or its behavior. Where a general identification is sufficient, this 'best guess' diagnosis can save significant time and resources otherwise expended if a second sample has to be obtained, prepared, resubmitted and re-examined.

Eggs and pupae are notoriously difficult to identify. Identification keys for eggs and pupae are scarce and often unreliable. Often larval or nymphal stages are also very difficult to identify. Resubmission of more advanced life stages or holding live specimens until after eggs hatch or immature forms develop may be required for positive identification.

COMPARATIVE ILLUSTRATIONS

For insects that look similar, comparative illustrations are the most effective method of separating them. Collecting or constructing illustrated comparative keys for groups of common pests is very valuable. For example, a key to identifying caterpillar pests of field corn, or a key to identifying moths caught at black light traps, or a key to the identification of household invading ants, can be of great worth to a diagnostician.

Text books and field guides containing insect comparisons are on the book shelves of most diagnosticians. Over time, a diagnostician will also collect a series of independent keys and photo comparisons that can be taken out and referred to quickly. Sometimes these are pictorial keys with selected distinguishing characters highlighted or magnified.

While it is impossible to provide all of these aids or even their references in this text, it is instructive to discuss comparative illustrations used to diagnose frequently confused pests.

Comparative illustrations are most often used to help separate insects and other pests that appear very similar. Diagnosticians use simple comparative illustrations to help clients recognize the difference between insect pests at the order level. For example, to help homeowners differentiate winged ants and winged termites, a diagnostician may point out the key distinguishing features, (elbowed antennae, slender waist, unequal wing length) of ants. Even over the telephone, explaining this difference to a client holding a winged ant can bring tremendous relief.

Sawflies

Slightly more complicated are explanations regarding the differences between sawfly larvae and caterpillars. Most people confuse a sawfly infestation with an outbreak of caterpillars.

FIGURE 5.148 Redheaded pine sawfly (*Neodiprion lecontei*).

Sawflies are not caterpillars however. They may appear to be similar in form and behavior but they belong to a completely different order of insects called Hymenoptera. They can be separated by observing the larvae. Sawflies have six or more pairs of prolegs or false legs, whereas caterpillars (Order Lepidoptera) never have more than 5 pairs. In addition, hymenopteran prolegs never bear crochets (gripping hooks in a small circle at the bottom of the proleg).

Separating the two, otherwise similar appearing pests, is crucial to assessing damage potential and recommending proper pest management techniques.

Using a photo collage or assemblage of photographs together, as below, is a very effective diagnostic tool. It is an example of a basic comparative illustration. In the same way, diagnosticians themselves use comparative illustrations to help separate very similar insects at the family or genus level.

Prolegs (false legs) Comparison

Sawfly larva

7

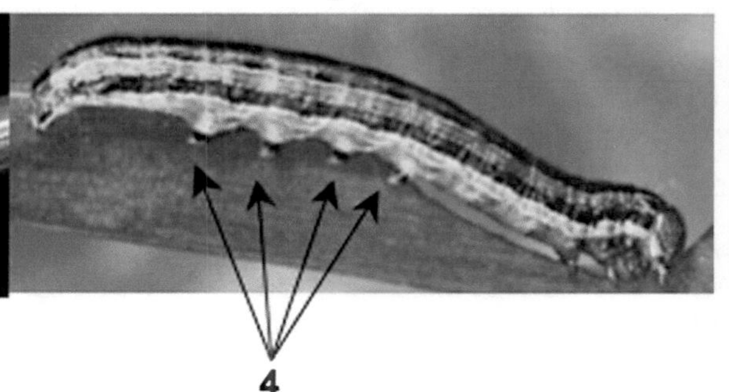

FIGURE 5.149 Sawfly larva with 7 pairs prolegs compared to armyworm caterpillar with 4 pairs of prolegs.

Corn Rootworm Beetles

Corn Rootworm beetle adults may be easily distinguished by using comparative illustration.

One can easily see and differentiate the three species by comparing the photographs. However, spots, stripes and colors are not always conclusive diagnostic features of insects. Many insect species exhibit various color phases.

FIGURE 5.150A,B,C (A) Northern corn rootworm (*Diabrotica barberi*). (B) Southern corn rootworm (*Diabrotica undecimpunctata*). (C) Western corn rootworm (*Diabrotica virgifera virgifera*).

Bean Leaf Beetles

Consider the bean leaf beetle photographs below. All are the same species (*Ceratoma trifurcata*). Color is most usually yellow but may also be orange or red. Wing covers typically have

FIGURE 5.151A,B,C Color and marking variations in bean leaf beetles.

four black rectangular marks, but they may also have only two or even no marks at all.

As demonstrated above, strict reliance upon colors or spots is not a reliable diagnostic feature for bean leaf beetles. A more consistent diagnostic marking is the black triangle behind the pronotum, illustrated below. All bean leaf beetles share this feature.

FIGURE 5.152 Bean leaf beetle, arrow showing the diagnostic black triangle.

Corn Ear Caterpillars

Corn earworm (*Heliothis zea*) caterpillars also range widely in body color from yellow, green to red or brown.

FIGURE 5.153 Color variations in corn earworm larvae.

Because several caterpillars, including corn earworm, can be found feeding on developing ears of corn, it is tempting to differentiate them based on color. However, color is not consistent, even over one species of corn ear infesting caterpillars.

To separate the corn ear feeding caterpillars one must consider the more important distinguishing features, lateral stripes, setae and head capsules. These are used as diagnostic differences that separate them from fall armyworms (*Spodoptera frugiperda*) and western bean cutworms (*Striacosta albicosta*).

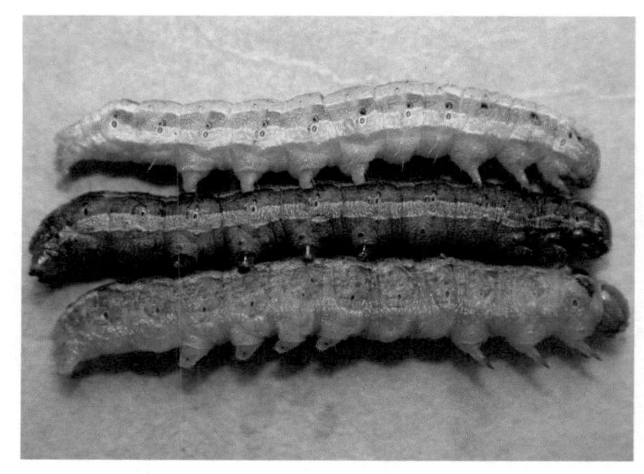

FIGURE 5.154 Lateral stripe differences between common corn ear feeding pests: Corn earworm (top) fall armyworm (middle) and western bean cutworm (bottom).

FIGURE 5.155 Head capsule differences between common corn ear feeding pests: Corn earworm (left) fall armyworm (middle) and western bean cutworm (right).

Diagnostic features can be further enhanced with arrows and words to help point out specific diagnostic character differences. The following illustrations are particularly valuable when placed side by side for comparison.

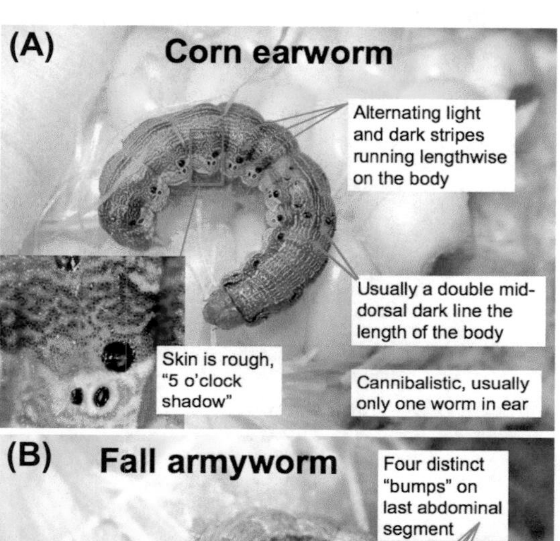

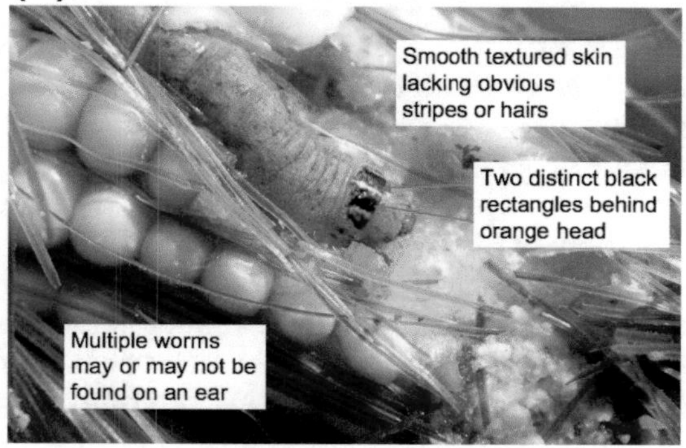

FIGURE 5.156A,B,C (A) Corn earworm; (B) fall armyworm; (C) western bean cutworm.

Lady Beetles

Separating Asian lady beetles (*Harmonia axyridis*) from other beneficial species of lady beetles also can be confusing if only colors or wing patterns are considered.

FIGURE 5.157 Color and marking variations in the Asian lady beetle.

Asian lady beetles are known for their range of color variations (from reds to yellows) and the variation of spots on their wing covers (some have many, some have few and some lack spots altogether).

The most apparent identifying character of Asian lady beetles is not color or spotting on the wings, but rather, a black, "M-shaped" marking on its otherwise light-colored thorax, just above the wing covers. Some M's are darker and more complete than others, but their presence is always an effective diagnostic tool.

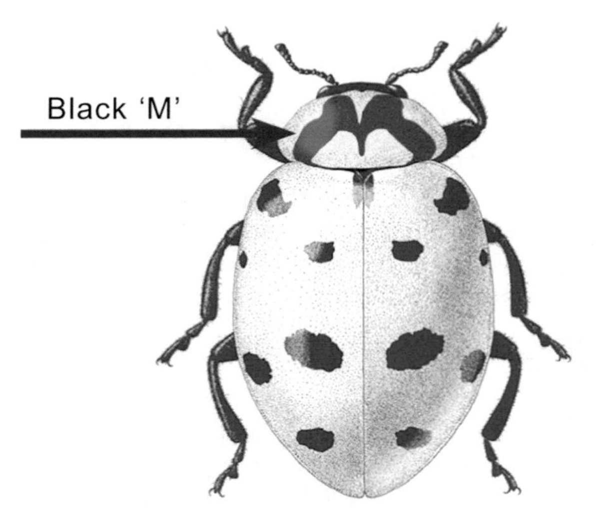

FIGURE 5.158 The most apparent identifying character of Asian lady beetles is a black "M-shaped" marking on the thorax.

Light Trap Insects

Some insects are nearly dead ringers for others, yet their biology and damage potential is quite different. When using light traps to monitor and predict the occurrence of pest insects, adult moths are frequently captured for identification. Comparative diagnostic keys are particularly useful in this effort.

The European corn borer (*Ostina nubilalis*) and the celery leaftier (*Udea rubigalis*) are both small, light-colored moths of the family Crambidae. Both are highly attracted to light traps. These two species are notoriously difficult to separate. Each holds wings in the characteristic triangle position typical of the family Crambidae and both have light colored wings with dark, irregular, wavy bands that cross them in exactly the same way. For most, the insects are virtually identical. However, when compared side-by-side, European corn borers are slightly larger and do not have the long 'snout-like' mouth parts that celery leaftiers do. These differences are difficult to explain without the use of side-by-side comparisons of a series of specimens.

FIGURE 5.159 European corn borer adults (top row) are slightly larger and do not have the 'snout-like' mouthparts common to celery leaftiers (bottom row).

More importantly, for those using light traps as an aid in management, European corn borers are highly destructive pests on corn. Celery leaftiers feed on beans, beets, celery, and spinach – but not corn. Making management recommendations without knowing for certain which of the two pests are present, can lead to undesirable consequences.

Bat Bugs and Bed Bugs

Prior to the late 1990s, most specimens of Cimicidae submitted to insect diagnostic laboratories were determined to be bat bugs (*Cimex pilosellus*). Even though bat bugs are nearly identical in appearance to bed bugs, they are very distinct biologically.

Bat bugs develop in colonies of roosting bats where they feed exclusively on bat blood. Occasionally, however, bats roost in attics or behind walls of homes. When bats either migrate or are eliminated from the building, the populations of bat bugs remaining in the roosting site are forced to move to find food. When they relocate into human living areas they may incidentally feed on people in much the same way that bed bugs do.

The important difference between the two species is that populations of bat bugs, in the absence of their bat hosts, cannot sustain and reproduce. They have to have bats.

Differentiating between bat and bed bugs is particularly important because control techniques are drastically different. Controlling bat bugs first requires the elimination of any bats that are present in the home or building. This is accomplished by exclusion techniques also known as 'building them out', i.e., sealing entrance holes. After the bats are excluded, efforts should be directed at preventing the bugs, residing in the roost area, from moving to other areas of the home where people live. Residual insecticides applied to all cracks and crevices near where the roost was located will help to intercept and kill them before they get to people. Light fixtures, chimneys and window casings should all be treated.

This treatment strategy is completely different than a strategy for treating bed bugs. Bed bugs are managed by applying pesticides, heat treatments and sanitation practices to areas where humans sleep. Bed encasements and clutter removal under and around beds and furniture are usually recommended. None of these techniques impact bat bugs. What works for one definitely does not work for the other.

Sadly, because bat bugs are so similar in appearance, they are often mistaken for bed bugs. Microscopic examination is needed to distinguish them.

Compare the length of the hairs on the head and thorax of the specimens below. Bat bug (left) hairs are twice the length of bed bug (right) hairs. A comparative illustration is most effective.

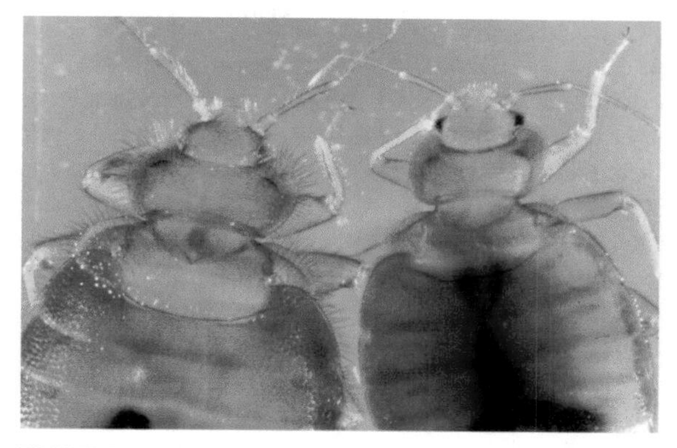

FIGURE 5.160 Comparison of fringe hairs: bat bug (left) bed bug (right).

Grubs

Immature stages of the beetle family Scarabaeidae can be very difficult to distinguish. As a group they are referred to as white grubs. All are white, C-shaped and have reddish colored heads and legs, and they all feed in the soil, most often on roots.

FIGURE 5.161 White grubs are white and C-shaped.

The specific identification of grubs is crucial to pest management because life history of each species is quite different. Some go through multiple generations per year, some only one and other species live two, three or four years. Grubs and adult beetles of certain species may feed and congregate on completely different plants than others. Some feed, some do not. Some are nocturnal, some diurnal. Some are newly introduced exotic species that have resistance to pesticides, others are native species that seldom cause economic damage even when present. In sum, important differences exist between the various species of grubs.

Knowing if, when, where and how to control grubs depends upon species identification. Fortunately, each species has differences in the pattern of the hairs or bristles present on the under side of the terminal abdominal segment, called the raster.

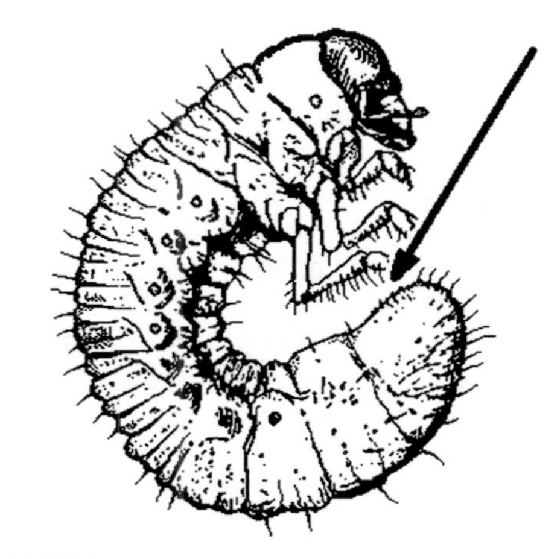

FIGURE 5.162 White grub: arrow showing raster.

Rastral setae patterns are diagnostic for each of the grub species that occur in agriculture or in turfgrass. For example, four of the common grubs found in agriculture are photographed below. Note the distinct pattern (or in the case of the masked chafer, absence of a pattern) of setae on the raster.

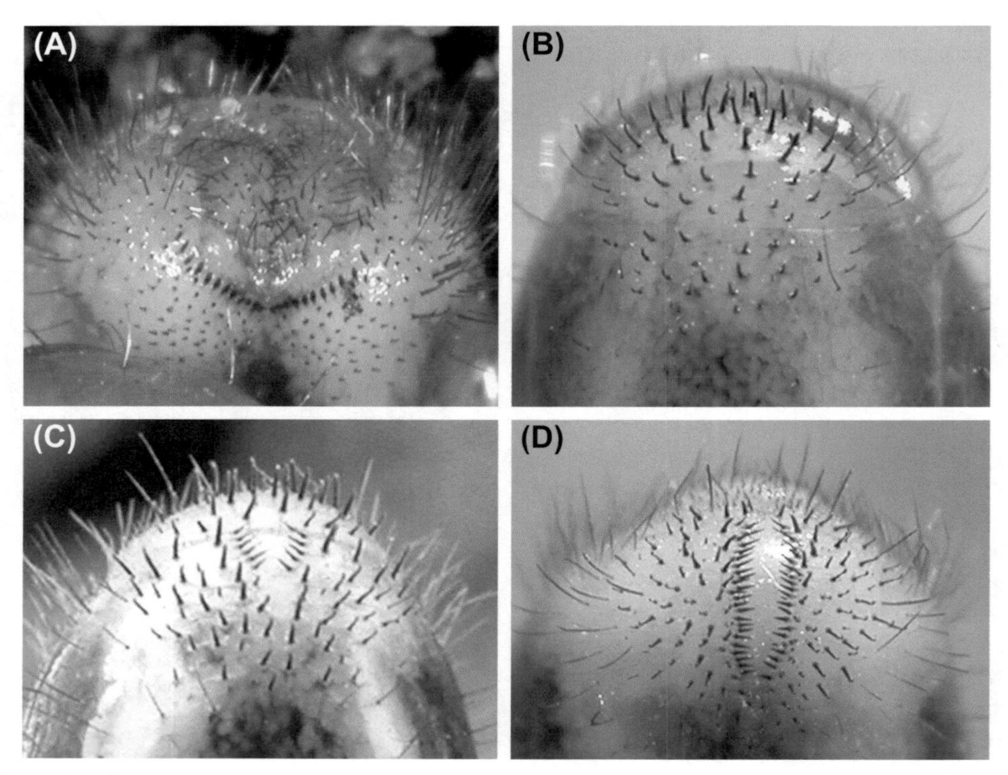

FIGURE 5.163A–D White grub rasters: (A) Asiatic garden beetle (B) Masked chafer (C) Japanese beetle (D) May beetle.

When assembled and compared as in the photo arrangement below, grubs can be diagnosed much more easily.

Know Your Grubs
Identifying Rasters

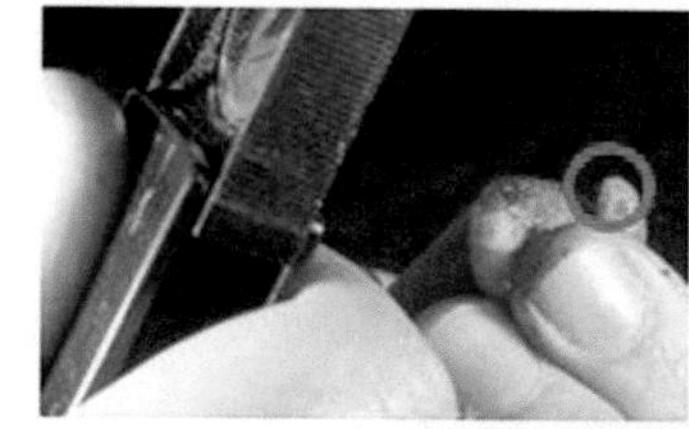

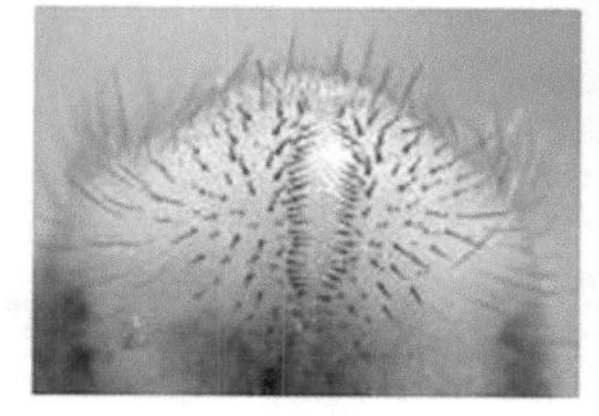

May beetle
(3-year grub)

Japanese beetle
(1-year grub)

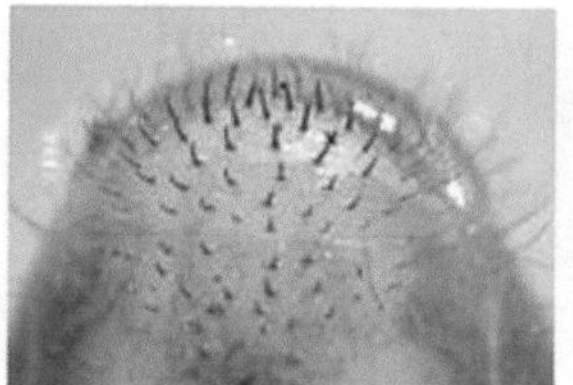

Masked chafer
(1-year grub)

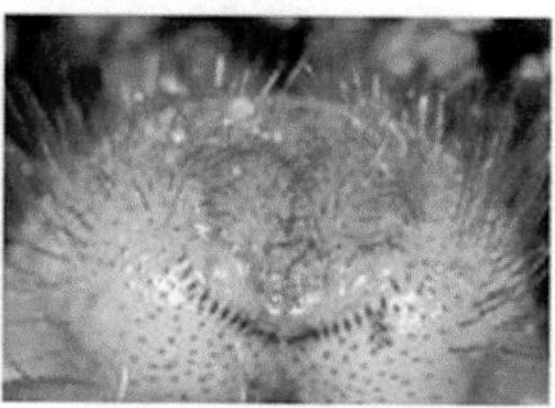

Asiatic garden beetle
(1-year grub)

FIGURE 5.164 Identifying common grub species based on rasters.

Cockroaches

Cockroaches are particularly troublesome pests in homes. Most homeowners associate them with filth and disease and will go to great lengths to control or 'exterminate' them. Understanding that not all cockroaches are the same, helps in both control and in assessing the need for management. Cockroach species can be distinguished best by using comparative illustrations.

The American cockroach (*Periplaneta americana*) is the largest of the above series and can reach 37 mm in length. The body is reddish-brown with a pale yellow border around the pronotum. Wings are as long as or longer than the abdomen. This roach feeds on a variety of foods but is most common where decaying organic matter is prevalent. It commonly inhabits its dark, moist basements, bathrooms and sewer systems.

The Oriental cockroach (*Blatta orientalis*) is very dark brown or even black in color, is not quite as large as the American roach (19–25 mm in length) and has reduced, non-functional wings that do not cover the body. This species often moves in and out of doors and frequents the same moist, dark habitats as does the American roach.

The German cockroach (*Blattella germanica*) is the most common cockroach species found in restaurants and kitchens in the United States. It is a general feeder but is normally attracted to fermented foods and to kitchen and pantry areas where food and moisture residues accumulate. It grows to about 12 mm in length and can be distinguished by its two dark lateral stripes along the pronotum.

The brown-banded cockroach (*Supella longipalis*) is often found throughout homes, hospitals and apartments and is comparable to the German cockroach in size. It does not require the close association with water that the German roach does, thus can be found anywhere in the building. It can be distinguished by the two light

FIGURE 5.165A–E Identifying cockroaches. (A) American cockroach (B) Oriental cockroach (C) Brown-banded cockroach (D) German cockroach (E) Pennsylvania woods cockroach.

FIGURE 5.165A–E Cont'd

colored, transverse bands that cross the base of its wings and abdomen.

The Pennsylvania woods cockroach (*Parcoblatta pennsylvanica*) most often occurs out of doors near wooded areas where it breeds in stumps, hollow trees and fallen logs. It is a strong flier and males are particularly attracted to lights at night. It is sometimes inadvertently brought into homes with firewood. Finding these inside a home or garage should not be of concern as they do not breed inside the home. Adults are dark brown with yellow margins along the thorax and forewings.

Confusing Flies

Flies can be a vexing diagnostic problem. There are many of them and when it comes to flies that occur in and around homes and farms where livestock are present, many of them look similar. Behavioral differences between flies may be the easiest diagnostic tool. When behavior differences can be pointed out within a comparative illustration, much more powerful tools can be made.

Diagnosticians seldom have difficulty distinguishing deer flies and horse flies because both are large bodied, dark gray or brown flies that have large heads, eyes and biting mouthparts. These are very fast flying flies that bite animals on the necks and rear ends.

FIGURE 5.166 Horse fly (*Tabanus* sp.).

FIGURE 5.167 Deer fly (*Chrysops* sp.).

When the two major genera of Tabanidae (*Tabanus* and *Chrysops*) are compared, *Tabanus* species are larger, approximately 25 mm in length and have clear wings. Chrysops species are much smaller, 12 mm in length and have a characteristic dark stripe across their wings. Females of both groups are strong fliers and vicious biters.

Musciod flies all resemble the housefly. There are many species of muscoid flies, however, and some can be considered pests. There are subtle apparent differences between them but fairly obvious behavioral differences.

House flies, blow flies and cluster flies are all nuisance pests that often get into homes and yards. These flies do not bite people but are considered filth flies because of where they breed.

FIGURE 5.168 House fly (*Musca domestica*).

The house fly Family Muscidae: (*Musca domestica*) is about 6 mm in length, has large red compound eyes and a gray thorax with four longitudinal dark lines down the back. It has sponging mouthparts, does not bite and is the most common fly near homes and buildings. It is often found in windows and on screens.

An adult cluster fly (family Calliphoridae) is slightly larger than the common house fly, is dull-gray with black markings and has golden-yellow hairs on the thorax. Cluster flies in their immature stages are earthworm parasites. Eggs are deposited on the soil and the maggots burrow into the bodies of living earthworms to feed. Adult flies are very lethargic and are often found congregating in wall voids and window sills of homes to pass the winter.

FIGURE 5.169A,B (A) Cluster fly (*Pollenia musca rudis*) (B) earthworm hosts.

FIGURE 5.170 Green bottle blow fly (*Lucilia Phaenicia* sp.).

The blow fly (family Calliphoridae) contains hundreds of species. These flies often sport metallic colors on the thorax and abdomen, have unique wing venations and distinctive hair-like bristles on their body. They are well known because they infest garbage cans and dead animal carcasses where they lay eggs. Blow fly maggot infestations are often used as evidence in forensic death-scene investigations.

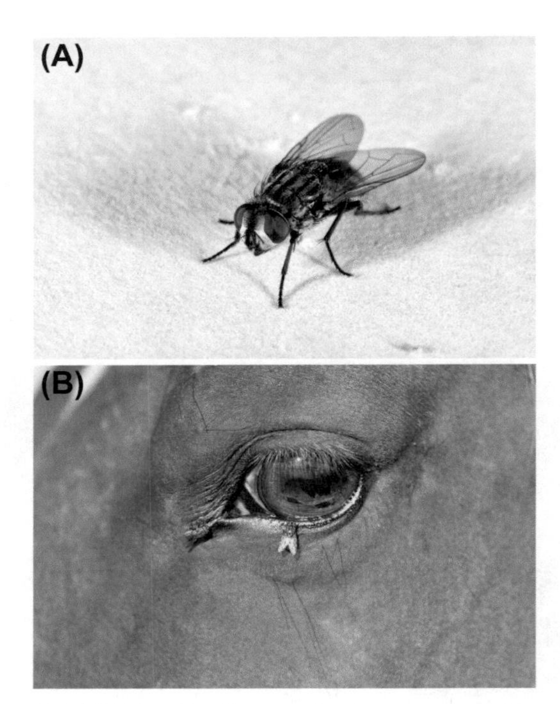

FIGURE 5.171A,B Face flies (*Musca autumnalis*).

Face flies (*Musca autumnalis*) strongly resemble house flies in size and appearance, but may be slightly larger and darker. They can be separated behaviorally, because they congregate around the eyes and nostrils of livestock, whereas house flies are more likely a general nuisance pest in buildings or around homes.

FIGURE 5.172A,B,C Stable flies (*Stomoxys calcitrans*).

Stable flies (*Stomoxys calcitrans*) are gray, 15 mm in length, resemble house flies except that they have a 'checker-board' pattern on the abdomen and have distinct biting mouthparts. They congregate on the lower legs of livestock.

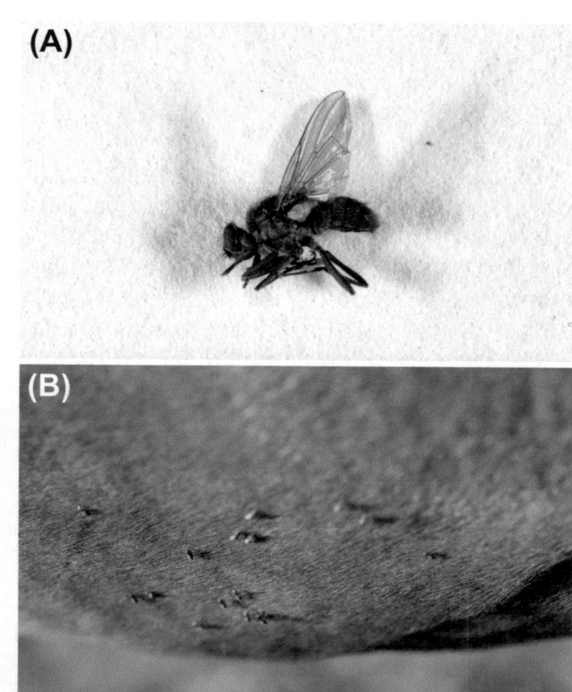

FIGURE 5.173A,B Horn flies (*Haemotobia irritans*).

Horn flies (*Haemotobia irritans*) resemble stable flies but are more slender and much smaller. They are gray-black flies that congregate on the backs, and sides of livestock and usually rest in a head-downward position.

Insect comparisons can be based on behavioral similarities as well as physical. Illustrative comparisons containing behavioral differences and similarities as well as physical, become very powerful diagnostic tools.

Behave Alike Pests

There is as much advantage in featuring groups of insects that behave alike as pests that look alike. Diagnosticians find great value in a composite of pests that can be discussed as a group. Extension specialists speak to clientele groups in this way. Most extension talks include a series of pests that affect a commodity of interest. For example when speaking to vegetable or fruit growers, an extension specialist may entitle the presentation, 'Insects That Feed on Fruits' or 'Pests That Damage Tomatoes'. Insects that feed on similar foods, such as mold beetles, for example, may be grouped together.

Grouping pests by the commodity that they damage or feed on is often the most effective way to teach about them. This allows for comparisons and contrasts of the pests and also facilitates more effective discussion about control and management because many of them can be treated together. Pests may be grouped together because they share other commonalities, however. Pests that behave alike may include insects that share the irritating behavior of invading homes. Other pests can be grouped by the habitat that they prefer, such as pests that are attracted to moisture, or to lights. Pests also may be grouped by how people perceive them, for example insects that bite people may be a valid way to group them.

Regardless of how the groupings are made, comparative photos of insects side by side enhance the discussions. Such photographs are commonly found on various web sites and informational outlets. Some of this availability is due to the fact that industrial affiliates who sell pesticides can profit by advertising that their products control a large group of pests. Other times, insect groups are listed together because their management strategies are similar.

Regardless of the reason, diagnosticians do well to study, and teach about groups of pests. By way of example, we will discuss:

- Pests that feed together (mold feeders)
- Key pests of a commodity (turfgrass pests)
- Pests that have a peculiar behavior (occasionally invading pests)
- Pests that require unique habitats (moisture loving home pests)
- Pests of human health (biting pests).

Photographs and descriptions of pests in a group can be created with a little photographic skill and a series of germane facts. When complied they are an effective way to teach principles or lead discussions, as well as identification techniques.

PESTS THAT FEED TOGETHER

Mold Feeders

The Food and Drug Administration (FDA) refers to insects in stored foods as 'natural contaminants.' A certain number of insects or insect parts are allowable because it is biologically impossible and economically impractical to grow, harvest, and process raw products that are totally free of non-hazardous, naturally occurring, unavoidable defects such as insects.

A family style photograph of mold feeding beetles is provided below. Even though the insects are quite different one from another, a common element in all of them is that they feed on molds. Most molds in the stored industry occur in grain products. The insects pictured below are all secondary pests of stored grains. They do not harm the kernels directly, but rather damage the commodity because their presence may exceed the FDA approved limits of foreign contamination placed on sale of the food. Thus it is helpful to have them all at a glance when discussing mold-feeding insects. Insect samples submitted for diagnosis from moldy grains are almost always identified as one of this group.

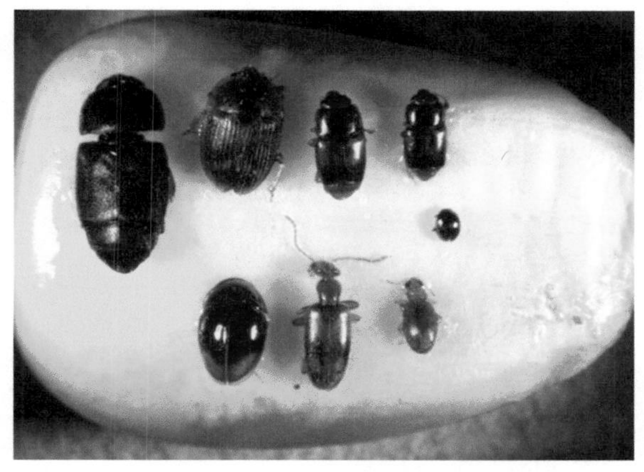

FIGURE 5.174 Common mold feeding beetles in stored grain.

KEY PESTS OF A COMMODITY

Turfgrass Pests

Key pests that damage a commodity may well be presented together. Even so, many pests may fit into this group. It may be profitable to further break down the pests based upon other commonalities such as when they occur, how they feed or exactly what part of the plant that they damage.

For example, turfgrass can be affected by many insect pests.

A very practical way of classifying key turfgrass insect pests is by separating them by where they live and feed in the turfgrass system.

Some pests are found in and feed only on the upper leaf canopy of the turfgrass. Common turfgrass pests that live and feed in this area include the caterpillar stages of cutworms, armyworms and sod webworms that have chewing mouthparts, as well as all stages of greenbugs (aphids) that feed with sucking mouthparts.

A second group of insects resides in the thatch area of the turfgrass profile. Stems and thatch harbor insects such as billbugs (adults and larvae) with biting mouthparts and chinch bugs that feed with sucking mouthparts. These insect pests are potentially more serious than leaf feeders because they feed in the portion of the plant where the growing point is located. In most cases, if the growing point is damaged, the plant will die.

The most serious insect pests of turfgrass grown in the U.S. occur in the roots and soil, the lowest level of the turfgrass profile. It is here that insect pests with biting mouthparts may destroy the root system of the plant. White grubs, the immature stage of beetles, are the biggest threat to grass roots but occasionally other insects such as mole crickets and some ants may cause damage.

Pest control techniques (including cultural, biological and chemical options) should be chosen based on an understanding of the pest they are intended to manage. By classifying turfgrass pests, a manager may be able to determine the most appropriate management technique. Interacting factors including; time of season, turfgrass value, turfgrass condition, where the pest resides, pest life stage, how it feeds, its life cycle and development, all influence how effective any given control method may be.

FIGURE 5.175A,B,C Armyworms (Noctuidae spp.) arrows depicting identifying characters.

Insect Pests that Feed on Leaves in the Turfgrass Canopy

Armyworm moths are brownish gray in color, and have a wing spread of about 37 mm. The distinguishing white-colored mark in the center of each forewing helps separate them from other moths. Caterpillars are the damaging stage of the armyworm in turfgrasses. They can grow to be as long as 37 mm and have characteristic white, orange, and brown stripes that run the length of the abdomen.

The fall armyworm's head has a predominantly white, inverted Y-shaped suture, usually visible between the eyes. (In the common armyworm, the suture is H-shaped.)

Armyworms feeding on grass blades may actually cut off the plants at crown level. Turfgrass stands may exhibit an overall 'ragged' appearance with brown patches. To be certain of an armyworm diagnosis, scout ragged turf areas from June to September, searching for the small caterpillars. Damage is most severe during hot, dry weather in midsummer.

Black cutworm moths are similar in shape and size to armyworm moths but are dark gray or black in color and have a unique black triangular mark on the forewing.

Black cutworm caterpillars develop through five larval instars. Compared to armyworms, black cutworm caterpillars are more uniformly colored on the dorsal and lateral surfaces, ranging from light gray or gray-brown to nearly black. On some individuals, the dorsal region is slightly lighter or brownish in color but ventrally the larva tends to be lighter in color. Black cutworm larval epidermis bears numerous dark, distinguishing, granules over much of the body making it appear granulated.

Caterpillars are nocturnal and seek cracks, crevices and aeration holes in which to hide

FIGURE 5.176A,B,C Black cutworm (*Agrotis ipsilon*) (A) Moth; (B) & (C) Larvae. Arrows depict identifying characters.

FIGURE 5.177A,B Sod webworms (Crambidae spp.) (A) Moth; (B) Larva.

during the day. During the evening they emerge, to feed on the crown areas of the turfgrass grass plants, often completely severing them before moving on. Feeding on the grass plant growing point kills the plant. Heavily damaged turfgrass appears as if it has been scalped.

Sod webworm adult moths are buff-colored and have characteristic snout-like projections extending forward from the head. They measure approximately 12 mm in length. At rest, webworm moths fold their wings around their body, giving them a cigar-shaped appearance.

Sod webworm caterpillars are brown or green in color and display characteristic, small, dark spots all over the surface of the body. Compared to cutworm and armyworms, sod webworm caterpillars are smaller, never more than 1 inch long when fully mature.

Larvae consume the leaves and stems of turfgrass just above the crown. As webworm larvae mature and continue to feed, large brown patches characterize the injured turf areas. Like armyworm damaged areas, sod webworm damage may resemble scalped or drought-stressed turf with scattered, irregular, brown patches and an overall ragged or thin appearance. A sure sign of caterpillar activity is a coarse, greenish, sawdust-like fecal material (frass) deep in the canopy around the perimeter of damaged areas.

A simple 'flush test' is often used to sample for turfgrass feeding caterpillars. Cut out both ends of a coffee can, drive it into the turf, and fill it with soapy water. After a couple of minutes the insects will float to the surface.

Caterpillar damage is most severe in fine fescues, Kentucky bluegrass, and non-endophytic grasses.

Green bugs are pale to dark green insects, very small (less than 3 mm), have pear-shaped bodies and may be winged or wingless. Immature stages are similar in color and feeding behavior to the adult but they are smaller and do not ever have wings.

Green bugs have many generations per year. They become active in spring and remain active for the duration of the growing season.

Early damage symptoms first occur in shaded areas. Infested areas, which range from several inches to several feet, turn pale green then rust color as the aphids suck the juices from the leaves. These symptoms may initially resemble drought injury.

FIGURE 5.178A,B Green bug aphid (*Schizaphis graminum*) (A); Green bugs feeding (B).

Look for masses of aphids lined up along the leaf mid-vein of discolored chlorotic plants.

Insects that Feed on Stems in the Turfgrass Thatch

Insects that reside in the thatch area of a lawn and damage the stems and crowns of the plant include billbugs and chinch bugs.

FIGURE 5.179A,B Bluegrass billbugs (*Sphenophorus* spp.) (A) Adult (B) Larva.

Adult billbugs are recognized as small (12 mm), black or brown weevils with long, pronounced 'snouts,' characteristic of the family Curculionidae. Adult beetles are present from late April through May on hardscape areas, such as sidewalks and driveways near turf.

Larvae are cream colored with a reddish head capsule and appear similar to white grub larvae, but are smaller (6–9 mm) and legless.

They can be found in crowns and stems for a time but eventually move into the soil to feed on roots.

Although both the adults and the larvae feed on the grass plant, it is the larval boring and tunneling damage to the stems and crowns that causes the serious injury to turfgrass. Evidence of feeding is small, clearly visible holes as well as stems that eventually turn straw-colored and can be pulled out easily. This often exposes very diagnostic, light brown, sawdust-like fecal material (frass) in the crown area of the stem.

Damage is most severe during hot, dry periods and resembles many turf diseases, such as dollar spot, that begin as small circular patches and expand and eventually coalesce.

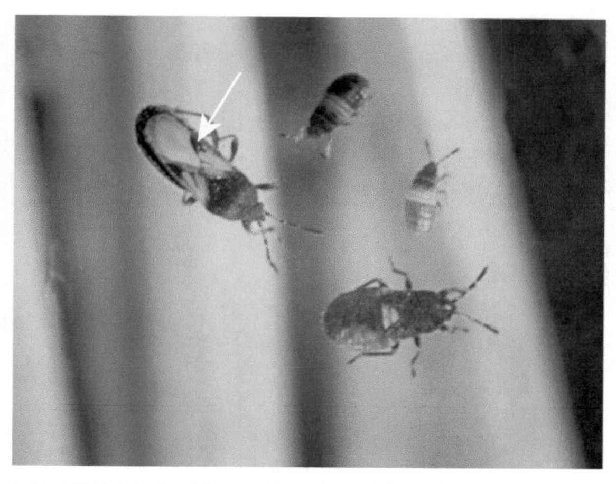

FIGURE 5.180 Hairy chinch bug (*Blissus leucopterus hirtus*).

Adult chinch bugs are small (5 mm), white and black in color, have flattened elongate bodies. The head, pronotum and abdomen are gray-black in color and are covered with fine hairs. The wings are white with a black spot, the corium, located in the middle front edge.

Immatures (nymphs) are smaller, bright red to purple, similar in shape to adults, but lack wings. A nymph also may have a distinct, cream-colored band midway across the body.

Both nymphs and adults suck sap from grass blades, causing a yellow, then straw-brown turf. Patches of damage grow and spread over time and are most severe in hot, dry microenvironments, especially during droughty years. Thatch-forming grasses and fine fescues grown in full sun are most susceptible to chinch bug injury.

Most cool-season grasses are susceptible to chinch bug infestations when stressed. A vigorously growing, well-maintained stand of turfgrass, with minimal thatch is less apt to be damaged. Endophyte-enhanced turf species also may be less prone to chinch bug damage.

While a site may have large populations of these pests, the bugs often remain unnoticed for some time because of their relatively small individual size. Look for them on the margin between damaged and healthy grass. Be aware that they will run for cover when the turf canopy is disturbed. Using a soap flush test (described above) helps to verify and sample for chinch bugs.

Insects that Feed on Turfgrass Roots in the Soil

FIGURE 5.181A–D White grubs (Scarabaeidae spp.) (A & B) Adults (C) Larvae (D) Close up of grub.

FIGURE 5.181A–D Cont'd

There are several species of Scarabaeidae that infest and damage turfgrass root systems, including Japanese beetles, masked chafers, European chafers, May or June beetles, green June beetles, and Asiatic garden beetles. Adults vary widely in color shape and size.

By contrast, immature stages are very similar in appearance. They are plump, white, C-shaped grubs, 6–50 mm in length and have many wrinkles or folds along the body behind the head. White grubs always have red-brown heads and six legs. Grubs have three larval instars and multiple grubs may be present concurrently. Grub species may be separated by examining the rastral hair patterns of grubs as discussed earlier.

Grubs begin feeding in the soil immediately after hatching. Older larvae burrow below the freeze line to overwinter, then return to the turfgrass root zone in the spring to resume feeding. As soil is mined, turfgrass roots are pruned and grass dies. Grubs may feed for a single year or multiple years – depending upon the species. Turf damage is particularly noticeable from late summer (August) through the fall, but may also be apparent in the spring.

Grubs prefer thatch-forming grasses, such as Kentucky bluegrass. Feeding damage resembles early drought stress, and consists of wilting, dead or dying, irregular patches of turf. Distinguishing symptoms include sod that pulls up easily, like a carpet, revealing the white grubs underneath, or large areas of severely disturbed sod, caused by the foraging activities of skunks or raccoons. Usually grub damaged turfgrass has a sponge-like feel to it as it is walked on.

Grub scouting should be done in early August, before severe symptoms become manifest, by cutting 0.1 square meter sections of sod, 50–75 mm deep from several locations and manually searching for larvae.

PESTS THAT HAVE A PECULIAR BEHAVIOR

Occasionally Invading Pests

Pests that occasionally enter homes and other buildings but do not breed or feed there are often called 'occasional invaders.' These are considered nuisance pests because they do not directly harm people, pets or the contents of a home. This is not to say, however, that they are minor pests in any way. Complaints of hundreds of insects in a kitchen, swarming over a table or around a person's head are anything but 'minor.' Homeowners and professional pest managers expend great energies to manage occasional invader pests every year. Diagnosticians receive a large number of occasional invader samples for identification and control recommendations.

Understanding how and why the invaders gain entrance into the building is the first step in managing them. Below are photographs of the 12 most common occasional invaders in the U.S. These are not particularly similar in appearance and can usually be identified by comparison to their photographs. Most seem to enter structures in an effort to avoid severe weather events that occur out of doors. Shelter from the elements is considered to be the motivation for occasional invaders to enter homes and buildings. This phenomenon normally occurs in the fall time, however, manifestations of invasion are sometimes not noticed until warm days during the late winter or early spring, when the insects become active and search for ways to get back out of the home. Occupants sometimes incorrectly assume that the pests have been breeding in the home because they see more of them in late winter than during the fall. This is understandable because occasional invaders tend to congregate over time and then become active more or less at the same time, giving the appearance of massive invasions.

Managing occasional invaders effectively largely consists of using preventative controls. Reducing attraction to a building, eliminating access by sealing them out and using perimeter pesticides barriers. Sealing occasional invaders out of a building includes techniques such as calks and sealants applied to cracks and crevices, utility entrance points, fixing windows and doors, using screens on windows, door sweeps on doors.

Consciously selecting, placing and using outside lights, selection of siding and roofing materials, colors and building locations in relation to habitats that are known to harbor high pest populations, are the first steps in reducing pest attractiveness to buildings. Selection and placement of outdoor plants (sometimes the removal of them) can affect occasional invasions of pests.

Sealing the building so that pests do not have an access to enter is a second step. Calking entrance points such as holes, crevices, outside utility ports, fixing broken windows and screens, repairing soffits, chimneys, installing door sweeps and door screens go a long way in reducing occasionally invading pests.

Application of outside perimeter pesticides that precisely coincide with pest occurrence at buildings is also very important. Timing depends upon an understanding of the pest cycles and upon close monitoring of the pest.

Selection of pesticides is also important because to be effective they must be active when the pest is present. Modern day pesticides are designed to break down with time, sunlight and water. This design is preferable to the long-lasting pesticides of the past that contaminate the environment. However, the comparatively short residual of insecticides today mandates that they be applied more precisely than ever. Selection of product, formulation and application technique all factor in to this decision.

Pesticides must be active upon contact with the pest and they must be placed such that they have the best chance of making this contact. Depending upon the pests, outside perimeter applications may be needed only on one side, only around windows and doors, or even only partially up the foundation of the building.

Finally, recognizing the pests that may potentially invade buildings is important and can change geographically, seasonally and over time. The following list contains the 12 most notorious 'occasional invader' pests in the United States.

The Obnoxious Dozen

Leaffooted bug
Box elder bug
Asian lady beetle
Brown marmorated stink bug
Elm leaf beetle
Clover mite
Potato leafhopper
Hackberry psyllid
Woods roach
Strawberry root weevil
Black vine weevil
Cluster fly

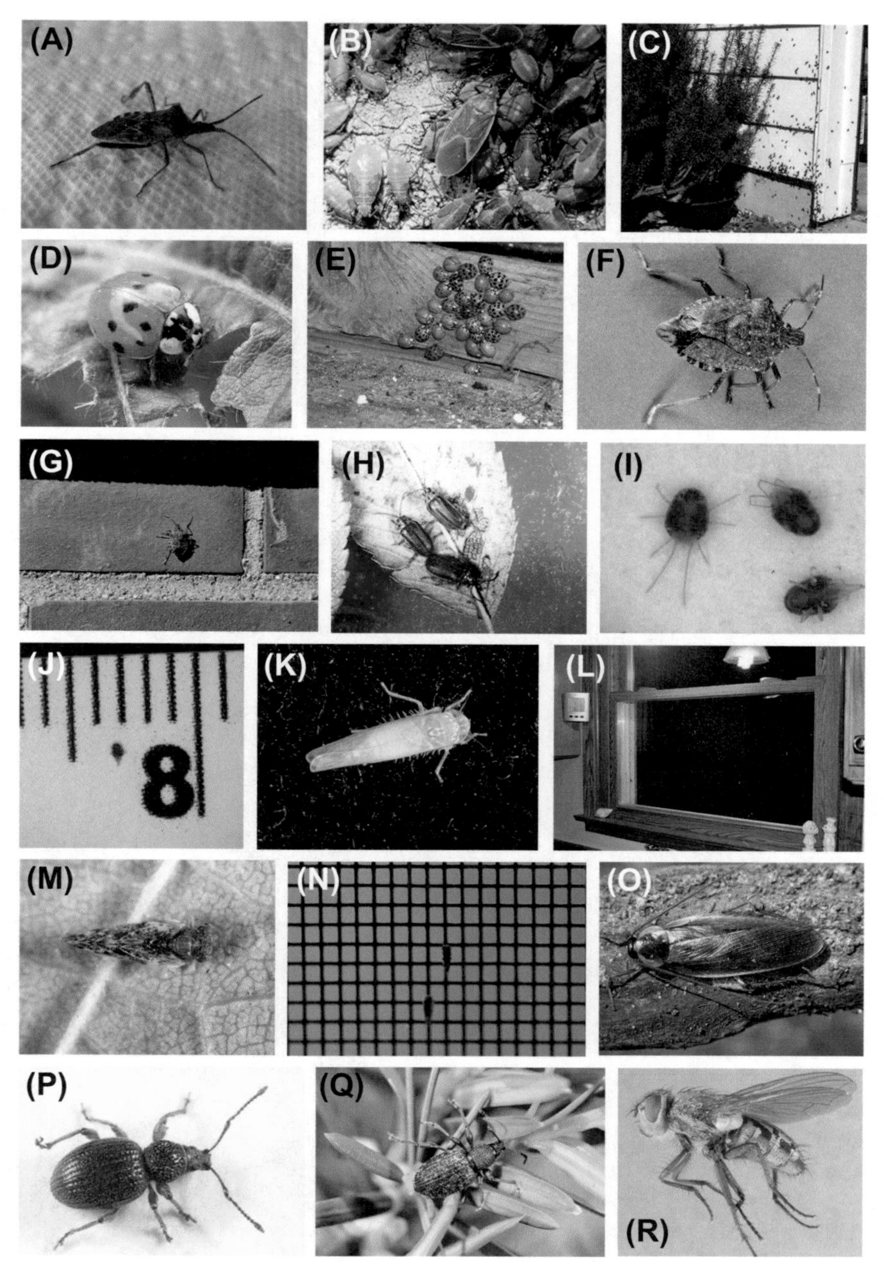

FIGURE 5.182A–R Twelve most common home-invading pests (A) Leaffooted bug (*Leptoglossus* spp.) (B, C) Box elder bug (*Boisea trivittata*) (D, E) Asian lady beetle (*Harmonia axiridis*) (F, G) Brown marmorated stink bug (*Halyomorpha halys*) (H) Elm Leaf beetle (*Xanthogaleruca luteola*) (I, J) Clover mites (*Bryobia praetiosa*) (K, L) Potato leafhopper (*Empoasca fabae*) (M, N) Hackberry psyllid (*Pachypsylla* sp.) (O) Pennsylvania wood cockroach (*Parcobatta pennsylvanica*) (P) Strawberry root weevil (*Otiorhynchus ovatus*) (Q) Black vine weevil (*Otiorhychus sulcatus*) (R) Cluster fly (*Pollenia* sp.).

PESTS THAT REQUIRE UNIQUE HABITATS

Moisture Loving Indoor Pests

Recent building trends of tighter, more energy efficient buildings naturally result in higher interior relative humidity. Along with all of the advantages, however, tight buildings lead to an increased opportunity for fungal and mold growth and thus mold feeding insect pests such as are described above (mold feeding pests).

In addition, other arthropods also favor damp buildings. Fungus gnats build up high populations when organic matter is wet.

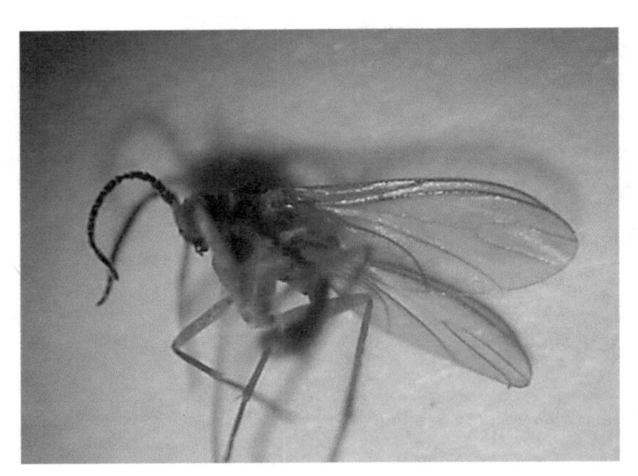

FIGURE 5.183 Fungus gnat (several genera of Sciaridae and Mycetophilidae).

In addition to standing water near the foundation of the buildings, and clogged or broken gutters, overwatered plants are a prime but often overlooked source for these pests.

In many cases pests get into buildings simply because they occur in high numbers close to the building. When plants are dense and grow next to a building they become a bridge for occasional invaders to enter.

Mulch is designed to retain water for plants and is very commonly used around homes and other buildings.

FIGURE 5.184 Slugs (Gastropoda).

Unfortunately, mulch also brings with it ideal habitat for moisture loving pests such as slugs.

Mulch on top of organic soils provides ideal habitat for a series of potential building invading pests including:

FIGURE 5.184A–D Soil originate home-invaders. (A) Springtails (Collembola) (B) Millipedes (Diplopoda) (C) Pillbugs (Isopoda) (D) Sowbugs (Isopoda).

FIGURE 5.184A–D Cont'd

The presence of these pests indoors usually indicates wet soils close to the foundation of the building. Correcting the problem may require the removal of the soil and mulch from around the foundation of the building or changing the conditions such that it can dry out thoroughly.

Crickets are well known for their affinity for moist shady conditions. Crawl spaces and damp basements provide ideal habitats for them. House, field and camel crickets in particular become pests in homes when conditions are just right.

FIGURE 5.185A,B (A) Camel cricket (several species of Rhaphidophoridae) (B) House cricket (*Acheta domestica*).

FIGURE 5.185A,B Cont'd

Insect pests such as silverfish and firebrats breed in moist, warm environments such as basements, boiler and utility rooms and sometimes in attics.

FIGURE 5.186 Silverfish (*Lepisma saccharina*).

FIGURE 5.187 Psocid (several species of Psocidae).

Psocids are interesting insects and are some of the first to inhabit moist foods such as grains, pet foods and paper products. Their populations build up quickly.

As discussed earlier, foreign grain beetles are common in homes where moisture has accumulated. Newly constructed homes in which the lumber has not yet dried sufficiently are susceptible to infestation of these insects.

FIGURE 5.188 Foreign grain beetle (*Ahasverus advena*).

Carpenter ants are common in landscapes around homes. They often nest in old mature trees where the center of the trunk is hollow and rotting. Although they do not feed on wood, they nest in it and it happens that water damaged wood creates the perfect conditions for them. If wood inside homes or buildings has become water-damaged in the past, it is a prime location for carpenter ants to build nests. As they expand their nests, adjacent wooden beams also can be damaged.

FIGURE 5.189 Carpenter ant (*Camponotus* sp.).

Insects associated with moisture problems, including high indoor relative humidity, can only be managed if the conditions that are conducive for the pest are changed. These include but are not limited to the following items:

TO REDUCE MOISTURE LOVING PESTS IN BUILDINGS

- Reduce any standing water near or in the structure
- Repair plumbing or roof leaks
- Be certain that rain guards, gutters, and down spouts are installed and are working properly
- Use dehumidifier or air conditioning units to reduce the inside humidity
- If contained areas such as wall voids, attics or basements have become wet, open them up to allow access and thorough drying with heaters and fans
- Make sure doors and windows are kept closed or use tight fitting screens
- Screen vents or other openings to the outside that may provide pest access
- Seal cracks and other openings in exterior walls around windows, utility ports and vents
- Use proper lights and lighting techniques on the outside of the building
- Use motion sensors to reduce light emissions
- Remove vegetation and mulch from contacting the structure
- Select and place trees, shrubs, ponds, drainage ponds with pests in mind
- Plan location of dumpsters, storage sheds and other outbuildings carefully

PESTS OF HUMAN HEALTH

Biting Pests

Diagnosticians may describe groups of pests based on the effect that they have on people. For example, people detest arthropods that bite. Such pests cause pain, itching and disease. Arranging information about the effects of biting pests is an effective means of organization. Disease transmission is the most important ramification of an arthropod bite. Chief among the disease carrying pests are ticks and mosquitoes. These two pests alone are responsible for the transmission of some of the world's deadliest diseases.

Requests to a diagnostic laboratory may begin with a description of the symptoms that a client is experiencing and end by asking for a diagnostician's appraisal of the cause. Even after assuring a client that medical diagnoses and recommendations are outside the area of an entomologist's expertise, clients normally persist by asking for a range of possible causes. This places a diagnostician at the uncomfortable interface between entomology and medicine. When it pertains to biting arthropods, when does a diagnostician's responsibility end and a medical professional's responsibility begin?

It is instructive to know at least enough about the potential problems and associated symptoms of insect bites in order for a diagnostician to rule out certain causations.

In epidemiology, a vector is any agent (including arthropods) that transmits a pathogen to another organism. The following human diseases, their symptoms and vectors should be understood by diagnosticians.

Ticks

Many species of the family Ixodidae (hard ticks) can vector disease pathogens. This is not to say that all ticks or tick bites result in disease transmission. In fact, most do not. Because of the serious nature of some diseases that are transmitted by ticks, diagnosticians should be aware.

FIGURE 5.190A,B Ticks (Ixodidae.).

Symptoms of tickborne diseases can range from mild symptoms treatable at home to severe infections requiring hospitalization. Although easily treated with antibiotics, many tick transmitted diseases can be difficult for physicians to diagnose.

The most common symptoms of tick-related illnesses are:

Fever/chills: With all tickborne diseases, patients can experience fever at varying degrees and times of onset.

Aches and pains: Tickborne disease symptoms include headache, fatigue, and muscle aches. The severity and time of onset of these symptoms can depend on the disease and the patient's personal tolerance level.

Rash: Many tickborne infections result in distinctive skin rashes. These are sometimes diagnostic for the disease. Like all reactions, however, they are quite variable from person to person.

The following tick transmitted diseases occur in the U.S. Each is vectored by individual species of ticks, although this is sometimes not well-documented. In addition to the above, general symptoms, a description of specific disease symptoms, known tick vectors and geographic occurrence follows.

Annaplasmosis is transmitted to humans by tick bites primarily from the blacklegged tick (*Ixodes scapularis*) in the northeastern and upper midwestern U.S. and the western blacklegged tick (*Ixodes pacificus*) along the Pacific coast. The first symptoms of anaplasmosis typically begin within 1–2 weeks after the bite of an infected tick. Severe clinical presentations may include difficulty breathing, hemorrhage, renal failure or neurological problems. Anaplasmosis is a potentially serious illness that can be fatal if not treated correctly.
Babesiosis is caused by microscopic parasites that infect red blood cells. Most human cases of babesiosis in the United States are caused by *Babesia microti* transmitted by the blacklegged tick (*Ixodes scapularis*) and is found primarily in the northeastern and upper midwestern U.S. Symptoms of babesiosis are similar to those of Lyme disease (see below).

Erlichiosis is transmitted to humans by the lone star tick (*Ambylomma americanum*), found primarily in the southcentral and eastern U.S. Ehrlichiosis can cause a rash, medically described as macular to maculopapular to petechial in appearance, and may present after the onset of fever.
Lyme disease is vectored by the blacklegged tick (*Ixodes scapularis*) in the northeastern U.S. and upper midwestern U.S. and the western blacklegged tick (*Ixodes pacificus*) along the Pacific coast. In most cases, a rash may develop at the site of the bite after 3–30 days. The rash is most often circular with a clear edge, known as erythema migrans. It is called 'migrans' because it grows in size, gradually expanding over several days from less than 50 mm across to very large, even covering an entire back. If the center of the rash clears, the rash may take on a characteristic bull's eye appearance. The rash is usually not itchy or painful. In early Lyme disease, patients often are diagnosed with the common flu and may suffer for many months with this infectious illness without treatment.
Powassan is transmitted by the blacklegged tick (*Ixodes scapularis*) and *Ixodes cookei* or *Ixodes marxi* ticks, in the northeastern U.S. and Great Lakes region. Signs and symptoms of disease caused by this virus can include fever, headache, vomiting, weakness, confusion, loss of coordination, speech difficulties, and memory loss. The virus infects the central nervous system and can cause encephalitis (swelling of the brain) and meningitis (swelling of the membranes that surround the brain and spinal cord).
Rickettsiosis is transmitted to humans by the Gulf Coast tick (*Amblyomma maculatum*). Most tick-borne rickettsial diseases cause sudden fever, chills, and headache (occasionally severe). Nausea, vomiting, and anorexia are common in early illness. Rash is common and usually appears two to four days after onset of fever, and typically presents as small, blanching, pink macules on the ankles, wrists, or forearms. These gradually evolve into maculopapules. They can spread throughout the body, including to the palms and soles. Clinically, tick-borne rickettsial diseases are difficult to differentiate from other diseases.
364D Rickettsiosis is transmitted to humans by the Pacific Coast tick (*Dermacentor occidentalis* ticks). This is a new disease that has been found in California.

Rocky mountain spotted fever is vectored by the American dog tick (*Dermacentor variabilis*), Rocky Mountain wood tick (*Dermacentor andersoni*), and the brown dog tick (*Rhipicephalus sangunineus.*) The rash varies greatly from person to person in appearance, location, and time of onset. Most often, the rash begins 2–5 days after the onset of fever as small, flat, pink, non-itchy spots on the wrists, forearms, and ankles and spreads to the trunk. It sometimes involves the palms and soles. A red to purple, spotted rash, usually not seen until the sixth day or later after onset of symptoms may occur in 35–60% of patients with the infection.

STARI (Southern Tick-Associated Rash Illness) is transmitted via bites from the lone star tick (*Ambylomma americanum*), found in the southeastern and eastern U.S.
The rash of STARI is nearly identical to that of Lyme disease, with a red, expanding 'bull's eye' lesion that develops around the site of a lone star tick bite. Unlike Lyme disease, STARI has not been linked to any arthritic or neurologic symptoms.

Tick-borne relapsing fever is transmitted to humans through the bite of infected soft ticks (family Argasidae). It has been reported in 15 states: Arizona, California, Colorado, Idaho, Kansas, Montana, Nevada, New Mexico, Ohio, Oklahoma, Oregon, Texas, Utah, Washington, and Wyoming and is associated with sleeping in rustic cabins and vacation homes. Tick-borne relapsing fever is characterized by recurring episodes of fever accompanied by other non-specific symptoms including headaches, muscle pain, joint pain, chills, vomiting, and abdominal pain. Symptoms tend to develop within 7 days after the tick bite and last an average of 3 days and are then followed by an asymptomatic period (no symptoms present) lasting anywhere from 4 to 14 days.

Tularemia is transmitted to humans by the dog tick (*Dermacentor variabilis*), the wood tick (*Dermacentor andersoni*), and the lone star tick (*Amblyomma americanum*). Tularemia occurs throughout the U.S. In the most common form of tularemia, a skin ulcer appears at the tick feeding site. The ulcer is accompanied by swelling of regional lymph glands, usually in the armpit or groin.

Tick Management

Preventing tick borne diseases is best accomplished by avoiding tick infested areas, inspecting for crawling ticks before they become attached and by using repellents. Measures to control them using pesticides are limited to small backyard areas and locations close to kennels or other areas where pets are kept out of doors.

Every state has publications with photographs of endemic ticks and descriptions of the diseases that they transmit. Ticks lay many eggs. Each species is slightly different, not only by way of the diseases it may transmit, but also by its preferred locations, intermediate hosts and the seasonal timing of its activity.

FIGURE 5.191 Female tick with eggs.

Mosquitoes

FIGURE 5.192 Blood sucking mosquito (Culicidae sp.).

It is estimated that mosquitoes cause more human suffering than any other animal. Mosquito bites cause severe skin irritation through an allergic reaction to the mosquito's saliva, but it is the danger of disease transmission that is of most concern to people.

In fact, more than one million people worldwide die each year and many millions more are infected and sickened by diseases that mosquitoes transmit. Mosquito vectored human diseases in the U.S. include viruses such as encephalitis, West Nile virus and dengue, as well as malaria, a protozoan disease. (It should be noted that these and other viral and filarial diseases such as dog heartworm may also be transmitted to dogs and horses.)

Encephalitis: Mosquito-transmitted encephalitis is a viral disease that causes serious infections of the central nervous system as well as brain inflammation. Symptoms may range from none at all to a mild flu-like illness with fever, headache, and sore throat. These progress to severe headache, high fever, chills, vomiting and in serious situations, disorientation, seizures, and coma. Several related viruses can be responsible for encephalitis including the following:

- **Eastern and western equine encephalitis** virus has a complex life cycle involving birds and several *Culex* species of mosquitoes. Mosquitoes feed first on infected birds and become carriers of the disease before feeding on humans and other mammals.
- **LaCrosse encephalitis** is thought to be transmitted by a specific woodland mosquito (*Aedes triseriatus*), sometimes called the tree-hole mosquito. It requires any of several small mammals as the warm-blooded, intermediate host.
- **St. Louis encephalitis** is transmitted from birds to man and other mammals by infected *Culex* sp. mosquitoes. This disease is rampant throughout the United States, but most severe along the Gulf of Mexico, especially Florida and affects the elderly and very young most severely.

West Nile virus is cycled between mosquitoes (*Culex pipiens*) and birds. It is transmitted to mammals (including horses) and man by mosquitoes that have fed on infected birds. Most common symptoms are fever, headache, tiredness, joint and muscle aches and often a rash. However, in a few cases the nervous system including the brain and the membrane surrounding it is affected, causing severe and irreversible complications (meningoencephalitis) and death.

Dengue is a serious arboviral disease even though it has a low mortality. Dengue symptoms include sudden, high fever, severe headaches, pain behind the eyes, severe joint and muscle pain, nausea, and vomiting followed by a skin rash three to four days later. Mosquitoes (*Aedes aegypti* and *Aedes albopictus)* are vectors. The recent spread of dengue can be directly attributed to the proliferation and adaptation of these mosquitoes. It is most common in Texas, Hawaii and Florida in the United States.

Malaria is a mosquito-borne disease caused by a parasite (*plasmodium)* transmitted by mosquitoes (*Anopheles* sp.). In the early stages, malaria symptoms include high fever, chills, headache, intense sweating, fatigue, nausea and vomiting. Often intense dreams and nightmares accompany this disease. As the plasmodium parasite develops in cycles, reproducing in the red blood cells and liver cells, associated symptoms also cycle. These may come and go at different intensities and for different lengths of time and are one of the major indicators of malarial infection. Longer-term effects include an enlarged spleen, impaired function of the spinal chord and brain stem, associated seizures, loss of consciousness and death. Anti-malarial drugs have been used to combat this disease for many years. Most recently, pioneering work has begun in developing a vaccine against this disease.

Mosquito Management

The best protection against any of the mosquito transmitted diseases is to prevent mosquito bites. Use insect repellent, wear long sleeves, long pants and socks or even stay indoors while mosquitoes are most active. Like for tick management, each state has produced copious amounts of information via fact sheets and web sites that list and describe the various mosquito species in the state and specific management practices.

Reducing pools of standing water (mosquito breeding sites) is one of the most effective methods or reducing certain mosquito problems. This is particularly effective in managing West Nile virus.

Chemically treating larger areas of water is still needed in some situations where large populations of mosquitoes develop. Using mosquitofish as predators and alternative materials such as oil or Bti (*Bacillus thuringienis* sp.) can also reduce high populations mosquitoes.

More controversial is the practice of chemically treating for adult mosquitoes. This practice is common in some areas, especially where the ability to manage mosquito-breeding habitats is limited and the threat of disease transmission is high.

Other Biting Pests

Other biting pests may afflict people. These include pests that do not commonly transmit disease but are a nuisance nevertheless.

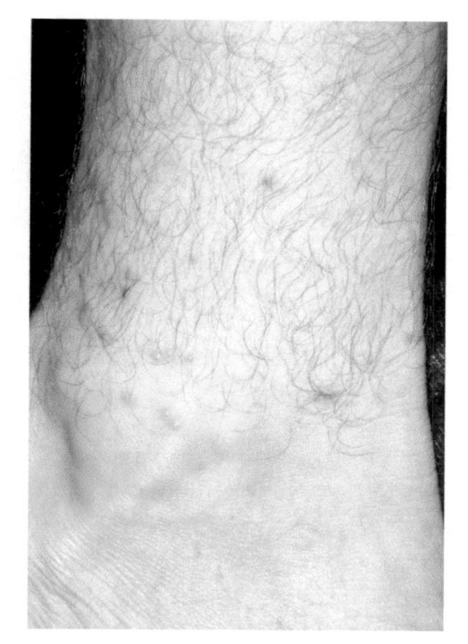

FIGURE 5.193 Bites on ankles caused by chiggers.

Symptoms occur as a direct result of a person's reactions to the insect feeding. When an insect feeds on a person it often injects its saliva into the skin, causing a person to produce a histamine response around the area of the bite.

Histamines are immune response chemicals produced by the body to assist in protecting against infections and diseases. It is the histamines that cause the intense itching, and the raised bump (wheal) typical of insect bites.

Unfortunately, the specific response, size and color of welt, intensity of itch and persistence of the wheal varies widely from person to person. As a result, it is not possible to state with confidence what pest may have caused the bite after the fact. However, identifying the casual agent can help a person avoid the bites in the future. Below is a list of the common arthropods bites that are often confused with mosquito bites.

FIGURE 5.194 Cat flea (*Ctenocephalides felis*).

Many wingless, biting, parasites of animals belong to the Order Syphonaptera. As a group we call them fleas. Most commom complaints to daignsoticians have to do with the cat flea which is the most common flea on domestic cats and dogs. Eggs and larve develop away from the host. If the primary host cannot be found, fleas can bite people. Severe itching and sometimes secondary complications due to people scratching are common.

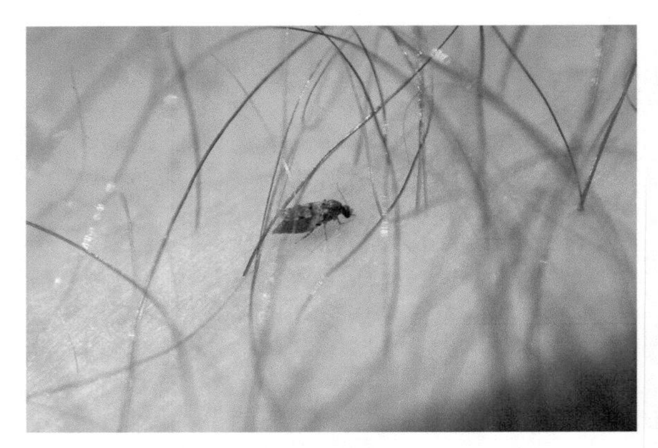

FIGURE 5.195 Biting black flies (Simulidae).

Biting black flies (also called buffalo gnats) and biting midges (also called punkies, or no-see-ums) are tiny flies belonging to the families Simuliidae and Ceratopogonidae respectively. They are found in moist habitats, both in running or stagnant water. They have piercing-sucking mouthparts and attack almost any warm-blooded animal in search of a blood meal.The bites of most species cause an immediate sharp pain. This is followed by the development of red wheals. For most people these lesions last up to a week or more and are reported as more painful than mosquito bites.

FIGURE 5.196 Head louse (*Pediculus humanus capitis*).

Lice are wingless insects that spend their entire life on their host. The common head louse is one of several species (including pubic louse and body louse) that feed primarily on humans. Lice are highly communicable and infestations, especially in children, often remain undetected for long periods due to the very mild itching symptoms that they cause.

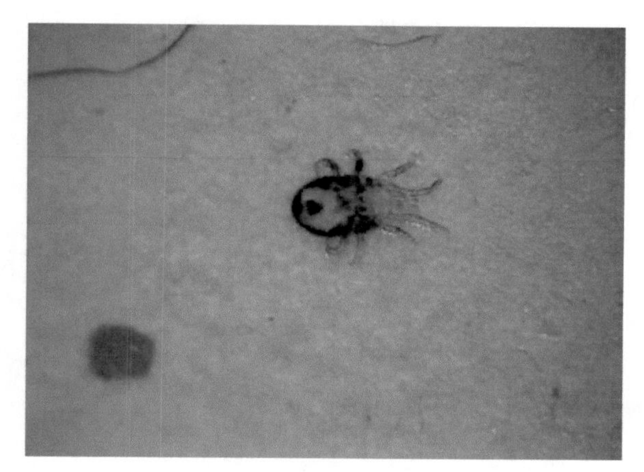

FIGURE 5.197 Northern fowl mite (*Ornithonyssus sylviarum*).

Several species of mites are bloodsucking parasites of birds and rodents. Under normal conditions they restrict themselves to their specific hosts. Problems occur when the primary host leaves and mites are left behind. They become hungry for a blood meal and start to wander in search of a new host. If a human is the first warm-blooded animal encountered, the mites will bite. The result may be a mild to severe dermatitis at the site of the bite. Chiggers are another mite species that causes many complaints.

Recently, cases of severe itching have been attributed to very small itch mites that normally feed in trees or other plants including straw (straw itch mite). When people work or

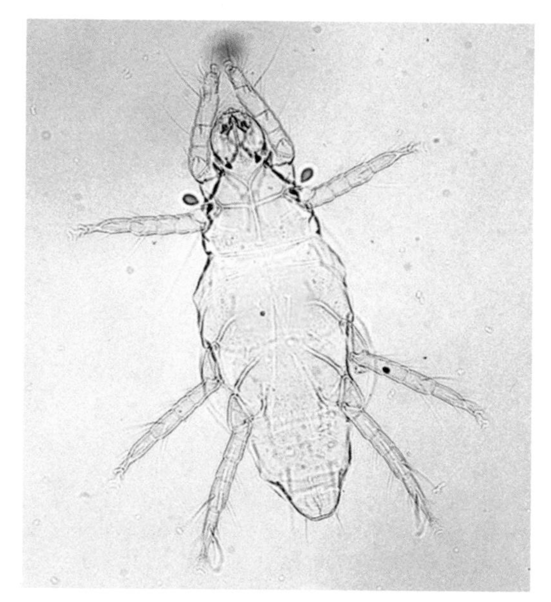

FIGURE 5.198 Itch mites (*Pyemotes herfsi*).

relax underneath infested trees or have direct contact with straw, they may come away itching. Often rashes also accompany mite biting.

Bed Bugs

FIGURE 5.199A,B Bed bug (A) Mouth parts (B) Whole body.

FIGURE 5.199A,B Cont'd

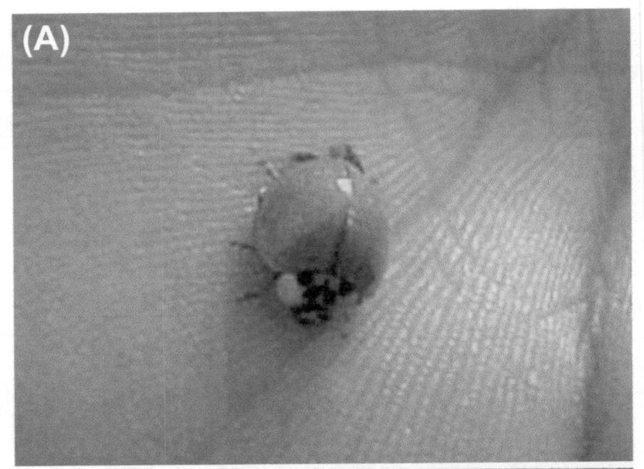

FIGURE 5.200A–C (A) Asian lady beetle (*Harmonia axiridis*) (B) Lace bug (Tingidae) (C) Thrips (Thripidae sp.).

Bed bugs hide in cracks and crevices in and around the bed during the day and come out at night to feed on people sleeping in the bed. The result of their bite is extremely variable. Some feel no effects of the bite at all. Most experience an itchy, red mark and an inflamed wheal at the site of the bite. Bed bugs are not known to carry diseases, but are undesirable because of their irritating bite.

Fall Time Nipping Pests

During the fall time, diagnosticians are asked about bites from very tiny insects that really should not be biting people. It is not really part of what entomologists recognize as their behavior. These reports are especially numerous on warm days when people are generally out of doors and perspiring slightly.

The general consensus from entomologists is that this group of nuisance pests does not actually feed on people but rather share the annoying behavior of biting or nipping anyway. The reason for the biting, nipping behavior is not known. Many believe that it is simply a probing or a tasting to determine food suitability. Once it is established that human skin is not a food source, they move on. However, mass interspecies communication is not really effective, resulting in the need for seemingly every last individual insect to sample for itself. The bite by itself is not severe enough to even break skin but the constant nipping and probing of hundreds

FIGURE 5.201 Insidious flower bug (Anthocoridae) *Orius insidiosis*.

of these pests becomes irritating enough to drive people indoors. In some people, a red wheal may develop with an itching sensation persisting for several days.

This peculiar behavior is shared by the following insects: lady beetles (Coccinellidae), several species but most common by the Asian lady beetle (*Harmonia axiridis*); lace bugs (Tingidae, several spp.); thrips (Thripidae sp.); minute pirate bugs and insidious flowers bugs (Anthocoridae) especially *Orius insidiosis* (Figs 5.200 and 5.201.)

CHAPTER

6

Understanding the Client

THE PURPOSE OF DIAGNOSTIC LABORATORIES

In chapter one, we discussed the need that people have for the management of insect pests. The purpose of an insect diagnostic service is to provide a rapid and accurate identification of an insect or mite, its signs or damage and in most cases, offer a conclusion or set of management recommendations.

The role of an insect diagnostician is to:

- identify the sample
- determine if it is a pest
- decide if it should be managed
- provide best management recommendations.

Most often, the first interactions between the general public and an expert entomologist are with insect diagnosticians. To be productive, a diagnostician must understand who is submitting the sample or question, their concerns regarding their sample and the expectations of the submitter. In this chapter we will explore the expectations that clients have when using an insect diagnostic laboratory and just how diagnosticians can best meet those expectations.

UNDERSTANDING THE NEEDS AND EXPECTATIONS OF THE CLIENT

The goal of a diagnostician upon receiving a sample or question is to completely resolve the concerns of the submitter. In order to accomplish this, a diagnostician must first understand *who* is submitting the sample and *why* they are submitting it. This is a challenge because often clients themselves do not know what questions to ask, nor do they know what background information to provide in order to help get appropriate answers. A diagnostician is therefore expected to anticipate what a client needs and expects and then to provide that information as part of the identification response.

For the purposes of this book, we will discuss clients that a public insect diagnostic laboratory might deal with on a regular basis. Undoubtedly, other diagnosticians will be able to relate to one or more clients in this list. We will identify those who most frequently use a public diagnostic laboratory, analyze what their real concerns are and postulate an educated guess as to what they expect to gain from their submission.

This can be made much easier by using a carefully crafted submission form such as the one presented shown in Figure 6.1.

Contemporary Insect Diagnostics
http://dx.doi.org/10.1016/B978-0-12-404623-8.00006-5

Plant & Pest Diagnostic Laboratory
LSPS – Room 101, Purdue University
915 W State St, West Lafayette, IN 47907-2054
765-494-7071 FAX: 765-494-3958
http://www.ppdl.purdue.edu

PURDUE UNIVERSITY

(PPDL-1-W) 1/14

Office Use Only: Date received:____________
Sample #: ____________
Account #: ____________

Date:____________

Submitter's Name ____________
Business ____________
Address ____________
City/State/Zip ____________
County ____________ Phone ____________
Fax ____________ Email ____________

Client's Name ____________
Business ____________
Address ____________
City/State/Zip ____________
County ____________ Phone ____________
Fax ____________ Email ____________

Please include a check or money order (payable to Purdue University) for $11 per sample ($22 out-of-state clients). DO NOT SEND CASH.
Send invoice to ☐ Submitter ☐ Client

☐ Perform only routine diagnosis ($11 in-state/$22 out-of-state)
☐ Please notify submitter if additional fees for advanced testing are needed
☐ Perform additional advanced testing if necessary (up to $50)

Mail reply to: ☐ Submitter ☐ Client
Fax reply to: ☐ Submitter ☐ Client
Email reply to: ☐ Submitter ☐ Client
☐ Copy Extension Educator

Information about Submitter/Client (please check one each for submitter and client)

Submitter	Client		Submitter	Client	(continued)
____	____	Extension Educator	____	____	Pest Control Operator
____	____	Homeowner	____	____	Nursery
____	____	Farmer	____	____	Lawn or Tree Care Co.
____	____	Dealer/Industry Rep.	____	____	Garden Center
____	____	Golf Course	____	____	Consultant
____	____	Landscaper	____	____	Purdue Specialist
____	____	Greenhouse	____	____	Other ____________

Check information desired:
____ Problem identification
____ Specimen identification
____ Control recommendations
____ Other ____________

Plant and Pest Information

Plant or Host: ____________ Cultivar/Variety: ____________

Location (choose one):
____ In dwelling
____ Tree/Shrub
____ Turf/Lawn
____ Golf Course
____ Flower bed
____ Vegetable garden
____ Field/Farm
____ Greenhouse
____ Nursery
____ Orchard
____ Animal/Human
____ Aquatic
____ Stored grain/Food products
____ Other ____________

Degree of Damage (choose one):
____ Heavy
____ Medium
____ Light

Insect Problem? (choose one):
____ Damaging plant
____ Biting/Stinging
____ Infesting food
____ Nuisance

for **Plant/Weed Identification Only**

Plant type:
____ Tree ____ Deciduous
____ Shrub ____ Evergreen
____ Vine
____ Groundcover
____ Herbaceous

Plant size:
____ Height
____ Width

Flowers:
____ Color
____ Month(s)
____ Size

Fruits:
____ Color
____ Month(s)
____ Size

Plant age:
____ Annual
____ Perennial (# years ____)

Unique features (bark, leaves, odor, thorns, etc.): ____________

Additional **Plant and Site Information**

Approximate age: ________ Height: ________ Number of years at present site: ________
Exposure: ____ Full sun ____ Partial shade ____ Full shade ____ Windy ____ Protected Irrigation frequency: ________
Root disturbance from: ____ sidewalks/driveway construction activities (describe): ____________
Size of planting: ________ % of plants affected: ________ Date first noticed problem: ________
Date planted: ________ Tillage practices: ________ Previous crop: ________
Chemicals/fertilizers applied (past 2 years)(include rates): ____________

Soil type: ____ sandy ____ clay ____ silt ____ loam ____ organic Soil pH: ________

DESCRIBE THE PROBLEM (Include symptoms, plant parts affected, pattern of occurrence, etc. Attach separate sheet if necessary):

Your tentative diagnosis/ID: ____________

FIGURE 6.1 https://www.pdffiller.com/en/project/13032792.htm?utm_expid=2952066-99.A_0_ThWlSD6YkSYRZhQUuA.0&utm_referrer=http%3A%2F%2Fwww.pdffiller.com%2F5287270-PPDL-1-Wpdf-PPDL---1---W-Physical-Sample-Submission-Form---the-Purdue-Plant--Other-forms-ppdl-purdue#red5287270 http://www.ppdl.purdue.edu/PPDL/pubs/PPDL-1-W.pdf

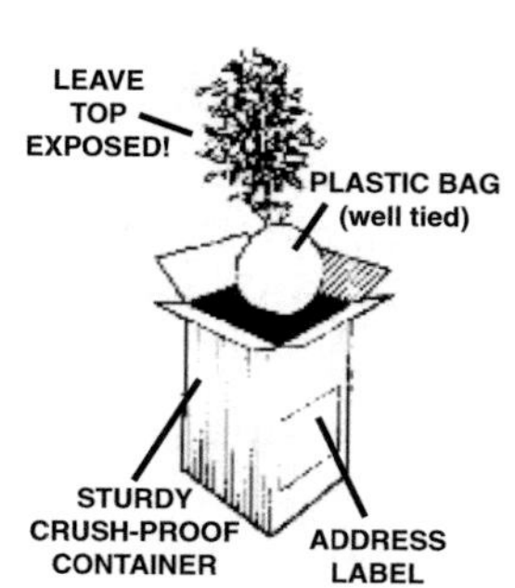

FILLING OUT THE FORM

1. Complete the form on the reverse side to the best of your ability. **Give complete information** pertinent to the sample, including background information.
2. State the problem clearly and indicate specific information desired.
3. Photographs or digital images of the problem site are helpful.
4. Attach an additional sheet if further explanation is necessary.
5. Submit white and yellow copies of the form, along with the specimen.

HOW TO COLLECT AND SHIP SPECIMENS

1. Collect fresh specimens. Send a generous amount of material, if available.
2. Ship in crush-proof container immediately after collecting. If holdover periods are encountered, keep specimen cool. **MAIL PACKAGES TO ARRIVE ON WEEKDAYS.**
3. Incomplete information or poorly selected specimens may result in an inaccurate diagnosis or inappropriate control recommendations. Badly damaged specimens are often unidentifiable and additional sample requests can cause delays.

SUBMITTING PLANT SPECIMENS FOR DISEASE/INJURY DIAGNOSIS

1. **HERBACEOUS PLANTS**: for general decline/dying of plants, send **WHOLE PLANTS**, showing **EARLY SYMPTOMS**, with roots and adjacent soil intact. **DIG UP PLANT CAREFULLY**. Send several plants. Bundle plants together and wrap roots in a plastic bag. Wrap the entire bundle of plants in newspaper and place in a crush-proof container for shipment. **DO NOT ADD WATER**.
2. **TREE WILTS**: collect branches 1/2 to 1 inch in diameter from branches which are actively wilting but **NOT** totally dead. Wrap in plastic to retain moisture. Collect a handful of feeder roots and place in a plastic bag.
3. **LEAVES/BRANCHES/FLESHY PARTS**: when localized infections such as cankers, leaf spots and rots are involved, send specimens representing early and moderate stages of disease. For cankers include healthy portions from above and below diseased area. Press leaves flat between heavy paper or cardboard. Wrap fleshy parts in dry paper.
4. **TURF**: samples should be at least 4" x 4" and include both the diseased and healthy portions of grass on the same sample piece. Place the sample on a disposable plate and wrap in newspaper for shipment.

SUBMITTING PLANT SPECIMENS FOR IDENTIFICATION

1. Include a 6-10 inch sample of the terminal (tip) portion of the stem with side buds, leaves and flowers in identifiable condition.
2. Place the sample flat between a layer or two of **DRY** newspaper, paper toweling or similar absorbent material. Try to prevent excessive folding of the leaves and place flowers so that you are looking into the center of the flower.
3. Pack the wrapped bundle in plastic, preferably with a piece of cardboard to keep the sample flat.
4. **NEVER PLACE ANY FRESH PLANT SAMPLE <u>DIRECTLY</u> IN PLASTIC!**
5. **NEVER ADD WATER TO THE SAMPLE.**
6. Shake excess water from **AQUATIC WEED SAMPLES** and place in plastic bag.
7. Wrap whole, uncut fruit specimens in paper, place in a strong box, and pack with additional paper to prevent crushing.
8. Package in sturdy crush-proof container and pack with additional paper to prevent shifting.

SUBMITTING INSECT SPECIMENS

Care should be taken to package insects so that they arrive unbroken. Be sure to separate and label the insects if two or more are included in the same package and provide appropriate information on each.

1. **TINY AND/OR SOFT-BODIED SPECIMENS**: such as aphids, mites, thrips, caterpillars, grubs, and spiders should be submitted in a small leakproof bottle or vial of 70 percent alcohol. Rubbing (isopropyl) alcohol is suitable and readily available. Do not submit insects in water, formaldehyde or without alcohol as they will readily ferment and decompose.
2. **HARD-BODIED SPECIMENS**: such as flies, grasshoppers, cockroaches, wasps, butterflies and beetles can be submitted dry in a crush-proof container. Do not tape insects to paper or place them loose in envelopes.

SUBMITTING NEMATODE SPECIMENS

If you suspect a nematode problem, and no other problem:

1. Collect at least one quart of soil from the root zone of affected plants. Include roots if plants are growing.
2. Place the entire sample in a plastic bag for shipment. Do not add water to the sample; do not allow it to dry out; and protect the sample from extreme heat (don't leave samples inside a parked vehicle in direct sunlight). It is often advantageous to collect a second, similar sample from a nearby area where plant growth appears normal.
3. To the **OUTSIDE** of each bag or package, attach a label, note ,or tag identifying the sample.
4. Send to Nematology Laboratory, Purdue University, Smith Hall, 901 W. State Street, West Lafayette, IN 47907-2089. If you have questions, please call 765-494-4611.

FIGURE 6.1 Cont'd

What Diagnosticians Learn from Sample Submission Forms

SUBMITTER CLIENT INFORMATION

The title (business) of the submitter or of the client can reveal important information about what should be included in a response. From the business name listed on the submission form, a diagnostician also may get a clue about the needs of the client. Knowing whether a response is to be written for a commercial grower or a hobbyist dictates how simple or inclusive to make the response. For that reason, it is very helpful if the submitter or the client can be identified based on job title. In the example submission form above, a checklist of the most common submitters to the diagnostic laboratory at Purdue University is provided. (While this list works well for the public diagnostician, a different list may be required by private diagnosticians.)

It is always safe to assume that the response should be written to the submitter. Even if a client is listed but the sample was submitted by a professional pest manager, the response should be written to the submitter. This rule makes it easier to know the audience level at which the response must be written. For example, little detail about chemical pesticides or applications is necessary if the sample is submitted by a professional pest manager. It may safely be assumed that a certified professional will be aware of those because of their training.

If specific controls are requested, knowing whether the submitter is a professional or homeowner influences what products should be recommended (professional use vs. do it yourself [DIY]), and what application recommendations to make.

CHECK INFORMATION DESIRED

It makes sense to clearly understand what information the submitter desires when submitting a specimen. A simple checklist including options (check one) such as:

- Problem identification
- Specimen identification
- Control recommendations and
- Other

helps to avoid misunderstandings and ambiguity.

When the 'information desired' is not indicated by the submitter, a diagnostician is forced to make assumptions based on what he or she may think the submitter needs to know. In cases where the submitter and the client are not the same person, the response should be created especially for the submitter and care should be taken to address the indicated or assumed needs of the submitter.

A certain level of identification and control detail is also dictated by the sample itself. For instance, if a brown recluse spider is submitted, it is incumbent upon the diagnostician to provide at least some basic warning of the potential harm that the spiders can pose in a home and how seriously they should be taken.

PLANT AND PEST INFORMATION

The more precise a submitter can be in describing where the specimen came from, its behavior, and the extent of damage it has caused, the more confidence a diagnostician can have in the diagnosis of the pest. Knowing where the pest was collected also helps a diagnostician anticipate what control recommendations to include in a response.

DESCRIBE THE PROBLEM

Allowing space for a submitter to describe the problem in their own words is extremely helpful. Often this section of the submission form is the most valuable to a diagnostician. It can dictate the direction an investigation may go and also adds valuable insight into how the submitter interprets or visualizes the problem.

In many cases a submitter will choose to attach a separate description or a letter at this point. Increased detail allows the diagnosticians to do their job more easily and completely. Likewise, additional background information aids the diagnostician in determining what to include in a response.

In a client's self-description of the problem there may be found an indication as to what controls they have already tried. This is helpful information to have in formulating an appropriate response.

If a client is already using bed bug encasements and residual pesticides, for example, then repeating that recommendation in a response would not be helpful. On the other hand, knowing that they have been used suggests that a comment about them is in order and even a further recommendation such as 'seek professional help' or 'use heat treatments.'

YOUR TENTATIVE DIAGNOSIS

An astute diagnostician can often tell how much a client already knows and what more they want to know about the sample by what they write in the tentative diagnostic section.

For example, if a submitter writes 'bug' in the tentative diagnosis line, it may be assumed they simply want to know if the sample is an insect rather than a piece of debris. An appropriate diagnosis in this case would be to confirm that it is an insect (or not). Chances are that keying out the insect to the family, genus or species level is not needed.

On the other hand, if the tentative diagnosis is *Blattella germanica,* then responding 'Insect' or even 'Cockroach' would be inappropriate. In this case, it is clear that the submitter wants to have an identification or confirmation at the species level.

Because composing a written response should center around educating the client, asking for a tentative diagnosis can help gauge the level of understanding that a client begins with and therefore at what level to write the report. A response can either simply validate the submitter's tentative identification or become a more extensive treatise on biology and management strategies.

For example, if a client submits a spined soldier bug (*Podisus maculiventris*) and asks if this is a brown marmorated stink bug (*Halyomorpha halys*), a proper response could begin with, 'Characters used to separate this specimen from a marmorated stink bug include the following ....'

Conversely, if the same insect is submitted and the submitter asks if it is a cockroach, then the above response would be entirely inappropriate. It would be better to begin with how bugs differ from cockroaches and why knowing these differentiations is important.

Assessing what is recorded in the 'tentative diagnosis' section not only provides a clue as to the level of identification that is required, but it also indicates what level of information should be included in 'management recommendations.' Depending upon the identification, some control information is mandatory, requested or not. In some cases, simply stating 'No controls are recommended at this time' is the most appropriate response. In other situations, very detailed management information and procedures are required.

WHO SUBMITS SAMPLES AND WHAT THEY EXPECT

There are potentially many people who may use insect diagnostic services. Each has a slightly different reason for doing so. A general discussion of the most common patrons is in order.

Extension Field Staff

All states have Extension programs that are part of the national land-grant university system. The mission of Extension is to pass along new or academically-gained information regarding agriculture, the environment, human health and the well-being of communities to the public. This is accomplished in many states by employing and sending Extension agents into rural areas to provide (extend) new and practical agricultural information to the end-users. Focus has since expanded to include human health and youth education, as well as food and agriculture-related programs in both rural and urban sectors.

Agriculture Extension agents are, by necessity, generalists. Though most have advanced college education, as well as significant practical experience working with the public, they cannot possibly know every detail about every crop planted. For this level of detail, they rely on Extension specialists who are experts in one smaller area of learning and are usually stationed at the university.

Extension agents are expected to teach those with whom they work. Education is implicit in the mission of Extension. In Indiana the name 'Extension Agent' was changed to 'Extension Educator' to reflect that important part of their job. Regardless of what they are called, Extension field staff must rely on specialists to support them in this important facet of their job.

One small portion of what Extension agents need to know concerns insects and their management. Insect diagnosticians are part of the support team. They back-stop Extension field staff by providing identification, confirmation and control recommendations concerning insects and other arthropods. Extension agents seldom require a diagnosis that goes beyond common name, but they almost always require a synopsis of the behavior, biology and pest potential of the insect. If it is a potential pest they also require a short synopsis of its potential threat, basic control recommendations and a reference to a publication, website, or other lead where they might learn more.

These facts are often simply passed along to their own clients. In some cases Extension agents submit samples on behalf of their client. In other cases they assist the client in submitting a sample. In all cases it is preferable to have the Extension educator involved at some level in both submitting the sample and in interpreting the results. Such involvement is also beneficial to the field staff as a means of keeping current, up to date, and generally aware of what is happening. This way they are sufficiently informed to field similar questions independently, saving specialists both time and resources.

Advantages of Extension agent participation in insect diagnostic services are that they often:

1. Recognize the problems and can provide a ready solution because they have seen it before. They can then circumvent the need to submit many samples to a diagnostician.
2. Understand exactly what the client needs to know, can ask the proper questions, and can best educate because of their close relationship.
3. Can provide the background information needed for certain diagnoses because they have been through it before.
4. Can interpret the results of the diagnosis and answer follow-up questions for the client.
5. Can better train their clients through public forums, seminars and the local media.

Professional Pest Managers

People who are employed in the pest management industry often submit samples to insect diagnostic laboratories. Depending upon what part of the industry they are in, they may submit samples from a wide range of settings, including agricultural, structural and urban, landscapes and turfgrass, health care or a variety of other specialty settings. Professional pest managers usually require a validation of their suspicions, although they sometimes ask entirely new or rare questions. Pest management/control is their specialty and so once they are sure what the pest is and where it came from they are usually quite adept at the control phase. Diagnosticians seldom need to expound on recommendations involving pesticides or their application to this group.

Homeowners

Generally, homeowners are the least informed of all the potential submitters. They are a difficult group to pigeonhole because homeowners are very diverse in terms of their entomological knowledge and background. In some instances diagnosticians need not provide an identification beyond the general 'bug, beetle, spider' level. What these homeowners want to know is if the pest has potential to cause damage: 'Will it hurt me, my family, my pets or will it damage my property?' If the response is yes to any of these, then a diagnostician should be prepared to offer details on how to prevent or manage the pest safely and effectively. Sometimes the correct diagnosis of a homeowner-submitted sample focuses less on answering questions asked and instead answers questions that should have been posed.

For example, if a sample containing a jumping spider (Family: Salticidae) was submitted with a note saying, 'I found this in my home, what is it?' a technically correct diagnosis would be as follows:

> 'This is a jumping spider, *Freya ambigua* (Araneae: Salticidae).'

However, a better diagnosis would give a more complete response:

> 'The sample was identified as a jumping spider. In nature, these spiders are beneficial predators of many insect pests. Because these are hunting spiders (wander about in search of prey) they may occasionally find their way into homes by accident. They are non-aggressive spiders and do NOT bite people, pets or cause damage to homes. Chemical controls are not recommended. They may best be simply removed to the outside or destroyed with a fly swatter or vacuum when found.'

So, even though the homeowner asked 'What is this?' it is unlikely that they really want to know the scientific identity of the spider. In fact, it is unlikely that even providing the common name 'jumping spider' will suffice. More likely, the real question on the mind of the submitter is, 'Will this spider hurt me or my family?' and 'What should I do now?'

As can be seen from the example above, by understanding a submitter's true concerns, even though they were not expressed, a proper response can be formulated that will eliminate need for a follow-up dialogue.

Farmers

Farmers, by definition, are generalists. They must know a great deal in addition to simply growing a crop. They must understand finance and business, mechanics, veterinary science, and marketing, in addition to agronomics, if they expect to stay in business. In addition they must understand crop production, how to prevent diseases, how to manage competitive weeds and how to protect against insect pests, both during production and during storage.

Beyond a confirmation or identification of their pests, what farmers really require is an idea of pest threat or potential injury. Most realize that they will always have a certain number of insects to deal with. What they need is an estimate of how many they should tolerate before taking action. Because farming is a business and profit is the bottom line, an action threshold (see Chapter 8) will often save them a lot of expense. A diagnostician that can provide recommendations of 'whether to' and 'when to' apply a control, in addition to an identification, is of great value to farmers and agriculturalists. In order to do this a diagnostician also must be quite familiar with the crop and its growth and storage practices, as well as the insect pest.

Businesses

Many clients that diagnosticians deal with represent large corporations or businesses. These may be agriculture-based or may be production or packaging businesses. Because insect pests can occur at nearly any point from the production of raw materials to processing, packaging and consumption, insect diagnosticians are required.

Diagnosticians are trained entomologists who understand both the biology and behavior of the insect, as well as best management principles. These often have to be site-specific, thus recommendations should be made on a case by case basis. Many food-related production processes and other industries such as pharmaceutical plants are regulated by the Food and Drug Administration (FDA) and other legislative bodies. Recommendations from the diagnostician must comply with these rules.

In many instances, large businesses already employ diagnosticians as part of a quality control team. These in-house diagnosticians must have more in-depth knowledge about a few key pests and much more specific knowledge about the system with which they are affiliated than other diagnosticians outside the company. Often these diagnosticians look to validate their findings with other practicing diagnosticians stationed elsewhere.

Legal Profession

When the legal profession becomes interested in a diagnosis it is usually because there has been a complaint. One party claims injury by an arthropod or pesticide and a second party is thought to be responsible in some way. Forensic entomology is a sub-science of entomology that deals with insects as part of legal investigations. We know that insects may reduce crop yields, contaminate foods, destroy buildings or products and may even harm human health. In order to get the correct facts, lawyers may seek out experts in diagnostic laboratories. Seasoned diagnosticians can recognize these cases simply by the questions that are posed. Questions such as:

'How long has this pest been present?'
'Where did this pest originate?'
'Should someone have known or done something differently?'

are most often posed by someone in the legal profession. These questions may be just an initial probing to determine if there is justification for a lawsuit, or they may be used to collect facts for litigation, as well as expert testimony for cases that are already underway.

Pest management contractors or others who have charge over managing pests can become party to a lawsuit when things do not go as planned. Pests that have contaminated foods, homes, hospitals and nursing homes are sometimes the basis for such inquiries. Pests that bite, sting or otherwise harm people are also the subject of litigation. Regardless of whether it is a seemingly simple identification or a management recommendation, any response from a diagnostician that may be used in legal proceedings must be very carefully thought out and accurately articulated.

Lawyers representing both the defense and the prosecution require the services of well-trained and meticulous insect diagnosticians. Involvement may vary from a simple written identification to an opinion, consultation, deposition or even testimony under oath in a court of law.

Professional Colleagues

Fellow entomologists trained in areas outside of diagnostics often need assistance with identification or control. Diagnosticians may be the first consultants that these experts go to with questions. Interactions can range from a simple confirmation of a tentative diagnosis to a precise and in-depth identification at the genus species level, in preparation for a peer-reviewed scientific publication. Insect diagnosticians are as indispensable to fellow entomology experts as they are to field staff.

Other Submitters

Insect diagnosticians find themselves at the interface between general public and entomology expertise. They are often the first entomologists consulted for an answer to most every entomological question.

RANGE OF QUESTIONS SUBMITTED TO INSECT DIAGNOSTICIANS

- I am in 4-H. Can you please identify my insect collection?
- Can you state unequivocally where this spider came from?
- Where do I get ants for my child's ant farm?
- I found a caterpillar and would like to watch it develop. What do you suggest?
- Where can I find pesticides for purchase?
- Are there more insects this year than normal?
- Why am I not seeing monarch butterflies? I usually see them by now.
- What pest control service do you suggest I use?
- I have a half bag of insecticide left over. Can I use it?
- Will we have more or less insects next year?
- I have invented a product that I believe will control mosquitoes. Can you help verify that it works?
- Last month while hiking I saw a big, black, prehistoric-looking insect that I have never seen before. I am 85 years old. What was it?

There are many different people from all walks of life, each with different reasons for submitting an arthropod for identification or a request for insect-related information from a diagnostic laboratory.

General Interest

Some submitters have a curiosity or general interest in facts, others such as news columnists might be looking for a story or for answers for their readers, or seeking a prognostication report. Requests are common from students in schools or from 4-H youth about where to find insects or how to improve a collection. Diagnosticians are expected to be able to answer questions as varied as vermiculture, butterfly rearing, fly tying for fishermen, horticulture planting, fertilizing, water quality matrices, and a host of other insect-related or sometimes unrelated topics.

One particular group of submitters possesses a very closely held belief that insects have infested their person. This is a category of submitters that has become so prevalent and yet is so unique that it justifies a section of its own.

Delusory Parasitosis

Delusory parasitosis, also known as Ekbom syndrome, is so common that anyone practicing general insect diagnostics is certain to encounter it sooner or later (Ekbom, 1938; Wilson, 1946; Hinkle, 2000). Established, public insect diagnosticians regularly see 5–10% of their sample load coming from clients who express symptoms of delusory parasitosis.

The disease is a form of psychosis whose victims hold a strong but delusional (false) belief that their person is infested with insects or mites. They express sensations of arthropods crawling on, burrowing into, or biting them. Typically, the psychosis worsens over time to include not only feeling the bites or crawling on the skin or scalp to perceptions of actually seeing the 'bug.' Such sightings are usually accompanied by bizarre behavioral descriptions, wherein the 'bug' bores in and out of the body, or is carried through the circulatory system. Often descriptions of bizarre metamorphosis are related in which the insect changes into a strange form, sometimes with wings or with non-insect-like abilities ranging from drilling into furniture, wood or even porcelain, in order to evade capture before going back into the person again (Hinkle, 2010).

Often this delusion of parasitosis extends to a person's dwelling in addition to their body. Furniture, automobiles, homes, and places of work can become 'infested.' It is common to receive detailed accounts, either written or oral, submitted with an enormous number of samples collected from various body regions or sites within the home. The sheer number of samples collected and submitted is often itself a sign of the disease.

The belief is so fixed that even when victims are made aware of the impossibility of their perceptions, are repeatedly told that their samples contain NO insects or mites or that NO insects or mites behave in that fashion, their belief is not shaken and they will continue to submit samples as if to convince by sheer persistence or by an overwhelming number of samples.

It is rare that two or more people from a single household suffer from this disorder at the same time, even when considering the tremendous peer pressure that is brought to bear. Even though illogical, the fact that only one member of the residence is afflicted does nothing to convince a person suffering from delusory parasitosis. Multiple opinions from different people also fail to convince a person suffering from this disease. An insect diagnostician may safely assume that they are only one of a large number of people that this person will contact, looking for validation as much as diagnosis.

Ethical pest managers never treat such cases with pesticides because the end result is always the same: no control. The best response by anyone handling these cases is a referral to a medical professional such as a dermatologist who may hopefully refer the client to a psychologist for help.

It is important that an insect diagnostician not provide a diagnosis outside of his or her expertise. Having no formal training in either medicine or psychiatry, a diagnostician or credentials cannot diagnose Ekbom syndrome or even delusory parasitosis but can only make a referral and hope that someone down the line is able to assist.

Unfortunately, delusory parasitosis is often not immediately recognized by medical professionals and thus is sometimes treated with creams or scabies-related solutions. This only exacerbates the problem. Delusory parasitosis lies on the fringes of current psychiatry and thus is often discounted even by professionals. Sadly, it is the cause of considerable suffering by the victim and the victim's extended family. Loss of sleep, pain, expense, and a loss of quality of life for the patient, as well as the family, is very real.

Ekbom syndrome spans educational, socioeconomic, racial, and age classes. It is not uncommon to witness a very well-spoken or

Delusory parasitosis, also known as Ekbom's Syndrome (Ekbom, 1938) is a presumed psychiatric condition ascribed to individuals who are convinced, in the absence of any empirical evidence, that they are infested with an insect or parasite (Poorbaugh, 1993; Webb, 1993a). These individuals experience itching, stinging/biting, and crawling sensations on or under their skin, which are often associated with excoriations, discoloration, scaling, tunneling or sores. Their conviction that they are infested is reinforced by their observation of particles described as sparkly, crusty, crystal-like, white or black specks and/or fibers. Typically, these individuals have consulted extensively with general physicians, dermatologists, and entomologists (Kushon et al., 1993) who could not find physical cause for their complaints. Despite findings ruling out lice, scabies or other medical causes, patients refuse to accept the diagnosis of delusory parasitosis (Koblenzer, 1993; Webb, 1993b), become extremely focused on eradicating the pests, and further compromise their skin by frequent scratching, excessive cleaning, and the application of various remedies such as prescription pesticides for lice or scabies, household cleaning products, and organic solvents or fuels. The symptoms are debilitating and the sufferer's distress is compounded by the lack of a concrete physical diagnosis (Lim, 2009; Hinkle, 2010).

highly successful person suffering from this disease. Experts have made some correlations with past drug abuse or recent personal stress such as loss of a loved one, a career change, loss of job or even a move. However, little additional research has been published concerning this mental disease.

The role of an insect diagnostician in cases of delusory parasitosis is to first examine the samples submitted, rule out any possible arthropod involvement and then to refer the sufferer to a medical professional who is able to help. It is incumbent upon all practicing insect diagnosticians to become familiar with the symptoms and signs of this syndrome so that a rational and explicit response is ready. The most desirable outcome is a speedy response so that the victim and the diagnostician can be distanced. Remember that regardless of how much sympathy is felt, an insect diagnostician will NOT be able to resolve this problem. Asking for additional samples or enduring multiple and lengthy conversations will at best prolong the distress and at worst validate an actual insect infestation in the victim's mind (Hinkle, 2010).

Actual Parasitosis

Having described Ekbom syndrome in detail, there are cases that may, at first glance, appear as delusory parasitosis but are, in fact, actual parasitosis. A diagnostician must rule out these possibilities before making any judgments. Certain mites such as scabies, chiggers, northern fowl and any of several itch mites can infest people or homes. Some small biting insects such as bed bugs, bat bugs, fleas, lice and springtails may also afflict people. Keeping these possibilities in mind as a diagnostician examines a sample will allow for the possibility of actual parasitosis. Questioning the patient is an effective way of corroborating a conclusion. When patients describe a truly bizarrely-behaving insect that they cannot catch, diagnosticians should beware. When the victim is very young, old or otherwise compromised, incorrect conclusions may result because communication lines are not optimal. In these cases extra thorough inspection is required.

Illusions of Parasitosis

Illusions of parasitosis refers to a condition in which dermal sensations caused by a change in a person's environment are attributed to insects. Dermatologists know that sensitive people develop itchy or sensitive skin in reaction to substances such as smoke, new carpets, dry air, static electricity, or environmental changes. These sensations can be misinterpreted as being caused by arthropods. However, one of the main differences between illusions of parasitosis and Ekbom syndrome is that those suffering from illusions of parasitosis can be reasoned with. With sufficient explanation, they can see and accept the reality of their confusion.

CHAPTER

7

Responding, Educating and Record-Keeping

RESPONSIBILITY TO EDUCATE

Responding to Clients

It is the responsibility of an insect diagnostician to provide more information than simply insect identification. Identification is the heart of what a diagnostician provides; however, by definition, diagnostics involves not only the identification of the specimen, but its placement into the context of the submitter's world. For a client this means determining the status of the specimen and – if it is a pest in need of control – what constitutes the appropriate controls. Diagnosticians must know the client as well as the insect.

Understanding the Client

Often a submitter is not able to determine what answers are needed or what questions to ask. An effective diagnostician knows what questions should have been proposed, but were not.

For example, consider a tick submitted by a mother who found it imbedded in the skin of her young child, accompanied by the simple question, 'What is this?'

An astute diagnostician will know that the mother does not necessarily want to know the family, genus and species of the tick, even though her question was 'What is this?' She almost certainly wants only to know if the tick is dangerous, if its presence compromises the well-being of the child in any way and if there are specific recommendations as to how she should respond right now to protect the child. Secondly, she may want to know how to avoid a similar situation in the future. Therefore, a diagnostician might spend the majority of the response time answering questions that have not actually been posed by the client.

A poor diagnostic response would be a simple though accurate answer to the mother's question 'What is this?' Imagine the response of a frantic mother who receives a report back that simply states '*Acaridae: Ixodes dammini,*' and nothing more.

A response report should be written for the client, not for a database or for peers. The amount and detail of the information required differs depending on who is submitting the sample.

Compare the needs of the mother above to the needs of a researcher who submits a similar sample while writing a paper about the distribution of deer ticks. The researcher may not benefit from a response report including the biology of and control measures for the tick; they may

Contemporary Insect Diagnostics
http://dx.doi.org/10.1016/B978-0-12-404623-8.00007-7

simply want a confirmed scientific identification of family, genus and species.

Obviously, needs vary widely between these two extremes. Professional pest managers may be in need of a general identification up to the family level accompanied by more detailed control recommendations. Medical professionals may only require an order level identification but require more details on the probability of disease transmission or other harm to the patient.

A diagnostician must also know the context in which the sample was obtained. For example, was it found in a garden? If so, a client may be interested in knowing if this pest will cause damage to vegetables or plants. If it was found on a person, diagnosticians may assume that the client would like to know its potential health concerns. In each case, a proper response to the question 'What is this?' varies based on the submitter's identity.

Effective responses require a diagnostician to understand the submitter nearly as well as the insect submitted.

Response Reports

Responding to a request for diagnosis is an art. Like all arts, improvement comes with practice. The basics of what constitutes an appropriate response may vary; in particular, private versus public diagnosticians have slightly different objectives when diagnosing an insect problem. However, it can be agreed that a response should be concise, timely, factual, and should satisfy the submitter. In addition, an appropriate response should provide enough of an informational lead to allow the client to further research the problem on their own. Most importantly, the goal of a response is to provide a satisfactory and inclusive response in one communication.

A poor response is one that is incomplete and requires a second or third query back to the diagnostician in order to satisfy the submitter. Having multiple exchanges is time wasted on the parts of both the client and the diagnostician.

Using a well-designed submission form (Chapter 6) is key to making this happen. When the directions for specimen submission are made clear and the needs of the client are manifest, a diagnostician can be confident both that the specimen will arrive in shape for identification and that the answers provided will best meet the needs of the submitter.

The goals of an insect diagnostician are to provide:

1. rapid and accurate identification of the specimen submitted;
2. clear and concise assessment of the problem, its relative importance and concern;
3. brief information about the biology of the arthropod, especially where it came from, why it is present and what the immediate future may hold as it relates to the submitter;
4. well-conceived and clearly articulated management recommendations and options;
5. leads on where to get additional information.

Diagnosticians refer to the period of time from when a sample is received at the laboratory until the diagnostic report is returned to the client as the 'turnaround time.' Most clients hope for turnaround time to be immediate but expect it to be within one week. Some laboratories are set up with the expectation of a much more rapid response. The nature of the sample also dictates, to some degree, the turnaround time expected. For example, the submission of an insect that has been feeding on a tree for several years in a row but has not significantly damaged the tree is of lower priority than a sample of a spider that has bitten a person and who is now in the ICU of a hospital. While these are extreme examples, they illustrate the necessity of prioritizing samples. Priority should always be human health first, economic injury potential second, and aesthetic or nuisance pests third, while samples that are for curiosity only should be lowest in priority for a public diagnostician. Diagnosticians working for private enterprises may have shortened turnaround times and priorities based more on the potential economic impact of the sample.

Diagnosis

An accurate identification of the arthropod serves as the basis of the remainder of the report. Insect diagnosticians are trained to use all available tools to identify insects and other arthropods (Chapter 4). Much of this expertise is acquired in graduate entomology studies and the rest is learned on the job. Training in taxonomy is a large part of what is needed. Problem-solving skills, which are primarily learned on the job, are required to deduce the facts of the case; a clear communication style, both written and oral, is also required.

Management Recommendations

As suggested in Chapter 1, identifying the sample is only the first step in a diagnosis. Most submitters are more interested in learning if it is a threat and if so, how to manage it. These two pieces of information are linked and both are critical to satisfying the needs of the client.

When a determination is made that the sample is a threat, a diagnostician is expected to offer advice on how to best manage it. Use of integrated pest management practices are always preferred because they are long-term, sustainable and are less hazardous to human health. These usually include multiple tactics. For example, changes in cultural management practices should always be considered in conjunction with chemical methods of control (Chapter 8).

Biological Information

It is incumbent upon diagnosticians to provide at least some biological information to the submitter. At the very least, clients will learn important facts about the arthropod and hopefully knowing these will affect subsequent sample submissions. Pertinent biological information for an insect includes:

- expected number of generations per year;
- a description of all life stages;
- what it feeds on;
- where it lives;
- its life cycle.

More or less information may be given, depending on the circumstances of each case.

Where to Get More Information

Not all information about a diagnosis can possibly be contained in a response report. Diagnosticians should direct a client to sources of further information. Referring the client to a written publication, a published text or a specific site on the internet is part of a well-written response (see for example http://extension.entm.purdue.edu/bedbugs/index.php, and Chapter 5, Bed bugs).

PURDUE
UNIVERSITY

Home | Make A Handout | Identification | Monitoring | Control Strategies | Prevention | Preparation For Treatment | Furniture Disposal | Finding Help

Bed Bug IPM
Technical Resources
for Educators

Information on the Monitoring, Control, and Treatment of Bed Bugs

What is Unique About This Web Site?

There are many very good web sites that all provide similar information about the basics of bed bugs. A few of these are linked throughout this site. This technical education web site differs in that it is designed to provide tools for those charged with educating others about bed bugs.

It includes technical resources such as power point presentations, printable brochures/handouts that can be custom designed for specific audiences, color photographs, specific question/answer write-ups, a 20 minute general video introduction to the bed bug problem as well as a series of small video (youtube type) clips that address specific techniques or questions about bed bug management. It also features an interactive video game that teaches about bed bug inspection, and is of particular interest to younger audiences. Finally, this site advertises a bed bug exhibit that can be taken to large conferences or other educational venues to provide general awareness information about bed bugs.

Purdue University Extension is pleased to provide these materials at no cost to you in an effort to increase the general awareness of the bed bug problem and to assist in the education of our citizens.

Also linked to this site are other resources, checklists, web sites and even a search program for bed bug products and services that have been developed elsewhere. Together, these are all provided as a buffet for educators teaching bed bug awareness and management to nearly any clientele group.

Know The Facts: Bed Bugs

Click here for the most common bed bug questions.

Videos

Bed Bug Prevention & Control Video - 20 min

Nat Geo Bed Bug Introduction
Bed Bug Biting Signs
Bed Bug Feeding Sequence
Hotel Room Inspection
Avoid Bringing Bed Bugs Home
Bed Bug Dog Detection
Bed Bug Infested Suitcase
Bed Bug Control - Steaming
Bed Bug Control - Spraying
Bed Bug Control - Encasement

Resources

Extension Entomology, Purdue University

Purdue University, Medical Entomology - Bed Bugs

EPA | Controlling Bed Bugs

NPMA Bed Bug Hub

Bed Bug Product Search Tool

Bed Bug Insecticide application target areas, key ingredients and formulations

Page 1 of 1

DIAGNOSTIC REPORT

Sample#	13-01217
Field ID	
Host	Household; Domestic dwellings
Received Date	8/5/2013
County	St Joseph
State	IN

Contact:

Tim Gibb
Purdue University
Department of Entomology
West Lafayette IN 47907

Phone **765-494-4570** Fax Email **gibb@purdue.edu**

Submitter :

Phone Fax

Email

Diagnosis and Recommendations

Host/Habitat	Household; Domestic dwellings (habitat)
List of Diagnosis/ID(s)	
Confirmed for Cimicid Bat Bug (Cimex adjunctus)	

Final Report

The sample submitted was identified as a bat bug. These insects are very closely related to bed bugs and for most people they are indistinguishable.
However, a few very important differences exist between the two species.

Bat bugs develop in colonies of roosting bats, which sometimes occur in attics or behind walls of buildings. Bat bugs may move into human living areas and incidentally bite people, when bats migrate or are eliminated from the building.

Can you find evidence of bats having been in your home?

The good news is that (unlike bed bugs) in the absence of the bat hosts, these insects cannot sustain and reproduce on people.

Control procedures depend upon sealing the bats out of the home and applying insecticides to the areas where the bat bugs are. In your case, concentrate on the second floor bath and laundry area where you are finding these bugs.

Tim Gibb
Department of Entomology
765-494-4570
gibbs@purdue.edu

Plant and Pest Diagnostic Lab Purdue University 915 West State Street West Lafayette IN 47907-2054 Telephone : (765) 494-7071 Fax : (765) 494-3958	Diagnosed By : Timothy Gibb (gibb@purdue.edu) Completed Date: 8/8/2013

Sample# 13-01217

IMPORTANT: Our report is based on the material and information that you have delivered to us. Purdue University cannot guarantee the accuracy of our report. Additional onsite information may indicate a different conclusion. Our evaluation is an opinion only and not a warranty of your results from your use of our report.

FIGURE 7.1 Example of bat bug diagnostic report.

Sample#	13-01146
Field ID	
Host	Household; Domestic dwellings
Received Date	7/29/2013
County	Elkhart
State	IN

DIAGNOSTIC REPORT

Contact:

Tim Gibb
Purdue University
Department of Entomology
West Lafayette IN 47907

Phone **765-494-4570** Fax Email **gibb@purdue.edu**

Submitter :

Phone Fax

Email

Diagnosis and Recommendations

Host/Habitat	Household; Domestic dwellings (habitat)
List of Diagnosis/ID(s)	
Not Detected for Mold mite (Tyrophagus sp./spp.)	

Final Report

The sample submitted was identified as mold mites. These are common in wet years like this and especially where pet food and litter boxes remain damp. Finding them near where condensation accumulates is common.

Mold mites usually go unnoticed except in occasions when they become abundant. They are usually an annoyance and nuisance and but not injurious. They are harmless to people and pets, furniture, house structures, clothing and so forth. Mold mites only develop where there is moisture or a high humidity.

As the name implies, they feed on molds and are common only where mold and fungi can flourish. They have been reported from a large number of items and places but the common characteristic of all infestations is the presence of a high humidity.

Control of mold mites may be difficult and is likely to require persistence. The first step should be to eliminate humidity or moisture that is producing favorable conditions for molds and mites. Spray or fog treatments with household insecticides will be of some benefit if the insecticides reach the source of the mites. Spraying without correcting conditions to eliminate the source will only provide temporary control, however..

Tim Gibb
Department of Entomology
765-494-4570
gibbs@purdue.edu

Sample# 13-01146

Plant and Pest Diagnostic Lab Purdue University 915 West State Street West Lafayette IN 47907-2054 Telephone : (765) 494-7071 Fax : (765) 494-3958	Diagnosed By : Timothy Gibb (gibb@purdue.edu) Completed Date: 8/2/2013

IMPORTANT: Our report is based on the material and information that you have delivered to us. Purdue University cannot guarantee the accuracy of our report. Additional onsite information may indicate a different conclusion. Our evaluation is an opinion only and not a warranty of your results from your use of our report.

FIGURE 7.2 Example of mold mite diagnostic report.

Page 1 of 1

Plant and Pest Diagnostic Laboratory PURDUE UNIVERSITY

Sample#	13-01762
Field ID	
Host	Household; Domestic dwellings
Received Date	11/13/2013
County	Milwaukee
State	WI

DIAGNOSTIC REPORT

Contact:

Tim Gibb
Purdue University
Department of Entomology
West Lafayette IN 47907

Phone **765-494-4570** Fax Email **gibb@purdue.edu**

Submitter :

Phone Fax

Email

Diagnosis and Recommendations

Host/Habitat	Household; Domestic dwellings (habitat)
List of Diagnosis/ID(s)	
Confirmed for Woodlouse Hunter Spider (Dysdera crocata)	

Final Report

The spider submitted was identified as a woodlouse hunting spider, Dysdera crocata, common throughout the US. The woodlouse hunter preys on pill bugs or sow bugs (order Isopoda) and derives its common name from the British common name for these crustaceans. D. crocata is known to feed on other arthropods as well.

The woodlouse hunter probably overwinters in its adult form. Mating is reported to occur in April, with the eggs being deposited shortly thereafter. The eggs are suspended within the female's silken retreat by a few strands of silk. These are likely what you are describing as the 'small cotton balls' found in the webs.

Medical Importance
D. crocata bites have very rarely been reported. Even when reported, the bites do not result in any systemic neurotoxicity or cytotoxicity.

By regularly removing the egg sacs (cleaning) you will reduce the population of spiders in your house.

Tim Gibb
Department of Entomology
765-494-4570
gibbs@purdue.edu

Sample# 13-01762

Plant and Pest Diagnostic Lab Purdue University 915 West State Street West Lafayette IN 47907-2054 Telephone : (765) 494-7071 Fax : (765) 494-3958	Diagnosed By : Timothy Gibb (gibb@purdue.edu) Completed Date: 11/18/2013

IMPORTANT: Our report is based on the material and information that you have delivered to us. Purdue University cannot guarantee the accuracy of our report. Additional onsite information may indicate a different conclusion. Our evaluation is an opinion only and not a warranty of your results from your use of our report.

FIGURE 7.3 Example of spider diagnostic report.

In some cases diagnosticians must refer clients to medical personnel such as physicians, dermatologists, or others. Other times, it may be advisable to refer clients to professional pest managers or those responsible for manufacturing and providing pesticides and other tools. In fewer cases, making contact with other specialists such as regulatory authorities, legal professionals, or others should be recommended.

Examples

Figures 7.1, 7.2 and 7.3 are actual examples of diagnostic responses for bed bug, mold mite and spider sample submissions. Personal references have been removed. Note the inclusion of the elements that we have thus far discussed.

Please refer to the following link for details on biology and control: http://extension.entm.purdue.edu/bedbugs/index.php

DIAGNOSTICIANS AND EDUCATION

Insect diagnosticians have an educational responsibility beyond simply teaching those who submit a specimen. A diagnostician's clientele group is often much larger and may include all employees of a company, staff and customers or, as in the case of university diagnosticians, it may include county extension agents, commodity-based clientele groups and large sections of the general public.

Educating groups of individuals can be done effectively by using photographs and written articles, including case studies of insects that have been submitted or are anticipated.

Providing a description of the insect's appearance and behavior, along with management recommendations, can be done in many ways. Depending upon the audience, a short fact sheet may be all that is required. Other audiences respond better if the facts are disguised in a general interest piece. A diagnostician must carefully weigh how much technical material to introduce and how much entomological jargon can be understood by the audience. If the audience is composed of peers or scientists, the diagnostician can use the scientific names of insects, entomological terms such as genus and species, instar, diapause and chemical names of pesticides to concisely convey a great deal of information. However, laypeople are largely turned off by technical information that they do not understand or feel it is not useful to them. In these cases, presenting too many raw facts can be counterproductive to the educational process.

Consider the following two educational articles written about chiggers. Note how the first is written in very technical terms and contains only the facts. By comparison, the second article is intended to educate a more diverse lay audience. Both contain the essentials of chigger identification and bite prevention. The first uses formal technical descriptions while the second uses informal language and humor to retain interest. Both are effective articles if they are given to the appropriate audience.

CHIGGERS

Identification

Chiggers are the larval stage of certain mite species in the family Trombiculidae. Common names include chiggers, harvest mites, and red bugs.

General Description

Adults are oval shaped, approx. 1 mm in length, with a bright red velvety appearance, and possess eight legs.

Larvae have 6 legs, measure 0.2–0.4 mm, are rounded in appearances and usually display a chrome-orange hue. Each has a single dorsal scutum bearing two sensillae and 4–6 setae.

Line drawing of a chigger, magnified many times.

Biology/Behavior

Trombiculid mites go through a life cycle of egg, larva, nymph, and adult. Gravid females lay three to eight eggs in a clutch, usually on a leaf or under the roots of a plant, and die by autumn. Eggs eclose soon after. Larvae typically develop in overgrown brush, grass, under or around trees or in berry thickets that are shady, humid or near stream banks. They are especially congregated where small rodents are abundant.

Larvae quest on plants until contact is made by potential hosts. Feeding on hosts is only done in the larval stage. After feeding on their hosts, the larvae drop to the ground and become nymphs, then mature into adults that possess eight legs. In the post-larval stage, they are not parasitic and feed only on plant materials.

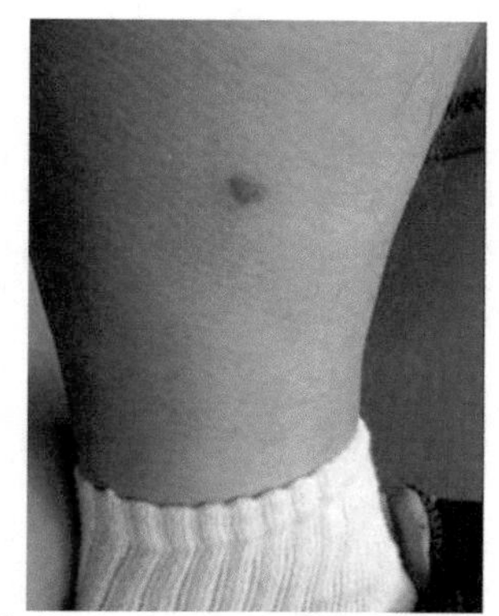

Typical appearance of a chigger bite on leg.

Trombiculosis

The larval mites feed on skin cells, but not blood, of a large variety of hosts, including humans, rabbits, toads, box turtles, quail, and even some insects. Most complaints describe severe itching resulting from the injection of digestive enzymes into the skin to breakdown skin cells.

Specific immunoresponse varies from person to person, but most common symptoms include severe itching within 3–6 h, followed by dermatitis consisting of macules and wheals.

Chiggers vector diseases such as scrub Typhus in some parts of the world but not commonly in the US.

Control Recommendations

Itching can be alleviated through use of over-the-counter topical corticosteroids and antihistamines. Hot showers or baths also will help reduce itching.

Personal Protection

Avoid walking through non-mowed fields, brush, and other overgrown areas. Avoid brushing up against vegetation where chiggers congregate.

When hiking or camping in potentially chigger-infested areas, wear long pants that are tucked into boots or socks and long sleeve shirts. Clothing made of tightly woven fabrics will tend to keep chiggers from reaching the skin as easily.

Apply an insect or tick repellent. Products containing diethyl toluamide (DEET) or permethrin (clothing treatment only) are most effective. Be sure to read and follow directions for use on the container.

Showering or bathing immediately after coming indoors effectively removes chiggers which have not yet attached. If not possible, thorough and brisk rubbing of the skin with a dry towel may remove many chiggers before they are able to attach and feed.

Reducing Discomfort from Bites

Apply over the counter anti-itch medication (hydrocortisone, Calamine lotion, etc). Physicians may recommend oral Benadryl or a prescription-strength steroid cream.

Controlling Chiggers Outdoors

Chigger infestations are less common in maintained turfgrass and landscaped environments. Occasionally, especially in sites recently cleared for development, chiggers may infest vegetation around a home. Hosts such as wild mammals, birds or reptiles can help sustain chiggers in backyard settings. Keeping grass cut short and vegetation well-trimmed can raise soil temperatures and lower humidity enough to make lawns less hospitable to chiggers.

Residual insecticide sprays, such as those containing bifenthrin, cyfluthrin, esfenvalerate or permethrin, can help suppress chigger numbers. Granular insecticides generally are less effective than sprays; however, among granular products, bifenthrin performed best against chiggers in recent studies.

For more information on the biology and control of chiggers, contact:

Timothy J. Gibb
IPM Specialist and Senior Insect Diagnostician
Purdue University

The information given herein is supplied with the understanding that no discrimination is intended and no endorsement by the Purdue University Cooperative Extension Service is implied. Any person using products listed assumes full responsibility for their use in accordance with current direction of the manufacturer. Purdue University is an equal opportunity/equal access institution.

Information listed is valid only for the state of Indiana.

CHIGGERS ARE DOWNRIGHT RUDE!

Americans should not have to tolerate rude behavior, especially from something as small as a chigger! And yet, that is just what we are exposed to every summer from May through September in Indiana. Chiggers are adolescent mites, so tiny that they are seldom seen. Several can actually fit on the period at the end of this sentence. It is only the 'teenage' mites that bite people. Apparently, once they mature to adulthood, they grow out of the immature and obnoxious behavior of biting people and live the rest of their lives feeding peacefully on plants.

Gangs of teenage chiggers all have the following MO: they wait on the tips of tall grasses, shrubs and weeds to drop off onto any larger animal that happens to brush by. Usually these animals are birds, amphibians or small mammals but the mites are just as happy with the odd human that passes by.

When chigger mites fall onto shoes or pant legs they begin climbing in search of tender, moist skin to bite. They seem to concentrate in areas where clothing fits tightly against the body, such as around the ankles, groin, waist or armpits. This is exactly the rude behavior that I am talking about. A simple bite on an arm or back of the neck may be acceptable, and can be scratched in public. But public scratching of the groin, armpits or under the bra strap is an entirely different matter. It is socially unacceptable, politically incorrect and may even be illegal in some countries.

But scratch you must!

Once chiggers bite, there is no alternative. Chiggers do not burrow into the skin but rather pierce skin cells with their mouthparts and inject their special chigger saliva. This saliva contains enzymes that breaks down

Line drawing of a chigger, magnified many times.

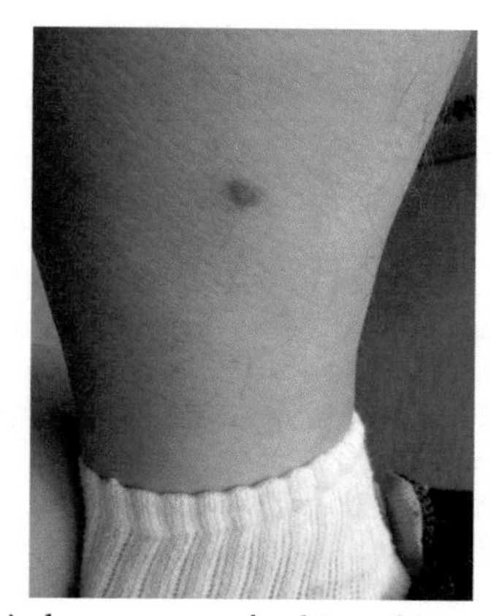

Typical appearance of a chigger bite on leg.

cell walls and causes the skin cells to liquefy. Meanwhile, human immune systems quickly react to this foreign enzyme, resulting in not only infuriatingly, and intense itching but also in the formation of a hard red wall at the spot of the bite. Chiggers capitalizes on this body reaction by using the round wall, called a stylostome, as a straw to suck up their meals of dissolved body tissues, and then promptly drop off immediately after feeding. The itching intensifies, however, over the next 20 to 30 hours even though the mite is no longer present. Depending on the person and individual sensitivities, the itching may continue for days or even weeks.

So, how does one stop chigger bites from itching? Well, aside from amputation, physicians can sometimes prescribe an antiseptic/hydrocortisone ointment. This may help ease the itch and reduce chances of secondary infections caused by the itching and scratching, but it is not a perfect answer.

One of the most effective methods of preventing chigger bites is to change clothes and take a hot soapy shower as soon as possible after being in potentially chigger-infested areas. The mites are so small that it may take them several hours to crawl from shoes to where they want to bite. Washing them away with a sudsy shower before they arrive is an effective preventative.

The best solution is to avoid getting into chiggers in the first place. Stay away from tall grasses and shrubs where chiggers are known to occur. Chiggers love to live in brambles, as most people who pick black raspberries know. They also inhabit grasses close to the ponds and streams where bank fishermen stand. (Both raspberry pickers and fishermen can easily be spotted due to their obsessive itching.) If you must go in those areas, tuck your pant legs into your socks and apply insect repellant containing DEET to the shoe and ankle area. This will stop many of the mites from gaining access to skin and beginning their climb to areas where clothing fits tightly. (Theoretically, avoiding tight-fitting clothes or even going naked might help. If nothing else, it will certainly confuse the little biters – not to mention friends and neighbors.)

Timothy J. Gibb
Insect Diagnostician

The information given herein is supplied with the understanding that no discrimination is intended and no endorsement by the Purdue University Cooperative Extension Service is implied. Any person using products listed assumes full responsibility for their use in accordance with current direction of the manufacturer. Purdue University is an equal opportunity/equal access institution.

Information listed is valid only for the state of Indiana.

Education Delivery Vehicles

Response Forms

The heart of a response report is education. The greatest service a diagnostician can provide a client is education. The response report that is prepared by a diagnostician should focus on helping a client see the insect or arthropod in its habitat with its unique biology, behavior and survival requirements, and then to properly place all of this in the context of human activities.

Public, as well as private industry diagnosticians, also have an obligation to provide enough direction to a submitter such that they will be able to recognize and deal with the same sample should it reoccur in the future.

Explaining what identification characters or behaviors to look for will help educate the client.

Public Messages

As a professional entomologist, there is an inherent responsibility to educate others. This is largely done through the preparation of written materials. The same information that is put in a response report can be communicated using printed materials in other forms, with a potentially wider impact. Extension personnel are known for producing one- or two-page fact sheets, sometimes called extension publications. These hit the same points as the response reports except that they also include a detailed description of the arthropod, something not needed if a specimen is in hand.

Generally, fact sheets include:

- descriptions of the arthropod;
- biology and distribution;
- potential concerns;
- management recommendations;
- where to get help.

The biggest difference is that (1) these are more detailed; (2) are prepared in advance of a sample; and (3) distributed to a wider audience.

The strategy of educating and helping potential submitters to be aware of and recognize a pest before they even see it enables them to be armed and ready with the information they need to deal with the pest.

When these groups are empowered as teachers themselves, the effect is magnified. They can educate others about what to look for in order to recognize a potential pest. This strategy effectively compounds the power of a diagnostician. At the same time, it may actually decrease the number of those particular samples and questions that are sent into the laboratory, freeing up the time for the diagnostician to do other things. It is a win–win situation.

Written Publication Delivery

Information can be delivered in a variety of different ways. Effective extension specialists use a combination of several traditional but still effective methods including:

- extension publications;
- fact sheets;
- newsletters;
- hot news blurbs.

Whatever the form of communication, the information itself remains standard. Below, we will consider examples of written diagnostic information and compare the advantages of the various means of delivery.

FACT SHEETS

Fact sheets, as the name suggests, should include the essential facts necessary to educate the reader. They usually provide both common and scientific names of the subject and a clear and precise description of its appearance. Use of color photographs is most appropriate as they can truly replace a thousand words.

Fact sheets are written to clientele groups most interested in and in need of the information (see the discussion of first detectors in Chapter 9). What is most important is that they receive facts about what to expect in terms of potential impact: what, where, when, and what can be done about it.

The end of the fact sheet normally gives a lead to find more information if desired.

This same information can be supplied as part of larger publications such as newspapers, newsletters or trade magazines. Let's face it, not every article in these publications is read by every subscriber. Basic fact sheet material must be slightly dressed up in a manner that captures the attention of a reader. This can be done in a variety of ways and geared to the talents, personality and comfort of the diagnostician.

A diagnostician must recognize what the audience requires and must write to the audience.

ELECTRONIC POSTS

Placing written materials on electronic sites, including websites and blogs, is an effective means of getting information to the public. Electronic newsletters are also commonly used to get time-sensitive 'hot' information into the hands of the users.

Google and other search engines are often used by the general public to find information on the internet. Unfortunately, a lot of garbage also can be found on the web. Much of it is designed as an advertisement for products or services. Some contain erroneous and outright false information.

Working with professional web designers to format and populate a site with appropriate material can go a long way in making a website user-friendly and also easy to find.

Even a very informative website is of little value if it is not used. Promoting a website or a blog among potential users is important. Various methods of advertising are commonly used. Regardless of the method, promotion is key to effectiveness.

Social Media

The term 'social media' refers to the use of electronic means (online) to engage in two-way or multi-way conversations with other individuals. The fact that social media can facilitate conversation or social interaction is what makes it different than simply broadcasting or posting materials to a blog or a website. Although some blogs are now set up to allow for posting and answering questions, they are still considered a much less advanced method of social media. Twitter, Facebook, and other social media channels are designed for more immediate interaction and ongoing conversation.

The effective power of using social media as a means of extending educational materials depends upon the number of people who 'follow' the account. As users become more familiar with a certain media channel and begin to follow it, the power increases.

Large companies use social media channels to communicate with their employees. In the same sense, diagnosticians and other extension specialists may use Twitter, Facebook, interactive blogging and other social media vehicles to communicate with large clientele groups, including extension agents, growers, consultants, producers and pest managers.

Social media is an effective communication tool for diagnosticians because it:

- allows for nearly instantaneous delivery of materials;
- encourages followers to ask questions of both the person who posts as well as others online (conversation);
- naturally creates groups of followers - real people who are interested in the subject;
- creates a thread of the conversation that can be retrieved and joined by followers who connect at a later time.

Social media is also an effective vehicle to deliver the same information as in printed documents, but in an electronic format. Communication via social media has the advantage of being able to make the information instantaneously available and, as more and more people become comfortable with these methods of communication, it can reach an even wider audience. Current social media include:

- e-mail groups;
- blogs;
- Facebook;
- Twitter;
- Pinterest;
- chat rooms;
- list serves.

Both the diagnosticians who send out information and the target audience using social media are usually of a younger generation and tend to be fairly tech-savvy. Many more mature patrons of a diagnostic laboratory still rely on the telephone as their sole means of communication, while the 'middle' generation seems most comfortable with using email. It is almost certain that newer forms of communication will become standard among the next generations to come.

Traditional Mass Media

The use of mass media is important to a diagnostician and can be used very effectively to get a diagnostician's message out. The major difference between the use of mass media and other publication methods is that informational facts are delivered via an interviewer. There is almost always a TV or a newspaper journalist who asks questions of the diagnostician. It is within the answers to those questions that the diagnostician's message must be delivered. This presents a slight difficulty in that it is up to the journalist to ask the 'right' questions.

Keep in mind that a journalist has slightly different objectives than a diagnostician. They need to add an element of human interest element to a story to make it newsworthy. They often lean towards the drama and emotion of the story to attract attention and capture the reader or listener's attention. In contrast, diagnosticians are scientists; they prefer to stick to the raw facts. Stories of newly introduced pests lend themselves very nicely to the objectives of both the interviewer and the diagnostician. In other cases, however, a diagnostician may desire to downplay the media hype of a situation. In these cases potential conflicts can arise. That said, however, mass media remains a very powerful vehicle to deliver a diagnostician's message.

Probably the best lesson to learn is that the interview is almost always at the invitation of the journalist. They usually call and say that they are working on a particular story and ask if they may ask you some questions. In the case of a radio interview, the questions may be asked and answered during that initial call. In the case of a newspaper writer, some questions may be asked then and follow-up questions later. Television crews usually make an appointment to come and interview the diagnostician, either at an interesting site or at the laboratory. In all cases, journalists are under a time crunch. They need the interview sooner rather than later, otherwise it is not news.

Diagnosticians must understand the objectives of the journalist in order to work with them most effectively. Diagnosticians are seldom trained in how to deal with the media. Most of what they learn comes from trial and error. Nevertheless, the following tips may be useful to diagnosticians working with the mass media.

TIPS FOR INTERVIEWING

- **Be calm.** It is natural for any person to become visibly nervous when being interviewed by a journalist. The microphone, camera and lights all combine to cause a diagnostician's mind to go blank. Even spitting out their own name becomes a challenge. Secondly, an interviewee is never absolutely certain what or how questions will be posed. If the questions come in an unexpected form, this may cause a mental lapse that is very apparent, especially in radio or television.
- **Be collected.** Remember that if a journalist has contacted you, it is assumed that you are the expert. You have the background, training and knowledge to comment on the subject. Your answers are what really make the story.
- **Make suggestions.** Do not be afraid to make suggestions as to what the salient points of the story are, what you think the audience should know, and what questions you have been receiving at the laboratory. These are valuable to a good journalist. You may suggest sites where the television interview might be done. If there are opportunities for some film footage of interesting live shots, it makes the story come alive. Photo options of insects or other props for newspaper stories likewise enhance a newspaper or magazine story. Journalists appreciate such suggestions, even if they are not always used.
- **Answer the question.** Politicians are famous for ignoring a posed question and answering a question that they would like to answer. On the other hand, a simple 'yes' or 'no' answer to a question does not lend to an interesting interview. A diagnostician must develop a technique to include an overarching message as they are answering the interviewer's questions. This can be done by prepping the interviewer in advance. One method is to discuss the essential message with the interviewer off-camera. This allows an interviewer to formulate and ask the right questions.
- **Be prepared.** Know what the essential elements of the story are and take advantage of this opportunity to get the message out. Never agree to an interview for a story that is beyond your expertise or one with which you are unfamiliar. It is better to decline, delay, or recommend another expert who can better add to the story.
- **Wait for the question.** Timing is everything. When the door opens, be very concise and confident with your answer. It is already assumed that you know much more information about the subject than the time allows. You cannot share all of what you know in a two-minute interview.
- **Be polite.** Always be polite and gracious. Never interrupt an interviewer. Remember that even though you may make suggestions, they are running the interview. Be friendly both on and

TIPS FOR INTERVIEWING—Cont'd

off the record. These are people that you will want to add to your network. If they have a good interview with you, they will likely remember you the next time an insect story arises. In addition, they will often leave their business card. When time-sensitive and newsworthy stories arise on your end, you can call and suggest to the journalist that a story is waiting. The more they work with you, the more trust will develop.

- **Educate the interviewer first.** Journalists are experts in journalism. They usually know little about the subject. It is up to you to educate them first. That way they will be able to ask the right questions on the air.
- **Remember that you represent your university, company and profession.** The way that you speak, dress and act will reflect upon the institution you represent.

Public Speaking

Diagnosticians are often sought after as public speakers for workshops, seminars, conferences and other events. These engagements may be anywhere from 30 minutes to day-long training. Diagnosticians often use PowerPoint as a tool to present their information because insect photographs are so riveting and descriptive.

Presenting public talks and seminars is not easy. The information must be audience-specific. One of the major mistakes that diagnosticians can make is to make their presentations either too technical or too general for the audience. It is critical for a speaker to know the audience and to appropriately adjust the material and the way it is presented.

Speaking must also be entertaining enough to hold the audience's attention. Humor, real-life cases, stories, photographs, hands-on activities, question/answer sessions, breaks, eye contact, pauses, audience participation, and moving about the room are all ways a speaker can capture an audience.

Professional public-speaking coaches can work to help develop a smooth and confident presentation style. However, public speaking skills must be learned, largely on the job. There is no way around that. The more speaking engagements, the more likely a speaker will develop their own personal style which works for them. Confidence is critical for a good speaker. This is an attribute that comes largely from preparation but also through practice. It is normal to be nervous. Public speaking is scary and is dreaded by many. Using and channeling nervousness, however, can make a speaker better because it helps a speaker to prepare.

There are thousands of lists containing tips for effective public speaking available either in print or on the web. Many of these are very good and a potential speaker would do well to consider them and incorporate those that work. On the other hand, speaking is an individual activity and each speaker must develop their own strategy for delivery.

That said, the following is a list that may help new diagnosticians or even old diagnosticians become better. Speaking skills are continually honed over time.

Effective public speaking requires that a speaker:

- **Be qualified.** Know the material inside and out. Be the expert that the audience expects. Research matters.
- **Be professional.** The way that you speak says much about who you are. Language and grammar is critical. Materials on screen must be accurate. Spelling has to be correct. Handouts must be professional.

- **Prepare the speech.** 'Winging it' will backfire. Know what you want to present. Know what words you will use and when you will use them. Sequence of thoughts and presentation can make or break a speech.
- **Know the presentation.** Do not read from the screen. Assume that the audience can read. If using PowerPoint, consider getting instructions on most effective design, and usage. There are many very good demonstrations of how to make effective PowerPoints.
- **Practice.** Do a dry run of the speech in front of a mirror, to colleges or a video camera. Critique and improve. Disruptive hand gestures, facial expressions and use of filler words such as 'um' and 'er' can be detected and avoided.
- **Dress appropriately.** Professional dress sets a speaker apart. Clothes that are not comfortable and are in need of constant readjustment are poor choices.
- **Avoid distractions.** Hands can be used effectively to point out or direct the audience's attention, but can also be distractive if used to fumble with jewelry, a pen or placed in pockets. Holding anything, unless it has a purpose in the presentation, is distracting.
- **Know the equipment.** Know how it works and be prepared for slight glitches – they are certain to occur. Good speakers work around them.
- **Know the layout of the room.** This dictates the best place for a speaker to base their presentation. Sometimes speakers are tethered to the podium, other times they are free to walk around the room.
- **Presentation is key.** Effective PowerPoint usage can lead both the speaker and the audience through the talk smoothly. Effective use of props can enhance a talk. These must be planned for and executed precisely.
- **Control nerves.** Speaking slowly is important. An obviously nervous speaker is generally one who races through the material.
- **Emphasize important points.** This can be done by repeating, highlighting, directing attention to or summarizing them. Help the audience to know the 'take-home' message.
- **Gauge your audience.** Speaking to a live audience will allow a speaker to pick up signals as to whether the audience is following, bored, in need of a break or uninterested. Adjustments should be made on the fly.
- **Humanize yourself.** This can be done by making eye contact, using personal stories, anecdotes or sample cases as appropriate.
- **Captivate.** Humor is an effective means of making the presentation interesting. Care must be taken to be sure that the humor is appropriate and not overdone. Either of these can ruin the presentation quickly.
- **Never exceed a designated time.** This shows respect. Never keep an audience past the designated stopping time. Audience members are generally unforgiving if this rule is violated. It is considered a cardinal sin to take someone else's time. A following speaker must be able to start on time.
- **Assess the impact of a presentation.** This can be done with pre and post surveys or in a variety of other ways. Seek feedback.

Remember that every speaker is an individual, with strengths and weaknesses, when it comes to making presentations. Diagnosticians must develop a strategy that works for their own personality. Remember that audiences truly want the speaker to succeed. Give them what they want.

If done well, return invitations will come.

RECORDING DATA AND DATABASES

Need for a Database

A database is created to keep a systematic record of data for particular activity. In the case of an insect diagnostic database, a large amount of information accompanies the submission of a sample for diagnosis. Many of these are fields relating to the submitter and the client, such as business, location (city, county, state), job responsibilities, and requested assistance. If recorded and entered into a database, these can serve as a historical record. Furthermore, when linked to specific pests, items of information such as identification (family, genus, species), whether or not it is a potential pest, location found, degree of damage, insect behavior, and management techniques recommended, such information can become a very powerful and searchable historical record.

What to Record and Why?

There are several reasons for insect diagnosticians to create or use an existing database. The first and most obvious is to create an automated method of receiving and responding to samples as they come in to an identification laboratory. A database can also serve functionally as a mechanism to respond to submitters and track accounting. Other information about submitters, such as address, phone, billing address, and occupation can be used to automatically generate reports and billing.

These can reduce redundancy in writing reports by:

- establishing a standard for related entities;
- providing suggested action plans and problem-solving strategies;
- making data accessible and intuitive to team members and database users.

In addition, a database is a method of organizing data that will provide an opportunity to trace patterns in submissions. For example, because the threat of new pest introductions is always present, a database can provide important information such as:

- the first record or first recorded incidence of a pest;
- spread of the pest in time and space;
- when the pest is first reported each year, when the population peaks and when it is gone;
- where (what habitat or eco-locations it was found in) and what general geographic location (city, county state) it comes from;
- what population levels are extant.

Although the database is not intended to be a quantitative survey, the number of laboratory requests or submissions can indicate general population growth trends.

A database is naturally a historical record. When a database is properly constructed, pest identifications by location, submitter, or dates can also serve as a platform for prognostications of insect occurrence and abundance. These allow for predictions of when certain pests will become problematic in various commodities times or locations.

Archiving Data

Diagnosticians from private laboratories will undoubtedly want to create their own database. This allows them to have sole control over what is contained and how and for what purpose it is used.

Public diagnostic laboratories have used independent databases in years past but now many have joined forces in utilizing a common database, the National Plant Diagnostic Network database (NPDN).

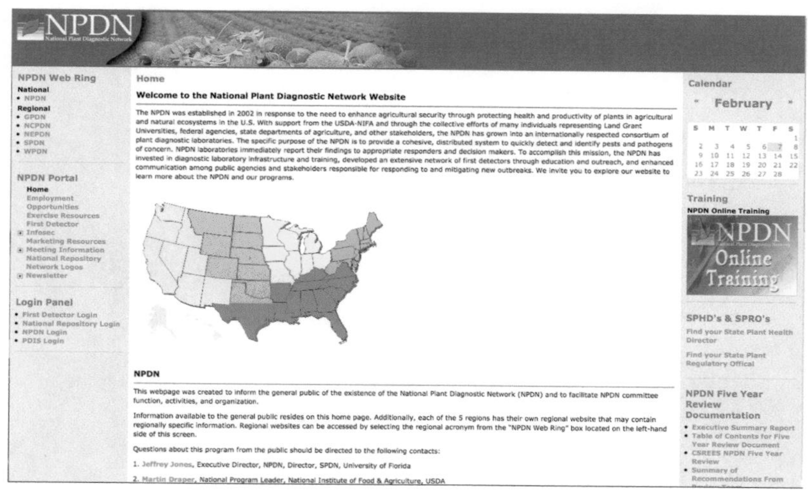

FIGURE 7.6 NPDN web site - http://www.npdn.org.

The Plant Diagnostics Information System (PDIS) was developed to facilitate plant diagnostic lab activities for a large organization of land-grant institutions, state departments of agriculture and the U.S. Department of Agriculture. These laboratories provide services for plant disease diagnosis, plant identification, and insect identification. PDIS is a system of web and database applications designed to facilitate record-keeping and communication needs of laboratory personnel and their clients.

The Animal & Plant Disease and Pest Surveillance & Detection Network was established to develop a network linking diagnostic facilities across the country. The network is a collective of land-grant university pest diagnostic facilities. Because so many otherwise independent diagnostic laboratories are working in concert, the collective power of the database is increased.

CHAPTER

8

Making Management Recommendations Using IPM

PEST MANAGEMENT RECOMMENDATIONS

A significant part of a diagnostician's responsibilities is making pest management recommendations. This responsibility is largely what separates a diagnostician from a taxonomist. Making appropriate pest management recommendations is a difficult task. No two situations are exactly the same; consequently, there is no standard set of management recommendations for every case. Each has to be made on a case-by-case basis by someone who is familiar with the existing facts and the many options available.

Any recommendation for management must be based first and foremost on an accurate identification of the pest and secondly upon an awareness of the environment in which it is found. An appropriate recommendation in one situation may differ completely in another. Site awareness is not only the responsibility of the person who implements the management, but is also the responsibility of the diagnostician who will make the recommendations. A diagnostician can ascertain much about the site based on the sample submission form. The form also provides indications about who will apply the control methods. When this information is lacking or unavailable, it becomes the duty of the diagnostician to find it.

Clients are largely unaware of all that goes into a control recommendation. This is understandable. Not everyone is trained in the principles of pest management. A diagnostician must take into account the capabilities of the submitter when making management recommendations. For example, if the person who will employ the recommendations is a professional pest manager, it may be safely assumed that the management recommendations can be more sophisticated and, therefore, that less detailed instruction is necessary than for a layperson. If it becomes clear to a diagnostician that the submitter has little knowledge about the pest or about how to apply controls, very simple and explicit management recommendations must be provided. A diagnostician must take this into account when making recommendations for the client.

Consequently, a large part of the diagnostician's duty is understanding the various management options that can be recommended and choosing from them based upon the specific pest situation. In order to offer the best management recommendations possible, an insect diagnostician must be up to date on current pest management techniques.

Contemporary Insect Diagnostics
http://dx.doi.org/10.1016/B978-0-12-404623-8.00008-9

PRINCIPLES OF INSECT PEST MANAGEMENT

To make human life tolerable, insect pests must be managed. People have been battling insect pests as long as humans and insects have shared our planet. Early control methods included cultural and physical techniques; basically, we avoided them or we swatted them. Since then many other methods of dealing with insects have been developed and implemented to manage insect pests. The many control methods used in pest management are worth further consideration because each may play a part in making pest management recommendations.

In earlier chapters we have pointed out that any pest management recommendation is most effective if we understand the identity, life history and biology of the pest. In this chapter we will describe the major methods or techniques of pest management and offer an example as to how they can be used in integrated pest management.

Major pest management techniques include:

- cultural control;
- biological control;
- alternative control;
- chemical control;
- mechanical control;
- and regulatory control.

Each of these techniques can be considered a control tactic in pest management. Most pests require a combination of several control tactics to manage them over the long term. Strict reliance upon a single control tactic is almost never adequate and often brings with it several unwanted consequences. We refer to a combination of control tactics as a control strategy. Strategies (multiple tactics used simultaneously) are the preferred way to manage insect pest populations.

A pest control strategy is a combination of two or more tactics used together.

Strategies involving simultaneous use of multiple tactics are the preferred way to manage insect pest populations. Even the use of chemical controls should be balanced with non-chemical approaches to maximize their effectiveness. Likewise, using sanitation techniques to remove available food sources may be most effective when coupled with exclusion techniques to minimize the reintroduction of the pests. All strategies are most effective if the life history and biology of the pest are well understood.

Integrated pest management (IPM) is a strategy that has been proven to be the most effective and sustainable solution for managing pest problems. In most cases a combination of several control tactics are combined into an overall strategy that will decrease the pest population status from 'pest' to 'non-pest.' It is important that a diagnostician understand the practice of integrated pest management.

All strategies are most effective if we understand the life history and biology of the pest.

IPM recommendations depend upon:

- an accurate identification of the pest;
- the environment in which the pest and controls will be applied;
- the capabilities of those applying the controls.

PRINCIPLES OF INTEGRATED PEST MANAGEMENT (IPM)

Integrated Pest Management is a decision-making process wherein observations, including inspection and monitoring, are used to make pest control decisions based on pre-determined management objectives. IPM takes an ecological approach to selecting control methods by combining a variety of chemical and non-chemical control tactics in a way that minimizes risk to people and the environment. The IPM process must include an evaluation and written records to document the procedure and results.

> Integrated Pest Management, or IPM, is an ecological approach to pest management that combines a variety of chemical and non-chemical control tactics in a way that minimizes risk to people and the environment. IPM requires accurate record-keeping and follow-up.

Although the basic components of IPM are always the same, the specific elements of an IPM program vary from one environment or situation to another. Diagnosticians must consider every pest situation independently. Evaluating the situation through inspection and monitoring techniques provides the necessary facts upon which to base decisions about whether or not to implement control procedures. After making the decision to implement controls, the various control options must be carefully selected, based on effectiveness and human and environmental safety. Finally, reviews and record-keeping promote informed decisions regarding how long to continue controls or whether or not to make changes.

In some ways, IPM can be viewed as a decision-making process in which observations (from inspection and monitoring) are used to make pest control decisions based on pre-determined management objectives.

> IPM always involves inspection, scouting, and monitoring. Pest identification, record-keeping, and evaluation are also basic to IPM.

An action threshold is the number of pests that can cause an unacceptable amount of damage if no action is taken to control them. This threshold is influenced by a tolerance level for the pest and by the pesticide's potential to cause harm. Prior to any control action, a pest must be correctly identified. IPM includes many potential treatment options, including human education, habitat modification, horticultural/agricultural design or redesign, and physical, biological, regulatory, chemical and cultural control methods. When using chemical controls, 'least-toxic' chemical control methods should be selected.

IPM COMPONENTS

The following five main components of IPM are considered essential:

1. Inspection.
2. Monitoring and establishment of tolerance level.
3. Situation-specific decision-making.
4. Application of pest management techniques.
5. Evaluation and record-keeping.

Inspection

The first step in the IPM decision-making process is a thorough inspection. This is done not only to identify the pest, but also the habitat in which it exists and the nature of its interactions with people. Identifying both the pest and its environment are key to determining the seriousness of the problem and the steps that must be taken to manage it.

Making sound decisions about pest control options requires:

- Accurate identification of the pest involved.
- Recognition of environmental conditions that support or have the potential to support the pest population.

Pest Identification

Pest identification is a step that is sometimes taken for granted. Nevertheless, accurate identification of the pest must always be confirmed. Although this process is one of the most basic elements of pest management, mistakes in identification are still common, especially when many pests are similar in appearance or behavior. Accurately identifying the pest and its damage, recognizing which life stages are present, and understanding the life history of the pest and how it interacts with people are all factors that help a pest manager anticipate damage and exploit the weak links in the pest's biology. Management efforts should never take place before the pest is properly identified.

The importance of correctly identifying the pest cannot be overstated, but this does not mean every pest manager must have a degree in entomology. Access to an insect diagnostician is the preferred way to get an accurate identification. Diagnosticians, in turn, are greatly assisted by pest managers and clients who have at least a basic understanding of pest identification skills. Diagnosticians are responsible for training and providing resource recommendations to prospective pest managers.

Accurate pest identification is not always a simple procedure. In many cases, the insects are either not present, are hidden, or are in a form (egg, pupa etc.) that is not readily identifiable. In such cases, identification must be made based on the presence of pest signs, such as holes or tunnels, fecal materials left behind, or damage done to a plant or product. Confirming pest identification is critical.

During inspections, pest managers should carry along some important tools to assist in accurate diagnosis.

A magnifying lens is important because many insects are small and the identifying characteristics that separate them from other insects may be almost invisible. Collecting the insects is almost always standard. Small collection vials can hold the insects while identification is being made, preserve the insect for future reference, or transport it to diagnosticians. Having field guides or other reference books handy is also recommended.

Purpose of Inspection

The purpose of the inspection is to gather information that will be used to make pest management decisions. Pest managers performing inspections must be knowledgeable about the types of pests that typically occur in their system, where and how to find these pests, and the conditions that favor pest activity.

Effective pest managers anticipate pest activity based on environmental conditions and past occurrences.

An important and common pest associated with a particular system is called a key pest. There are usually four or five key pests in each system. For example, one key pest found in kitchens throughout much of the world is the German cockroach. White grubs are key pests of turfgrass throughout the Midwest. Codling moths and corn rootworms are key pests in Midwestern horticulture and agriculture. Mosquitoes and deer flies are key health-related pests. Examples of key pests are provided in Chapter 5.

Pest managers as well as diagnosticians must be familiar with the key pests in the systems they manage. In order to be alert to signs of key pest activity, they should be cognizant of the yearly activity periods of the pest, when infestations are likely to occur, and what specific environmental conditions favor pest development.

Inspecting

An important component of pest management is inspecting for pests. Routine inspection is usually directed toward the key pests in the system and is designed to detect the presence or absence of those pests. Inspections also note any environmental conditions that may favor the introduction or the development of a pest.

Inspection is hard work. It requires a pest manager to get into places that only pests usually go. Finding pests means that a pest manager must think like a pest: 'if I were a cockroach, where would I like to live?' Knowing the basic requirements and preferences of each key pest allows a pest manager to determine how, where, and when to inspect, as well as what to look for.

Keen powers of observation are the basis for pest inspections; however, there are other tools that can greatly assist pest inspectors. Often they are so valuable that a proper pest inspection cannot be performed without them. For instance, when inspecting for pests in buildings and structures, a reliable flashlight is essential; without one, an inspector would not be able to see into the dark recesses where pests live. A shovel or a trowel is essential to inspect for pests underground. When inspecting for wood or stem borers, a sharp knife is required to open stems or branches.

Sweep nets are considered essential when inspecting for potato leafhoppers. These tools are basic requirements for inspectors. Even more specific tools and methods can be required for certain pests. For example, inspecting for the presence of spider mites (tiny arthropods that can threaten plant health) is often done by placing a piece of white paper or an index card beneath a branch and briskly hitting the branch to dislodge the mites. The pest manager counts the number of mites on the paper to decide if a pesticide treatment or some other control measure is necessary.

Inspection Process

The process for inspecting for pests requires that a pest manager:

- know the pest;
- identify pest signs;
- recognize the conditions that favor the pest;
- utilize proper inspection tools and techniques.

A valuable inspector knows how and where to find the pests in their various stages and sizes.

Other evidence of pest infestations include specific signs. In structural pest management, for example, bed bug signs include not only live bugs, but also cast skins and fecal specks. Dogs can detect bed bug and termite populations by smell. Some listening devices are used to discover termite or other wood borers at work even though they are hidden in wall voids.

Finally, good inspectors will look for pest-conducive conditions that might make it easy for pests to become established. When these are identified and eliminated in advance of a pest infestation, significant damage and health risks are avoided.

Environmental or Habitat Assessment

Specific environmental conditions can contribute to pest invasion or buildup by providing the habitat that pests need to survive. Part of the inspection process is to assess the local environment in relation to the needs of a pest. Every pest requires food, water, and shelter (harborage) Fig. 8.1).

We know that eliminating any of these three essential resources will also eliminate the pest. This can be likened to the legs on a three-legged stool. If any one is removed, the stool – or, in this analogy, the pest – cannot stand. Pest managers must inspect for these and other conditions that may favor pest outbreak. During any inspection, favorable conditions should be noted even if a pest is not detected.

Specific conditions may favor certain pests. Often these are related to weather conditions. For example, some insects do better in warm, dry conditions, while others thrive in moist, cool conditions. Conditions that might favor a specific agricultural pest may include soils that are

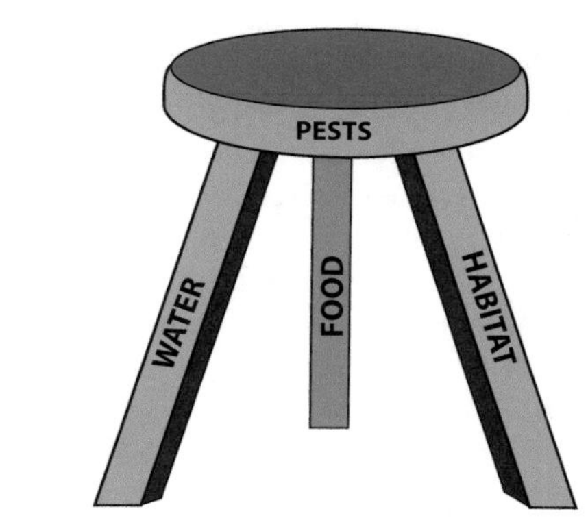

FIGURE 8.1 Pests require three basic elements in order to infest.

saturated, have a high pH, or are sandy. Other insect pests may require just the opposite. Factors such as differences in leaf litter, proximity to other food plants, tillage or cropping practices, clutter, or thatch all may favor different pests.

Professional pest managers understand the specific environmental requirements and preferences of the pests that they manage. This is why knowing the biology and life history of the pest is so important.

For example, a restaurant that has inadequate kitchen sanitation, a leaky faucet, excessive clutter in the storeroom, a filthy dumpster, and gaps under the doors make it ripe for pest infestation. If pests are not already taking advantage of these conditions, it is only a matter of time before they do. Identifying these pest-favorable conditions is part of a proper inspection.

An important advantage of IPM, as compared with traditional pest control management (chemical-only), is that it identifies and eliminates the conditions that lead to pest infestations before they happen. It is the pest manager's responsibility to identify habitat conditions that are favorable to pests and tell the client what steps are needed to rectify such situations.

ASSESSING ACCESS

If insects have access to a habitat that affords them the basics of life (food, water, and habitat) they will become pests. Access is a factor that may be used to a pest manager's advantage. For example, recognizing that insect pests may enter a building from the outside via utility ports, or under a door, will go a long way in making management recommendations. Insect pests may be unwittingly introduced from a contaminated field via machinery that has not been cleaned. Infested products, intentionally brought into a warehouse or a kitchen, may become a source of a pest infestation.

Pest managers must take 'access' into account when they make an assessment. Often, denying pests entry can be the best and most long-term pest management technique available. We refer to this as exclusion: keeping pests out.

An ounce of exclusion is worth a pound of cure.

Monitoring

Insect pests are usually thought of in terms of numbers or population size. Monitoring usually follows inspection and measures or gauges the size of the pest population. However, the size of the population alone does not mean much unless it is referenced in some context. Population estimates are given relative to some constant, such as space or time. For example, how many insects are in a lawn is usually measured as the number of insects found per square foot. In horticulture and agriculture, it is the number of insects per plant, per leaf, or per acre. Sometimes populations are estimated as the number collected per time unit (hour, day, or week). Still other populations are measured as the number of insects per sweep of a net or per shovel of soil. This will be covered in greater detail later in this chapter, but for now it is sufficient to understand that pest managers must estimate pest population size as a function of some standard parameter.

Monitoring may consist of pest observations made by the pest manager or by other people. For example, in a school, custodians, teachers, and staff members should be encouraged to regularly record pest sightings in a log so this information can be communicated to the pest manager. The same is true in homes, factories, farms, gardens, golf courses or wherever people spend time. Asking those who spend time there what they have seen is basic monitoring.

Many common pests are not active during the daytime or when people are present to actually see them. Other pests hide either underground or in places where they are difficult to see. Such pests often build up very high populations without being discovered. By the time they are noticed, their populations are sometimes so high and so entrenched that controlling them becomes very difficult. One principle of integrated pest management is devising a monitoring method so that if a pest invades, it can be discovered and eliminated quickly – well before it has time to reproduce and build up high populations.

Insect pest monitors are tools or devices essential to IPM. They are used to:

- detect early infestations of pests;
- estimate pest population numbers to aid in the decision-making process (for example, to decide whether to apply a control);
- determine how widespread the pests are;
- determine if a control strategy is effective and how well it is working;
- eliminate a potential pest before it gets started.

WHAT ARE INSECT PEST MONITORS?

Pest monitors are tools that can be very complex and full of technology, such as infrared sensors, motion detectors, and video surveillance.

Other very effective monitors are simple traps that can be homemade, inexpensive, portable, and disposable. Sticky traps are one version of such a pest monitor. These are often used indoors to help manage structural and nuisance pests in a home or school, but variations of these are used in conjunction with pheromone attractants (chemical lures) outdoors. They usually incorporate a plastic or cardboard base covered with a very sticky, glue-like substance. When pests walk or fly into the glue, they become stuck and cannot extricate themselves. Many such traps are available to the homeowner and professional alike (Fig. 8.2). The advantage of such monitors

FIGURE 8.2 Miscellaneous traps.

is that they work 24 hours a day, 7 days a week, and 365 days a year. They never need a vacation!

Other monitoring tools include devices that capture insect pests as they wander into and are caught in a trap, such as passive pit-fall traps. Most often, however, attractions such as light, scents, or food are used to entice pests to enter the trap. Common pest-monitoring attractant-based devices include black light traps, chemical-attractant (pheromones) traps, food-baited traps, or even traps that utilize attractive colors or shapes.

Pest managers monitor pest activity to gather information necessary for making pest management decisions. Monitoring relies on tools such as sticky traps (Fig. 8.3), pitfall traps, and light traps to collect data about pest activity between visits by the pest manager.

Other monitoring methods employ nets to catch insects, either used actively, such as with a sweep or aquatic net, or passively, such as with a malaise trap. Sweep nets are commonly used to monitor for agricultural and horticultural pests. A pre-determined number of sweeps of the net are made, passing over plant leaves and stems. Then the number of pests caught in the net are counted, which allows a pest manager to determine the number of insect pests per sweep.

Berlese funnels are another variation of a monitoring trap that use heat and light to repel, rather than to attract. A sample of substrate, such as leaf litter, soil, or grain, is placed into the

FIGURE 8.3 Sticky trap for brown recluse spiders.

holding container on top of a fine-mesh screen. The lights are turned on and left for a time. All small insects or mites in the sample are forced to move away from the light, causing them to fall through the screen and into the collecting jar below.

HOW MANY MONITORS SHOULD BE USED?

There are few established recommendations or even a formulae that dictate the number of traps to employ. The number of monitors to use often depends upon the precision of the desired estimate, the size of the area to be monitored, and the intended use of the monitors involved. For example, more traps should be used in a food-handling establishment than in a hardware manufacturing plant because pest management is much more critical where food is involved. In addition, a large building with many different areas containing food and water will require more traps than a smaller building. For horticultural or agricultural fields, enough monitors should be utilized to provide a reliable and consistent approximation of the population size. Sometimes, recommended numbers of monitors are available from product dealers or from Extension specialists in the area.

WHERE SHOULD MONITORS BE PLACED?

It is important to place monitor traps in pest-vulnerable areas (PVAs), which are any areas where conditions are right for a pest to establish and thrive. PVAs are normally those areas where water, food, and shelter are available. In buildings, kitchens and food-handling areas are prime PVAs. Any area where trash is handled is an automatic PVA because of the food availability and the many pest hiding places it provides. In outdoor areas, monitors should be located where they will provide the best information for the area under consideration.

HOW OFTEN SHOULD MONITORS BE CHECKED?

Traps can be checked at whatever time interval is deemed appropriate. Often, this is determined by the time resources that a pest manager can devote to monitoring as well as by the urgency of the pest situation. For example, if brown recluse spiders are discovered as a health threat in a school building, intense and frequent monitoring is required so that a pest management plan can be adopted quickly. On the other hand, where monitoring is being used exclusively to determine if a pest will arrive, traps may only require checking on a weekly or biweekly basis.

Remember that traps must be checked on a regular basis. Checking a set of traps is often referred to as monitoring. It simply involves looking at the kinds and numbers of insects that are captured and recording the gathered information on a form or in a logbook. Light traps capture a large variety of insects, not all of which are pests. These require significant sorting time. Pheromone traps are much more selective in the pests that they attract and can be monitored relatively quickly. Traps that are placed but left unchecked for extended periods lose their value quickly. They certainly cannot provide advance warning of a pest infestation if they are not checked regularly.

WHAT CAN BE LEARNED FROM INSECT MONITORS?

There are many things that can be learned about a pest population from checking monitors, most importantly, the identity of a pest. Correctly identifying the captured insect or insects can be very revealing. It is important to remember, though, that many insects found in a home or landscape or even in an agricultural crop are not pests. Many are actually beneficial, most are innocuous (neither harmful nor helpful), and a few are serious pests. The status of a pest is relative, since some pose a much greater threat than others.

It is important to understand that identifying the pest and researching its level of threat are critical to integrated pest management.

Other important pieces of information that monitoring traps may provide are relative population numbers and distribution. Traps placed in pest-vulnerable areas for 24 hours can give a good indication of the population distribution (how widespread the pests are) and the relative size of the infestation (minor, moderate, or severe).

A word of caution: having zero counts on sticky traps for a short period (24 hours) does not confirm that pests are absent. However, traps that are well placed for prolonged periods but which are consistently empty provide a reasonable indicator that there are no pests or that a very minor infestation level is present. The same can be said for sampling with a sweep net or another means. A zero count in one sweep or in one area of a field may not mean that there are no pest infestations. To validate population estimates, it is necessary to take multiple samplings that are evenly distributed in time and space.

Keep records of the number of pests trapped; accurate records are important. Professional pest managers utilize prepared log sheets (details will be discussed under the record-keeping section of this chapter). Date, time, specific location, pest identity, and numbers of insects found are the main items to be recorded.

Without a system of surveying a population of pests, managers would be forced to make uninformed decisions that may have unfortunate consequences beyond just wasted resources. For example, without actual evidence, a vegetable grower may assume that a pest is present on the vegetables and, therefore, apply a pesticide when no pests are present. Not only does this waste resources, but it also creates unnecessary risks to people and the environment. Another incorrect assumption might be that because the pest was once present or because it has been found in a neighboring field, it is present currently and requires control strategies. It is better to base decisions upon established facts.

One of the greatest benefits of monitoring is that it can be used as an evaluation tool to assess the effectiveness of any pest control strategy. For example, comparing the populations of pests before and after a treatment allows a manager to make informed assessments about how effective the treatment is. Monitoring should be considered one of the most basic tools of proper IPM.

Monitoring is essential for determining the population size and distribution of a pest, as well as to evaluate the effectiveness of control techniques.

Decision-Making Using Pest Tolerances

There are several important inputs into a decision-making process. Decision-making is based largely on pest tolerance levels. Other important factors such as site analysis, potential control methods, and safety concerns also play an important part.

Establishing a pest tolerance level is the first step in pest management decision-making. A single insect in a building or a cropping system seldom warrants a treatment, but a large number of pests might. That is why assessing the size of the pest population – the number of pests per building, per plant, per acre, or per animal – is so important. However, it is also true that all pests are not created equal; some pests may be potentially much more damaging than others. Both a termite and an ant may be pests, but while an ant is a nuisance, a termite can cause structural damage to a home, so the termite is the more significant pest. In this case, one termite is more troublesome than a colony of ants. Thus, the potential for a pest to cause damage must be assessed in order to determine if and when the pest must be managed.

The estimated number of pests (population size) can usually be an indicator of the amount

The likelihood that a pest will cause damage, or pest potential, must be a part of the IPM decision-making process.

of damage that may be expected. Therefore, an estimate of the damage potential can also substitute for the pest population size. Either of these estimates can be used to determine a pest tolerance level, which is the maximum amount of damage or the maximum number of pests that can be tolerated in any particular situation.

For example, a person may be able to live with, or tolerate, three flies in a home, but any more than that will make them uncomfortable. In this case, the tolerance level is set at three. Another example might allow 500 grasshoppers per acre as acceptable in a grass pasture, but when the population exceeds that 500, the pasture begins to sustain serious damage.

The graph in Figure 8.4 depicts the relationship between population size or damage potential and pest tolerance.

As the pest population increases, it may eventually reach a point where it is intolerable. That intersection is considered the tolerance threshold.

> Integrated Pest Management is not geared toward complete eradication of a pest. The goal is to maintain the pest population below the level of pest tolerance. Depending on the system, the tolerance level may be a function of economics, aesthetics, or human health and comfort.

Tolerance levels depend upon the system for which they are calculated. The tolerance threshold concept in IPM was developed for

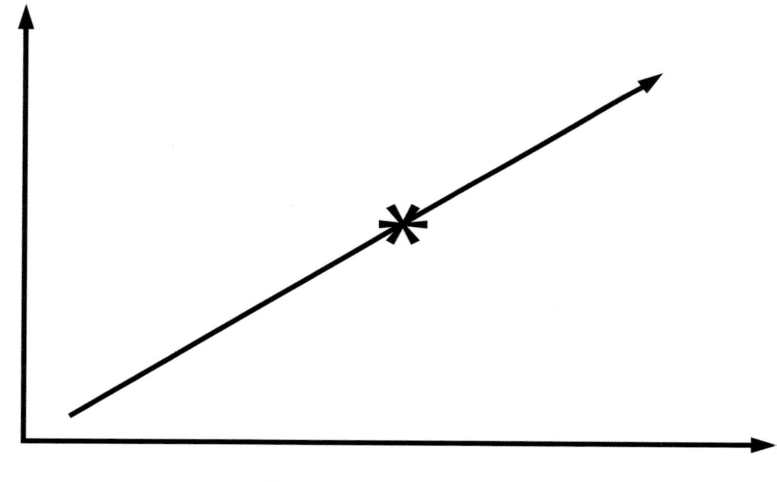

FIGURE 8.4 Tolerance threshold.

agricultural pest management, but it also applies to turf, landscape, and structural IPM. Each system has a different goal that will determine the number of pests that can be tolerated. In agriculture, the management goal is typically to maximize profit by producing the highest yield with the least production costs, including pesticide application costs. In turf and landscapes, the management goal is generally to maintain a pleasing appearance in an economically reasonable way. In buildings, the goal is to prevent pests from threatening human health and peace of mind, or harming contents or structures. In all of these systems, the pest manager must determine the level of pest population that can be tolerated without having a negative impact on the system's management goal.

Establishing a tolerance threshold is a simple concept in theory. However, in practice it is difficult to establish and may vary, depending upon several factors. Consider, for example, the following:

- Not all pests represent an equal threat.
- Not all environments, people or plants are equally sensitive.
- Pests may be more damaging under certain conditions than others.

To understand what tolerance standard to use, we must first recognize what system and what management goal we are striving for.

In agriculture, our goal is to maximize profit, thus the standard to use is economics.

Similarly, in construction of buildings or other commodities, minimizing losses is the management goal and so an economic tolerance is applicable.

In systems such as turfgrass and landscapes, it is not possible to put an economic value on things such as shade or beauty, so the goal becomes based on sensory-emotional levels and the tolerance is aesthetic tolerance.

When people are considered, health, comfort and safety must be the goal and become the tolerance levels.

ECONOMIC TOLERANCE

Establishing an economic injury level, or the point at which the cost of the expected damage outweighs the cost of the treatments, is critical to IPM in agriculture and horticulture production, as well as in building and product pest management. This concept is most easily understood by considering the cost that damage will cause to a building or product. If replacement or repair costs will be more than the cost of pest control, the economic tolerance level is surpassed and the pest should be controlled. Likewise, in agriculture or horticulture production, if applying a treatment will save money, the economic injury level has been surpassed.

> Economic Injury Level (EIL) is defined as the number of pests that cause monetary loss greater than the cost to control the pest.

Threshold levels have been painstakingly developed for a number of pest/crop systems; when available, they are provided by state IPM Programs in their pest management guidelines for specific crops. However, not many insect pest thresholds have been developed for other systems.

Cost is a factor that must enter into the decision to apply pest management options in any system. All pest management efforts cost something. Equipment, personnel time, and insecticides all have a readily calculable cost in dollars. However, the various control options also carry an inherent potential risk, such as cost to the environment or to human safety, should something go wrong. These are not easily quantifiable and thus are less easily calculated but nonetheless represent real potential costs.

AESTHETIC TOLERANCE

The true value of landscape plants, such as ornamental flowers, shrubs, lawns, and trees, cannot be easily translated into dollars. This complicates the development of the tolerance level. In such cases, assessment hinges on the 'relative worth' of the plant in its surroundings or on its 'aesthetic worth.'

In landscape pest management, the relative value of a planting can depend on such factors as its location in the landscape. For example, in a residential setting, some level of pest damage might be tolerable in a backyard, where relatively few people see it, while the same amount of damage in the front yard often is unacceptable. In golf courses, higher levels of pest damage and invasion are tolerable in the rough areas because the playability of the course is unaffected, while much lower levels of the same damage are acceptable in highly maintained areas. Still less pest damage can be tolerated on the tees and greens, especially on those closest to the clubhouse, where they are played on and seen by more people more frequently.

In sum, pest tolerance in non-economic areas is subjective and varies considerably from one specific situation to another. The aesthetic injury level is the point at which the appearance of damage to a landscape or building begins to be apparent and causes humans to be uncomfortable. In most cases, this is much lower than the economic injury level.

HUMAN HEALTH AND COMFORT TOLERANCE

Perhaps even more subjective than determining treatment thresholds in landscapes is determining pest tolerance levels for different people. As in landscape IPM, tolerance levels for people are determined by the potential of the pest to cause harm or annoyance and by the residents' personal tolerance levels. These are both influenced by the site or environment in which these two factors interact.

When personal health or safety is at stake, pest tolerance is comparatively low. For example, workers in an office building might tolerate a greater number of ants before implementing a costly management practice than they would brown recluse spiders. This is simply because of the different potential negative consequences associated with each pest. Pest tolerance levels are inherently very low in hospitals, daycare centers, or nursing homes because the sick, the young, or the elderly are highly sensitive to pests. Even if all pests were harmless, tolerance levels would still differ because each person has his or her own individual comfort zone. The reality is that some people don't mind a few insects in their environment; others lose control at the presence of one bug. Some people would accept the presence of an insect if they were convinced that it could neither bite nor sting; others would never be comfortable with a bug in their environment regardless of its danger. Everyone responds differently.

In summary, tolerance levels are at the heart of the decision-making process. These must be established in advance of a pest management program. When working with people, education is key to helping establish realistic tolerance levels. Educating clients about if, when, how, and where pests harm humans is the first step. Urban pest managers must recognize that tolerance levels are situation-specific. In some instances, very sensitive situations might dictate that even the presence of a single pest, if it can bite or sting people, may exceed the highest tolerance levels; in others, even great numbers of harmless insects may not be a serious threat, and thus a 'no action' policy may be the correct response.

Sites also affect tolerance levels. A common pest such as a house fly has a different tolerance level when encountered in a park than it does in the operating room of a hospital.

In short, a decision to manage a pest population depends on two major interacting factors: the size of the pest population and the tolerance level. Always remember that tolerance levels are based on a pest's probability to cause harm, either economically or by affecting human health, comfort or quality of life. As a result, tolerance levels differ depending on the person, the pest, where it occurs, and its population size.

SITUATIONAL ANALYSIS

It is clear that tolerance levels affect many pest management decisions. But due to the wide range of possible pests, the specific situations in which they may occur, and the variety of people who might encounter them, a list of universally acceptable pest-tolerance levels cannot be established. So, by definition, IPM must also be situation-specific.

In addition to considering the various tolerances that affect a situation, pest managers also need to analyze the site as well as the potential control tactics and their safety considerations.

SITE ANALYSIS A site analysis takes into account factors such as what protection is needed and the various pest tolerances. It also includes additional details that will affect the decision to manage a pest. For example, if the pest affects a plant or a crop, a site analysis notes whether the pest is attacking a growing plant or a stored product. It would also include details such as the plant stage of growth, where applicable, and the location of the plant.

For plants, pest managers must determine factors such as pest life stage as well as plant variety and development, health, and susceptibility to the pest. Environmental factors such as irrigation and soil compaction, as well as the interaction of many other possible stresses potentially affecting the plant, also must be accounted for in a formula predicting when a population of pests should be controlled. In some instances, specific numbers have been generated to give a general idea of pest thresholds and acceptable damage tolerances. Remember, though, that these are general guidelines only and should be adapted to the specific circumstances of the site and pest in question.

Sometimes managing pests in landscapes depends on several factors known only to those who manage the grounds. Factors such as upcoming events, expected traffic-use differences, specific weather conditions, irrigation stresses, soil types, and seasonal changes all play roles in IPM decision-making.

CONTROL METHOD ANALYSIS Potential advantages gained by applying pest management tactics must be weighed against implementation costs. This comparison is always at the heart of IPM. Direct-treatment expenses, such as chemical costs, personnel time, and resources, are relatively easy to calculate. Indirect costs, such as reentry or harvest restrictions, are more difficult to calculate, but they are often very important. Downtime for golf courses or athletic fields as well as access restrictions in buildings while pesticides are being applied must be considered. Even less easily measured costs, such as strained public relations, the potential for negative environmental side effects or associated lawsuits, must be taken into account (Fig. 8.5).

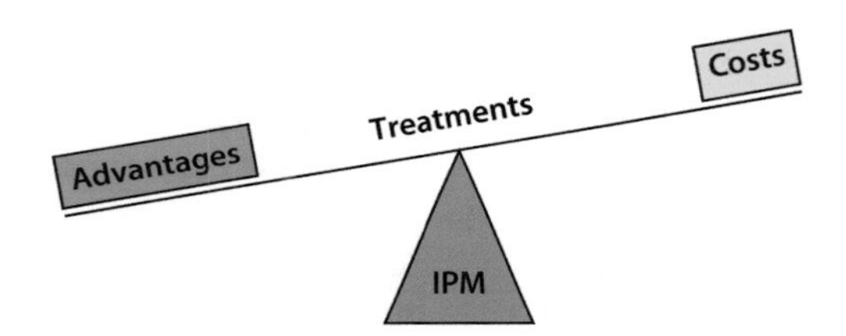

FIGURE 8.5 IPM decision-making.

SAFETY ANALYSIS Insect pests may damage plants, structures, and furnishings, cause annoyance, or create serious human-health concerns. Insect feeding may reduce a plant's ability to produce fruits or grains, or insects may feed on the produce directly, thus contaminating human food supplies. Insects can sting, bite, spread germs, and transmit diseases. At the same time, we know that chemical pesticides can also negatively impact people and can cause irreparable damage to the environment.

Over the years, society has come to depend on chemical pesticides as the primary tool for managing insects. Strict reliance on and overuse of chemicals have, in some instances, created a pesticide treadmill, requiring more and more chemicals in order to achieve consistent results. The potential for direct human exposure to pesticides, as well as for indirect exposure through environmental pollution, has become an important and volatile social issue. The dilemma faced by pest managers is that such concerns have not lessened public expectations for pest-free buildings and for high-quality, damage-free products. Fortunately, new pest management technologies, as well as the integrated pest management philosophy, are helping make this possible.

Unrealistic expectations regarding pest-free environments often confound IPM efforts. Realistic IPM expectations recognize that some level of pest presence may be acceptable, provided the long-term health of individuals or plants is not at risk. IPM requires a great deal of public education. Changes in public perceptions and tolerances of pest presence and damage must be adopted before the full range of IPM benefits can be achieved.

In summary, IPM encourages pest management practices based on a comparison of the advantages gained with the costs and disadvantages of a decision to treat. Site-specific judgments must be made in each instance and must take into account any potential perils, such as direct exposure to chemicals for humans and pets. Pest managers also must work to prevent environmental contamination by considering the fate of the pesticide through time and space as it is affected by factors such as proximity to surface or groundwater, type of soil, and density of plants. Experience and sound judgment while conscientiously considering each of the above factors are at the heart of site-specific decision-making.

Integrated pest management has the same objective in both the urban and the agricultural environment. Protecting human interests from damage – and not necessarily killing pests – is the basis of IPM. Decisions should be made on a situation-specific basis and must be consistent with an overall pest management plan.

Implementing IPM Control Options

The terms 'pest control' and 'pest management' often are interpreted as synonyms. However, there is a crucial, though subtle, difference between them. Pest control has traditionally relied only on chemical remedies for pest problems. In contrast, pest management involves a deliberate evaluation process resulting in one or more carefully chosen controls – pesticides being one control – to fit each situation. Pest management requires an understanding of pest population levels and the possible applications of different control tactics in a pest management framework. In pest management, pest tolerance levels are established and used as decision-making guides to clarify whether action against a certain pest is desirable.

In IPM, the word 'integrated' implies a multidisciplinary or combined approach, whereby several management options can be brought to bear on a single problem. This combined approach is considered the most environmentally healthy and the most viable long-term strategy available. Choosing from a variety of possible management strategies ensures that the best management fit is achieved for a particular site. The better the fit, the smaller the chance of undesirable consequences.

Once the decision is made to implement a management procedure, the next step is determining which tactics are appropriate. Arriving at such a decision is seldom a simple process. It is made simpler, however, when the pest manager has a thorough understanding of the biology, life cycle, and ecology of the infesting pest; the environment, and the available management options. Often the best method for choosing a specific tactic is to compare all evident advantages to all possible limitations. Factors such as effectiveness, ease of application, environmental impact, and a host of other on-the-job experiences will be of value to decision-makers. IPM involves many potential actions, including human education, pest habitat modification, horticultural/agricultural design or redesign, and physical, biological, regulatory, chemical, and cultural control methods.

When using chemical controls, 'least-toxic' chemical control methods should be selected. The major control tactics used in pest management are worth further consideration because each may play a part in the control of certain insect pests, depending upon where, when, and how they occur. Pest managers must recognize and understand the merits of each. Below is a basic description and an example of how each can be used in integrated pest management.

CULTURAL CONTROL

Once established, pests become part of the environment. They affect and are affected both directly and indirectly, by every practice occurring in their environments. Controlling pests by changing living habits or environments is known as cultural control. Many cultural practices can be manipulated to the detriment of pests. Making the environment less favorable to pest survival and reproduction is the goal of cultural pest control. In urban buildings, sanitation often is the most important cultural control.

Denying pests access to food, water, and a place to live through increased sanitation makes it difficult for pests to enter or to persist once they are present.

In landscape management, it is generally accepted that healthy, vigorously growing plants can withstand more pressure from pests than can plants that are stressed. In this way, cultural practices such as mowing, fertilizing, pruning, mulching, and irrigating all indirectly affect pest populations.

Inattention to these management practices can create stressed plants that attract pests.

SANITATION/PHYSICAL CONTROL

Proper identification and alleviation of these stress factors through cultural management changes are some of the longest-term and most environmentally-conscious methods of pest management in the landscape. For example, trees under stress due to any cause are relatively attractive and susceptible to wood borers and other insect pests. Landscape managers must understand the interactions that irrigating, fertilizing, mowing and pruning, soil compacting, and other human activities have on potential plant-pest problems.

BIOLOGICAL CONTROL

Biological control occurs in nature all around us. Animals preying on or becoming parasites to other animals, as well as diseases that kill animals or plants are all agents of biological control, more commonly called biological control agents. When we refer to the 'balance of nature,' we are often talking about natural biological control – one organism in the environment controlling another. In IPM, biological control is a method of managing pest populations by manipulating parasites (parasitoids), predators, or diseases (pathogens) in the pest's environment to the detriment of a particular pest population.

PARASITOIDS Parasitoids are often called parasites, but the term parasitoid is more technically correct. By definition, insect parasitoids have a free-living adult stage and an immature life stage that develops on or within an insect host and ultimately kills it. After feeding on host body fluids and organs, most parasitoids leave their hosts to pupate or emerge as adults.

> Parasitoids have a free-living adult stage and an immature life stage that develops on or within an insect host and ultimately kills it. They are effective control agents in pest populations.

Most beneficial insect parasitoids are wasps or flies, although other insects may have life stages that are parasitoids as well. Parasitoids usually complete their life cycles much more quickly and increase their numbers faster than predators, even though a parasitized host does not die as quickly as those eaten by predators.

Parasitoids are often the most effective natural enemy of pest insects, even though they may not be readily visible. Sometimes pest populations actually increase after pesticides are applied. This can happen when a chemical insecticide poisons the parasitoids before they can kill the pest insect host. Once free of the natural parasitoids, the pest insects are then free to increase in number.

PREDATORS Insect predators can be found in almost all agricultural and natural habitats. They are usually bigger than their prey and consume them very quickly. Predators are known for their ability to actively search out and destroy their prey.

Some, such as the praying mantis, are generalist predators, meaning that they will feed on nearly any insect that they can catch. Others are specialists, meaning that they will search out only a certain species or a certain stage of insect on which to feed. Their ability to search and destroy makes them very effective natural enemies of insect pests. Predators should be credited with controlling many would-be pests. Common predators include beetles, bugs, lacewings, flies, midges, and wasps.

> Predators, in either the larval or adult stage, search out and eat insect pests. Some have a voracious appetite and can consume many pests, ultimately reducing the pests' population.

Most beneficial predators will consume many pest insects during their life span, but some predators are more effective at controlling pests than others. Many predators may have only a minor impact on a pest population by themselves. But they still contribute to overall pest management when their efforts are combined with the effects of other predators and/or biological control agents.

PATHOGENS Pathogens are naturally occurring biological agents – such as bacteria, viruses, fungi, protozoans, and nematodes – that infect and kill insect hosts. Like humans, insects can be infected by pathogens.

Under favorable conditions, such as high humidity or pest abundance, pathogens may quickly multiply to cause disease outbreaks or epizootics that can decimate an insect population.

Entomologists and agriculturists are just beginning to truly recognize the value of naturally occurring pathogens in the prevention of insect outbreaks. Insect pathogens are one of the reasons that pest populations do not become problematic every year. Many pathogens have been identified, while many more are at work in the environment but have yet to be discovered.

Of those that have been identified, some have been mass-produced in laboratories and are now available to pest managers for use against insect outbreaks. Some of these microbial insecticides are still experimental, while others have been available for many years. Formulations of the bacterium *Bacillus thuringiensis*, or Bt, for example, are widely used by gardeners and commercial growers. Several species of nematodes can be purchased to control many different pests in landscapes, turfgrass, greenhouses, and urban structures.

As a group, these products are referred to as biorational or microbial insecticides because they are applied in a manner similar to conventional pesticides even though they contain a microscopic life form. One main advantage of insect pathogens is safety. They are relatively specific to select groups of insects and sometimes even target specific life stages. As such, they do not harm the environment or non-target animals, such as beneficial insects, pets, wildlife, or humans.

But unlike chemical insecticides, microbial insecticides can take longer to kill their target pest. Additionally, they are usually more expensive to use, and must be applied to an environment in a way that will allow them to survive. To be effective, most microbial insecticides must be applied to the correct life stage of the pest, so pest managers must understand the life cycle of the target pest.

Microbial insecticides are compatible with the use of predators and parasitoids, which may help to spread some pathogens through the pest population. While beneficial insects are not usually affected directly because of the specificity of these microbial products, some parasitoids may be affected indirectly if their hosts are killed. Insecticide applicators should note that although microbials are non-toxic to humans in the conventional sense, safety precautions should always be followed to minimize exposure.

> Pathogens can be important management tactics for many insect pests and are important components of an IPM strategy.

Biological control, the practice of using living organisms to control pests, is not a new science, but it is a control tactic that is beginning to see greater acceptance in insect pest management. Evidence of biological control efforts exists from ancient times and significant control was achieved through the introduction of biological controls in the early part of the twentieth century. However, with the introduction of modern synthetic chemical insecticides, especially during the latter half of the twentieth century, biological control became a largely forgotten science. New chemical insecticides were cheap and appeared to be 'cure-alls.' But this perception was unfounded, and as the disadvantages of synthetic chemicals became more evident and public demands for more environmentally-friendly methods of pest management grew, biological control agents were sought after once again.

Biological control offers many advantages over conventional chemicals. The most important advantage is that biological control methods can be less hazardous, both to people and to the environment. Once introduced, they can continue to be effective without further human intervention. Some instances of introduced biological control agents have been successful, while others have been disappointing. Of course, care must be taken to avoid introducing biological control agents into the environment only to realize later that these 'controls' have become pests themselves. For instance, Asian lady beetles were introduced into America to help manage aphids in agricultural crops. Now they have become a common nuisance pest in homes throughout much of the United States.

Predatory or parasitic organisms feeding directly on pests have shown promise for controlling pest populations. In nature, many different parasites or predators can be found, each with a unique ability to seek out and kill potential hosts or prey. Pest-specific diseases also have been used with varying degrees of success in the management of pest populations. Successful organisms, mostly fungi and bacteria, have been propagated in the laboratory and are now commercially available.

Naturally occurring biological control agents probably are more important than most people imagine. Insect pests usually are controlled naturally by beneficial arthropods and diseases, which can moderate or often prevent outbreaks of pest populations. It is telling that landscapes that are not intensively managed are seldom prone to serious pest outbreaks. Society has learned, through hard-won experience, that when the natural balance is upset by chemical or cultural interference, pests readily move in. Therefore, careful consideration must be given to how each management input can affect a system's beneficial organisms.

Identifying and conserving naturally existing plant and animal biological control organisms are logical steps in IPM implementation. Maintaining a balance between pests and their natural controls will diminish the need for pesticides, save money, and greatly benefit the ecosystem.

Overuse or misapplication of non-selective chemicals interferes directly with the potential of naturally occurring beneficial organisms. Pesticides have been shown to have the opposite consequences of their intended use if they kill the natural controls that hold a pest population in check. Releasing a pest population from its natural biological checks and balances may set the stage for a dramatic and often devastating resurgence of a pest population.

ALTERNATIVE CONTROL

Toxins such as azadirachtin, rotenone, nicotine, and pyrethrin all are derived from plants. Although these are not technically either biological controls or synthetic chemical pesticides, they are materials with a botanical origin that offer some of the same pest management advantages as true biological controls and chemical pesticides. These are part of a group called alternative controls.

> Alternative controls are botanical pesticides.

Alternative controls also include biological microbes (or microbe derivatives). Several such materials have been developed for use in pest control. Discovering new and better alternative control products is an active area of research that promises to have a significant effect on the future of pest control.

New technology and scientific advances have given IPM new tools for controlling pests. Some newer pesticides have been designed to be taken up into the plant, which make them more effective and less likely to harm insects other than the targeted pests. Advances in biotechnology have allowed 'pieces' of genetic code that are toxic to insects to be inserted and expressed in plants, making them resistant to insect feeding. Additionally, scientists have worked to grow plants that are resistant to specific insect pests. These are areas that will continue to see significant and exciting developments for alternative controls.

CHEMICAL CONTROL

The word 'pesticide' is often used as a synonym for 'insecticide,' but this is actually incorrect. Pesticides refer to a broad group of chemicals designed to control pests. While insecticides are one type of pesticide, there are many other types of pesticides as well. Some other examples are herbicides, which are used to control plants, and rodenticides, which are used to control rodents.

There are many different kinds of insecticides. Some, such as sulfur and arsenic, have been used for more than 3,000 years. Others are just being developed today. Because there are so many insecticides, they can be grouped in many different ways, including by chemical makeup, by how they work, by their form, and by their targeted insects. For example, insecticides are often placed into groups or classes of chemistry, depending on how they are synthesized. They are given common names that can relate back to the specific chemical name or chemical structure.

Sometimes insecticides are grouped by the method in which they kill insects, or their mode of action. These insecticides may kill by interfering with a specific part of the insect's nervous system, growth and development, or digestion.

Still other ways to describe insecticides involve their mode of entry, or how they get into an insect's body. Some are ingested while the insect is feeding, some are taken in as the insect respires, and still others rely on contact with the insect's epidermis.

In practical applications, pest managers may group insecticides by what insects they control, while in other situations, they group them by the formulations in which the insecticide is purchased and used (liquids, granules, fumigants, etc).

> Pest managers must understand the insecticides that they use, including the common and chemical names, modes of action, modes of entry, formulations, and target pests affected.

Use of synthetic chemical pesticides is an important component of most IPM strategies. Chemical treatments are especially effective because they can be applied with relative ease and can quickly bring a high pest population down to an acceptable level. When used in this manner and for this purpose, chemicals can be an integral part of an IPM program.

When pesticides are continually and exclusively relied upon to maintain pest populations at low levels, problems inevitably arise. Pesticides are toxic and are designed to kill. As a result, there is always a potential threat of pesticide poisoning to people or other animals that are not the target.

Chemical pesticides are notorious for their ability to move from an intended target zone to another. They may move through water or air into an unintended zone, where significant damage may occur to people, non-target animals, plants, or the environment. Chemical pesticides may also change forms – from liquids to gases or from solids to liquid – and pose very significant risks.

Sound IPM requires selecting only those insecticides that have the correct formulation, concentration, and proven result for the pest and site being targeted.

It is unrealistic to expect to eliminate all pests. Notwithstanding the broad-spectrum, long-residue pesticides used in the 1960s and 1970s, it has become clear that complete eradication of pests is impossible. In most situations, small pest populations that are monitored carefully over time and managed in such a way that they do not increase beyond certain tolerance levels become acceptable. When compared with the potential negative human and environmental health effects that complete pest elimination would cause – not to mention the costs of total reliance on chemical pesticides – a low, well-managed population is preferable.

Recently, great strides have been made in developing 'low-impact' chemistries. These environmentally-friendly, narrow-spectrum (more targeted), least-toxic pesticides have been developed by the chemical industry. As the U.S. Environmental Protection Agency registers these for use, other older 'high impact,' and 'less environmentally compatible' pesticides are being removed from the marketplace.

Using the best pesticide handling and delivery methods is sound IPM.

The solution to a specific pest problem does not always involve a new or better pesticide. Often the difference between success and failure in managing a pest population lies in knowing where, when, and how to apply a selected pesticide.

Targeting pesticide applications to only those areas where monitoring has determined a need for control (spot treating) decreases the total amount of pesticide applied and conserves natural biological controls already in place. IPM dictates that spot treatments replace blanket treatments wherever possible.

Using new technologies to deliver pesticides directly into the area where pests are present – that is, the target zone – diminishes the probability of exposure for non-target organisms, decreases the quantity of pesticides needed for treatment, and increases pesticide effectiveness. For example, replacing whole-room or even baseboard treatments with crack and crevice applications is one way of targeting pesticides in buildings and structures. Injection techniques may be used to deliver even smaller amounts of pesticides into soils or trees in the urban landscape. Select pesticides also may be absorbed and distributed throughout the plant, ultimately killing only those insects feeding directly on it. This is another example of targeting. Spot-treating crops where inspections have indicated that pests are present rather than treating the entire field is a method of targeting pesticide applications in agriculture.

Targeting is sound IPM.

In addition, IPM dictates that pesticides only be applied when they can prevent pest damage. Timing of insecticide applications is critical. In some instances, only one stage of an insect is damaging. Therefore, knowing when the pest does its damage is important. Application timing must be a part of an IPM strategy.

Proper timing of pesticide applications is sound IPM.

Professional IPM continually incorporates new procedures and technologies into pesticide-application methods, thereby decreasing both the populations of pests and the negative risks associated with pesticide applications. Development of pesticide-laced baits has greatly improved the ability of structural pest managers to control rodent and insect pests, such as mice, rats, ants, cockroaches, and termites. Baiting techniques significantly decrease the amount of toxic materials applied in the urban environment and yet achieve pest population controls equal to or better than traditional methods.

A sound IPM strategy requires the use of the most effective and least-toxic pesticide applied at the best possible time and in the best possible manner to only those areas that are known to harbor damaging levels of pests.

MECHANICAL CONTROL

Mechanical controls are also often referred to as physical controls. These are commonsense control techniques that are almost always used in combination with other management methods. Mechanical control methods range from tools as simple as a flyswatter, a window screen, or a vacuum to very complex and highly-engineered methods such as insect electrocuters, high-frequency energy devices (microwaves), or specifically designed, high-pressure air curtains which prevent entry.

Techniques using atmospheric or temperature extremes provide pest management professionals alternative tools for resolving situations where other controls cannot be used or are not effective. Applications of gaseous changes such as ozone, CO_2 or others have been shown to kill insects.

Mechanical controls are very important options in Integrated Pest Management. These include any physical or mechanical device used for controlling insect pests.

Insect traps are also a form of mechanical control. Examples of low-temperature controls include systematic freezing of library books, other archived materials, and taxidermy mounts of animals in large, walk-in freezers. High-temperature controls include heating of stored grains and green lumber, as well as the recent use of whole-house heat treatments to destroy deeply entrenched bed bug populations. A wide variety of traps can be employed as effective methods of control, as well as their usage for monitoring.

REGULATORY CONTROL

Preventing pest entry or spread is an essential objective of IPM. On a national and a statewide basis, regulatory agencies help prevent the spread of pests by enforcing rules on shipping food products, as well as sod and other landscape materials, from one area to another. A great number of insect pests are held in check by these regulations. Regulatory pest managers must understand how pests move and are introduced into a new area. The spread of pests can be limited by routinely inspecting, cleaning, and disinfecting equipment, transportation vehicles, and other materials that could transfer pests from one site to another.

Additionally, it is desirable to take steps to eradicate a small pest population before it becomes established. Government agencies actively monitor for invasions of suspected pests and stand ready to perform localized, intensive eradication procedures in an effort to avoid permanent establishment of new pests. Urban pest managers using similar tactics can also help prevent the movement and establishment of pests from an infected area to a new site.

PUTTING IT ALL TOGETHER

IPM combines multiple strategies for controlling pest problems. One of the biggest challenges for pest managers is deciding which strategies to use in a particular situation. The guiding philosophy of IPM is to use the most effective control measures that will have the least potential for negative impact on people and the environment, including beneficial organisms such as honey bees and pest predators.

A significant part of pest management is done well before a building or a landscape is constructed. Designing with pest management in mind is one of the most effective urban IPM techniques available. Both structures and landscapes can be designed that will make it difficult for a new pest to enter and thrive, making subsequent pest management tactics much more efficient.

Simple and commonsense control methods, such as sealing up entry holes, placing screens on windows, or cleaning off equipment when moving from one field to another, will all help prevent pest problems. Educational efforts can be effective as well. One recent campaign has focused on informing the public that transporting firewood is risky because the emerald ash borer can use the wood to hitchhike from infested to non-infested areas.

Professional pest managers know that if a pest does not arrive, they will not have to worry about controlling it. An ounce of prevention is worth a pound of cure!

Selecting the best treatment strategy for a particular pest requires:

- accurate identification of the pest(s);
- complete understanding of the pest habitat and treatment site;
- knowledge of pest biology;
- familiarity with the management system;
- knowledge of available treatment options and their effects.

Within this framework, IPM favors treatment options that have the greatest potential for providing long-term solutions to pest problems while minimizing negative impacts. The following IPM tactics are presented in the order of general desirability:

1. Education.
2. Habitat modification.
3. Horticultural/agricultural design (or redesign).
4. Mechanical controls.
5. Biological controls.
6. Least-toxic chemical controls.

The pest manager is the person responsible for making pest management decisions in an IPM program. Depending on the situation, the pest manager may be a farmer, private consultant, landscape manager, hospital administrator, school employee, or professional pest manager/contractor.

The pest manager inspects and monitors for pests, identifies pest problems, determines why the pests have become a problem, devises a solution for the problems, implements the solution, keeps thorough records on pest levels, documents actions taken and results, and, based on this information, reevaluates the IPM program on an ongoing basis.

Not all options in each of these categories will be available or desirable in every situation. The types of available strategies vary considerably from one system to another. The pest manager must always consider:

- the pest involved and the risk it poses;
- the resources and options available to the client;
- the relative risks and benefits of available control options.

Pest managers may use one category or method of treatment exclusively or in combination with other methods. Integrated Pest Management is always situation-specific, but it usually involves using two or more options or techniques in a logical way to achieve pest management goals.

Evaluation and Record-Keeping

Sound decision-making in IPM must be based on the best available information on pest activity and the conditions that promote infestations, as well as on the success of previous control measures. After a pest management practice is used, pest managers must follow-up to evaluate its effectiveness. Decisions to continue, increase, decrease, or suspend a pest management practice should be based on its past performance.

RECORDS

Records provide the historical data that can help a pest manager evaluate control techniques over time. In IPM programs, the pest manager collects data through monitoring and scouting activities, and keeps records on:

- pests that are encountered;
- control strategies used, including pesticides and non-chemical controls;
- the effectiveness of each control method.

Records must include site-specific details, such as the size of the pest-infected area and its exact location, population estimates, damage amounts, symptoms, dates, and where appropriate, weather conditions leading up to pest infestation. Records maintained during previous years allow pest managers to make informed, science-based decisions when planning strategies for future monitoring, scouting, and control activities.

Thorough record-keeping provides critical information that can be used to make future pest management decisions.

When information about pesticide applications is combined with data on pest activity levels, the true success and duration of pest suppression can be measured. This information should always be used as a basis on which to make future pest management decisions.

WHAT RECORDS SHOULD PEST MANAGERS KEEP?

With so much information to keep track of, pest managers need a way to organize the information that they collect. They do this through various reports and logs. Depending on the situation, pest managers will prepare and maintain the following:

- inspection reports;
- pest-sighting logs;
- pest-monitoring logs;
- pesticide application records.

INSPECTION REPORTS

Pest managers generally conduct detailed inspections at the start of an IPM program and continue to provide monitoring throughout the program. Pest managers use inspection reports (Fig. 8.6A) to document the results of these site inspections and to inform clients about the presence of pests or pest-conducive conditions. The inspection report is a thorough, room-by-room, plant-by-plant, or field-by-field assessment of the building, landscape, crop, or other site being evaluated.

In many cases, inspection reports include specific recommendations about how to correct a problem.

PEST-SIGHTING LOGS

The pest-sighting log (Fig. 8.6B) is a useful tool in structural IPM programs, particularly in schools, hospitals, nursing homes, and industrial settings. The first entry in a pest-sighting

(A)

Date:	Integrated Pest Management Report				
Part of Building Check	Station	Activity	Assessable	Satisfactory	Comments for
I. Building Exterior					
1. Garbage storage area					
2. Garbage handling system					
3. Parking lot and drainage areas					
4. Weeds and surrounding landscape					
II. Classroom/Hallways/Offices					
1. Ventilation/air ducts					
2. Doors and windows					
III. Food Storage					
1. Dry food storage					
2. Damaged/spoiled dry food					
3. Empty container storage					
4. Drains/pipes/sinks					
5. Overall sanitation					
Recommended 3-part NCR	White – Log		Yellow – Technician		Pink – Secondary IPM Contact

(B)

Integrated Pest Management Pest-Sighting Log					
Date and Time	Pest Seen	# of Pests	Person Reporting	Recommended Actions	Date Completed

FIGURE 8.6 Integrated Pest Management Forms (A) Inspection Report - used for recording the results of a building inspection. (B) Pest-Sighting Log - used to record observations by building residents and followed up actions to be taken by the pest manager.

log is simply a record of pest sightings made by building occupants, including the following:

- location of sighting (as precisely as possible);
- types of pests sighted;
- numbers of pests sighted;
- date.

Pest-sighting logs are kept in a convenient location and consulted by pest managers during routine inspections. The pest-sighting log identifies the problem areas where the pest manager needs to follow-up and ensures ongoing communication between clients and the pest control professional. More importantly, it involves building clients in the process of pest control, which is the first step toward educating them about how their daily practices influence pest populations.

PEST-MONITORING LOGS

Monitoring is an important part of any IPM program. A pest-monitoring log is a simple record of the number and type of pests encountered by the pest manager during scouting or visual inspections (Fig. 8.7).

Monitoring logs serve both a preventive and an evaluative role in the IPM program. They are preventive because information obtained through monitoring can indicate the need for an immediate control action. They are evaluative because a review of long-term records reveals general trends in pest populations that can be used to evaluate the success of specific control measures in particular, or the success of the IPM program in general.

It is standard practice among commercial pest management services for the pest manager to provide the client with a report or invoice that describes:

- the action performed;
- the cost of the service.

In IPM programs, the service report is expanded to include:

- information about the pest species encountered;
- documentation of conditions that promote pest activity;
- non-chemical control recommendations, such as repairing leaks or removing clutter.

PESTICIDE APPLICATION RECORDS

Many states have laws that require documentation for pesticide use. In most cases, the

Date:

Address:

Technician:

Trap #	Location	Date	Trap Catch

FIGURE 8.7 Pest-monitoring log.

following records must be maintained for some set period of time (it is important to find out what laws pertain to your state):

- pesticide applied (brand name and active ingredient);
- U.S. Environmental Protection Agency (EPA) registration number;
- formulation;
- rate of application;
- location of application;
- time and date of application.

For most pesticide applications it is a good idea – and in some states, a legal requirement – for the pest manager to provide the client with a copy of the EPA pesticide label and Material Safety Data Sheets (MSDS). Pesticide applicators, landscape managers, and facility managers should consult federal, state, and local regulations with regard to keeping pesticide application records to be sure that they are in compliance with all legal requirements.

EVALUATION

Evaluation is an important and ongoing part of any IPM program and occurs at multiple levels:

- evaluation of individual pest control actions;
- evaluation of program success.

EVALUATION OF PEST CONTROL ACTIONS In an IPM program, the pest manager must follow-up to evaluate the effectiveness of each pest management action. For example, if a pest manager treated an ant infestation by removing a food source and sealing a crack in the floor, he or she should conduct a follow-up inspection to determine if ants are still active in the area. If so, additional actions may be necessary.

Continual evaluation of each pest control action allows the pest manager to accumulate a helpful database of information that can improve pest management decisions in the future.

EVALUATION OF PROGRAM SUCCESS The record-keeping aspect of IPM provides an opportunity for ongoing evaluation of program success and can provide the following benefits:

- documentation from pest inspection and monitoring reports to determine general trends in pest populations;
- a means of assessing the success of pest management programs over the long term;
- comparisons with data from previous pest control practices to determine the relative success of IPM versus other methods.

Regular IPM program evaluation is recommended so that programs may be continually improved. It is up to pest managers and clients to determine the most appropriate schedule for evaluation of a particular program (Fig. 8.8).

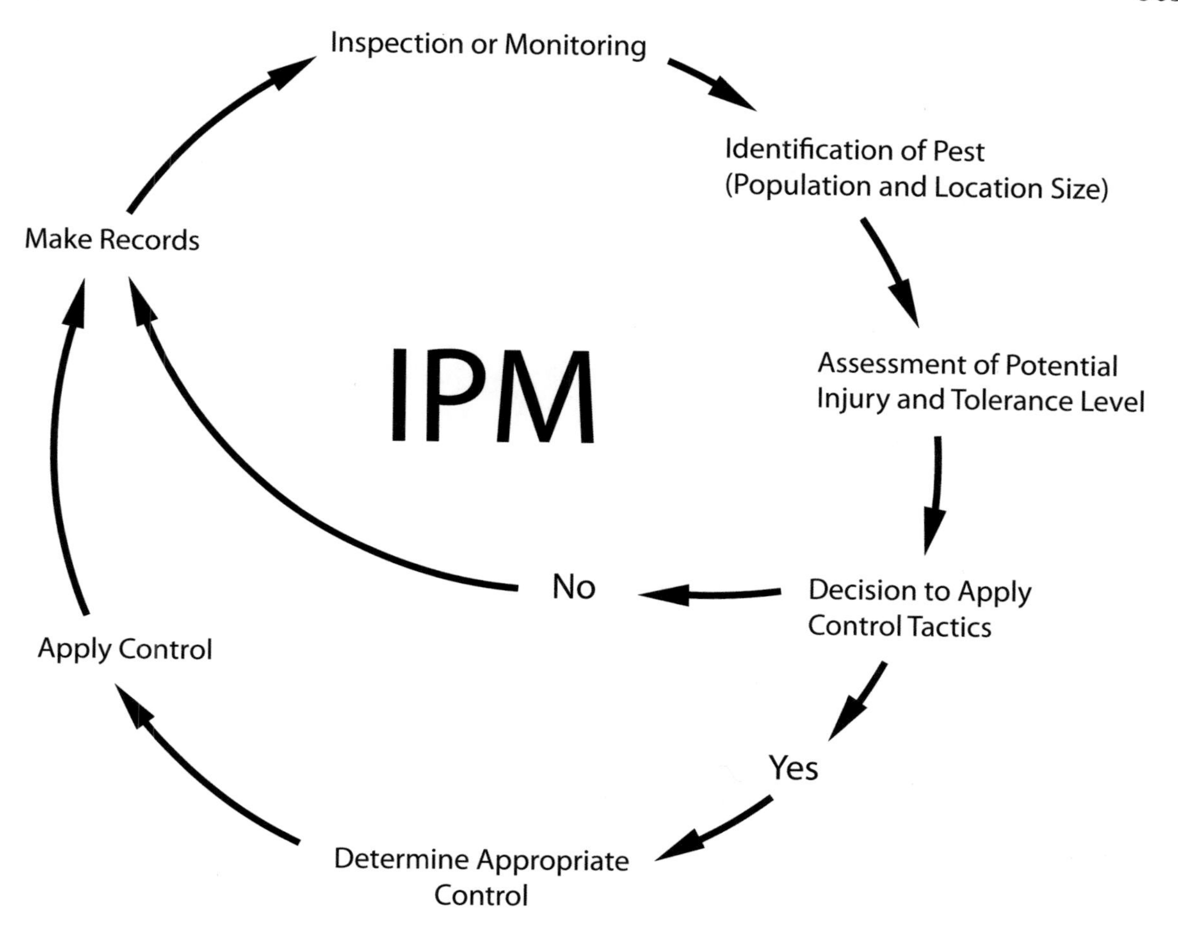

FIGURE 8.8 IPM action and evaluation flow chart.

CHAPTER

9

Networking

THE IMPORTANCE OF NETWORKING

It can be truly said that 'no man is an island.' Every person is connected to his or her surroundings in many ways. The same can be said of an insect diagnostician. The job responsibilities of a diagnostician mandate interconnectivity between many people, agencies, services and the public. No one can know everything about all insects. It is also impossible for a single laboratory to provide services for every entomological need.

Networking can be defined as a beneficial exchange of information or services among like-minded individuals, groups, or institutions. In the case of insect diagnosticians, networking may be referred to as the cultivation of productive relationships to aid in the identification and control of potential insect pests. This includes forming synergistic relationships with persons, institutions, businesses and laboratories that may aid in assisting people with pest problems.

Insect diagnosticians must network with colleagues, federal and state counterparts, specialists, identifiers and diagnosticians, state agencies, and local entities, as well as other diagnostic services and laboratories. Linking such contacts together in a professional network is critical to the success and power that a diagnostician can bring to bear on an entomological concern.

Networking facilitates positive synergy. This is especially true when pest management is considered. Federal and state agencies monitor U.S. borders for plant pest introductions and watch for pest outbreaks throughout the nation. Even so, each year new pests find their into our fields, factories and homes. To provide even the most rudimentary management recommendations, pests must be accurately identified. Regulatory agencies depend very heavily upon specialists with diagnostic skills to determine the exact identification of new or invasive insects. Often, regulatory agencies employ insect diagnosticians or identifiers. Insect diagnosticians, whether they are based at public land-grant universities or private diagnostic laboratories, are more effective if they can access professional contacts who also provide identification and management recommendations.

An insect diagnostician is an essential resource to many individuals, programs, agencies and companies. It is generally accepted that having a reliable and accurate identification and readily available pest management recommendations can save time, frustration, and money, as well as provide protection for people, food and property. When working in pest management, it is always beneficial to have the special talents of a diagnostician on the team. Insect diagnosticians are desired as part of many others' networking resources. Likewise, insect diagnosticians benefit by interacting with others who have talents, abilities, and expertise beyond their own.

The importance of networking cannot be overstated. It is critical for an insect diagnostician to develop a network not only of people, but also positions, responsibilities, institutions, services, agencies and laboratories that may assist in insect identification, as well as making

Contemporary Insect Diagnostics
http://dx.doi.org/10.1016/B978-0-12-404623-8.00009-0

management recommendations. Some of these contacts are used very frequently, others only when a specific need arises.

General networking for insect diagnosticians can be framed using four concentric rings as follows:

Diagnostician Networking

Tier 1 Colleagues	Teachers University Specialists Fellow Diagnosticians
Tier 2 Cooperative Extension	Extension Agents Master Gardeners Extension Specialist
Tier 3 State Officials	National Plant Diagnostic Network Cooperative Agricultural Pest Survey Animal and Plant Health Inspection Service Systemic Entomology Laboratory US Department of Agriculture
Tier 4 Local Entity	Educators Mass Media Legal and Medical First Responders

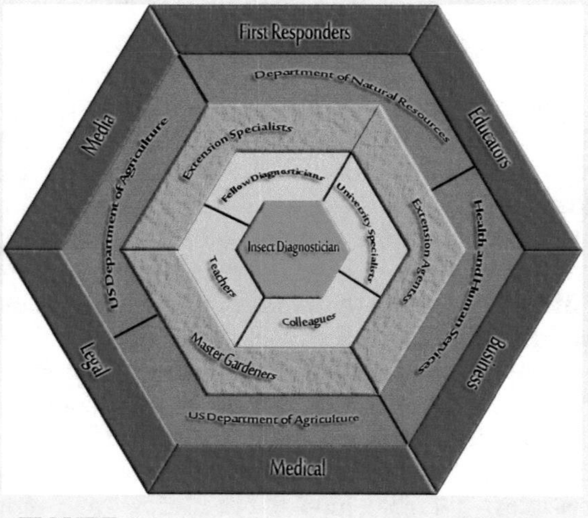

FIGURE 9.1 Insect Diagnostician networking tiers.

Tier 1: Colleagues

Peer Diagnosticians

No single person can be an expert in everything. Most diagnosticians have particular expertise, interest or training in certain groups of insects, or a particular crop, commodity, or pest management practice; they may feel most comfortable in making identifications and management recommendations in those areas. It is important that diagnosticians find and develop a working relationship with specialists in areas where they may not be strong. Shared expertise is the goal of networking.

Diagnostician networks can be formed and cultivated by attending professional conferences, working together in training meetings, sharing newsletters and sometimes co-authoring research or other publications. Conferring on samples and diagnostic issues by telephone or electronic communication strengthens networks.

DIAGNOSTICIANS IN OTHER DISCIPLINES

An insect diagnostician usually works within a laboratory, clinic or department in which others also work. Sharing the talents and training that diagnosticians from other disciplines may bring to the table is a vital first layer of networking. An insect diagnostician who has the luck to work closely with a plant pathology or a weed scientist diagnostician, for example, is extremely fortunate. If such a relationship is not afforded by proximity, these individuals must be sought out from within the university, state or at institutions elsewhere.

University Specialists

In academic, and to a lesser extent business institutions, many colleagues can be found with whom a diagnostician can network. Other entomologists, for example, can offer a new and beneficial perspective because of their unique training and specialty. These local colleagues and peers should become part of the first level or tier of networking.

Diagnosticians often consult with university faculty specialists regarding both identification and management recommendations. If diagnosticians serve in a department of entomology they may be blessed with an automatic network of contacts in various specialties. Scientists from departments outside of entomology are also excellent collaborators on a networking team.

Reaching out to other universities and colleges can yield yet another set of potential collaborators for the formation of networking alliances. It is important that diagnosticians find and develop a working relationship with other specialists and working professionals so that expertise can be shared.

Teachers

A student with graduate training in entomology naturally brings a set of experts and potential collaborators from academia to his or her new job, which includes teachers, professors, mentors and fellow students. Strong working relationships develop when both parties gain from the other and may endure for many years. Partners from a diagnostician's academic past make for strong and lasting network contacts.

Tier 2: Cooperative Extension

The Cooperative Extension Service is a nationwide, educational network that employs a very large cadre of entomology experts throughout the country. County extension agents reside in and serve most every county. Together with extension specialists, they become an important second tier of networking contacts.

To understand the Cooperative Extension system, it is instructive to briefly review the history of the land-grant university system in the United States.

Prior to the mid-1800s there were no public universities in the United States. Only private institutions existed, and tuition was often too expensive for the average family. In 1862 President Abraham Lincoln signed the Morrill Act into law, which gifted 10,000 acres of federal government land to each state. This was to be sold and the proceeds were to be used to create a public university (land-grant) to teach agriculture and the mechanical arts (engineering).

Today every state has a land-grant university. Although there are many public universities in most states, it is the land-grant university that has the responsibility for agricultural research, education and a major 'outreach' or extension education program directed towards the public.

In the land-grant system, there is much pride and satisfaction in being able to provide practical knowledge and information, based on unbiased scientific research, to both rural and urban citizens everywhere.

Most major land-grant universities also support diagnostic laboratories, including insect diagnostic laboratories, to meet the mission of the institution and to serve the public.

Extension Specialists

With the implementation of the Hatch Act in 1881, each state also was authorized to create an Agricultural Experiment Station with the purpose of conducting quality agricultural research. (Other countries also utilize Agriculture Research Stations in this manner and thus may be valuable networking components.)

In 1917, the Smith-Lever Act established the Cooperative Extension Service nationally with each state extension service administered through the state's land-grant university. The original purpose of the Cooperative Extension Service 'to disseminate useful and practical information' to the public has not changed significantly from then until today. Thus, the land-grant system encompasses three major missions:

1. Objective or unbiased research (done by the Experiment Stations);
2. Non-formal education and information dissemination (carried out by Extension Services); and
3. Classroom or college instruction (taught at each land-grant campus).

Today's modern land-grant university offers many programs other than agriculture and the mechanical arts, but the original mission is still unique relative to that of other public universities. The land-grant university has a major responsibility for 'outreach' to the general public.

Extension specialists are hired as experts to focus in one or more areas of agriculture teaching and research.

Extension specialists provide academic instruction via the classroom but also serve the public by providing non-formal or 'continuing education' through extension programs and applied research. They may be assigned to oversee the education and training of special clientele groups, usually commodity-based, such as field crops, fruits, horticulture or animals. Faculty extension specialists at the state campus also provide backup or supporting expertise to county agents. Extension specialists are critical associates in a networking environment.

Extension Agents

In addition to extension specialists and programs based at the land-grant university, local county extension offices are established to serve all 4,000 counties in the United States. Advisory committees at both the county and state level help to determine the priorities, programs, and educational needs of the local public.

Extension agents are employed in each county. These work specifically in agriculture, home economics, youth development or community development and generally become the first responders to educational needs within their counties. This vast army of extension agents provides useful, practical, and research-based information to agricultural producers, small business owners, youth, consumers, and others in their respective counties.

They rely on insect diagnosticians to provide expertise in areas where a generalist cannot. County extension agents are one of the primary users of a diagnostic laboratory. The lab is a vital resource to them. At the same time, county extension educators can be of great advantage to diagnosticians as they are often closest to the real-time happenings and needs in the state. A diagnostician actively networking with county extension agents makes for a very powerful alliance.

Below are the U.S. land-grant universities (listed by state) that receive Smith-Lever Act funds to operate their own Cooperative Extension program.

Alabama
- Alabama A&M University
- Auburn University
- Tuskegee University

Alaska
- University of Alaska Fairbanks

Arizona
- University of Arizona

Arkansas
- University of Arkansas
- University of Arkansas at Pine Bluff

California
- D-Q University
- University of California
- University of California, Berkeley

Colorado
- Colorado State University

Connecticut
- University of Connecticut

Delaware
- University of Delaware
- Delaware State University

Florida
- University of Florida
- Florida A&M University

Georgia
- University of Georgia
- Fort Valley State University

Hawaii
- University of Hawaii

Idaho
- University of Idaho

Illinois
- University of Illinois at Urbana-Champaign

Indiana
- Purdue University

Iowa
- Iowa State University

Kansas
- Kansas State University

Kentucky
- University of Kentucky
- Kentucky State University

Louisiana
- Louisiana State University
- Southern University and A&M College

Maine
- University of Maine

Maryland
- University of Maryland at College Park
- University of Maryland, Eastern Shore

Massachusetts
- University of Massachusetts Amherst
- Massachusetts Institute of Technology

Michigan
- Michigan State University

Minnesota
- University of Minnesota

Mississippi
- Mississippi State University
- Alcorn State University

Missouri
- University of Missouri
- Lincoln University

Montana
- Montana State University (Bozeman)

Nebraska
- University of Nebraska-Lincoln

Nevada
- University of Nevada, Reno

New Hampshire
- University of New Hampshire

New Jersey
- Rutgers University
- New Jersey Agricultural Experiment Station

New Mexico
- New Mexico State University

New York
- Cornell University

North Carolina
- North Carolina State University
- North Carolina A&T State University

North Dakota
- North Dakota State University

Ohio
- The Ohio State University

Oklahoma
- Oklahoma State University
- Langston University

Oregon
- Oregon State University

Pennsylvania
- The Pennsylvania State University

Rhode Island
- University of Rhode Island

South Carolina
- Clemson University
- South Carolina State University

South Dakota
- South Dakota State University

Tennessee
- University of Tennessee
- Tennessee State University

Texas
- Texas A&M University
- Prairie View A&M University

Utah
- Utah State University

Vermont
- University of Vermont

Virginia
- Virginia Polytechnic Institute and State University (Virginia Tech)
- Virginia State University

Washington
- Washington State University

West Virginia
- West Virginia University
- West Virginia State University

Wisconsin
- University of Wisconsin–Madison

Wyoming
- University of Wyoming

ASSOCIATED TERRITORIES

American Samoa

- American Samoa Community College

District of Columbia

- University of the District of Columbia

Guam

- University of Guam

Northern Marianas

- Northern Marianas College

Puerto Rico

- University of Puerto Rico at Mayagüez

Virgin Islands

- University of the Virgin Islands

Master Gardeners

The Master Gardener Program is very simple in concept but very powerful in practice. Experts in several crucial areas of gardening and horticulture, including entomology, provide intense training to interested individuals who then volunteer in their communities, giving lectures, creating gardens, caring for demonstration plots, and answering gardening questions from the public.

Since 1972, Master Gardener programs are active in all 50 US states as well as four provinces in Canada (McAleer, 2005). The most recent survey (2009 Extension Master Gardener Survey) estimates nearly 95,000 active Extension Master Gardeners, who volunteer over 5,000,000 community service hours per year. A total annual personal contacts total nearly 5 million and with media (TV, radio, print and internet), reach 102 million people. All said, this is a very powerful outreach program.

Insect diagnosticians often become involved in their local Master Gardener Program where they serve as professional trainers in insect identification and integrated pest management. Perhaps the greatest advantage of this involvement to a diagnostician is the instant network of thousands of gardening enthusiasts who are interested, willing and trained. Insect diagnosticians benefit from networking with master gardeners.

Tier 3: State Officials

USDA

State agencies are created at a national level with the commission to oversee and support specific public functions. In the United States, the United States Department of Agriculture (USDA), followed by the Environmental Protection Agency (EPA), the Department of Natural Resources (DNR) and the Department of Health and Human Services (HHS) are the agencies most closely aligned with the mission of an Insect Diagnostic Laboratory.

SYSTEMATIC ENTOMOLOGY LABORATORY

The USDA oversees an important role in insect diagnostics. This is largely done through a laboratory called the Systematic Entomology Laboratory (SEL). The SEL is home to some 37 scientists. A quick search of their staff expertise reveals that many of these are entomologists, highly trained in the systematics of one or more specific groups of insects. These scientists are considered as experts in their particular areas and can be a critical contact for consultation on specific identifications when needed by diagnosticians.

USDA Systematic Entomology Laboratory:

10300 Baltimore Avenue
Bldg. 005, Barc-West
Beltsville MD 20705, USA

The mission of the SEL is to conduct research to:

1. Develop comprehensive classifications, hypotheses of relationships, and identification tools for insects and mites on a world basis in support of U.S. agriculture and natural resources;

2. Provide identifications and associated taxonomic services to federal, state, and private organizations involved in research and action programs;
3. Develop and maintain, in cooperation with the Smithsonian Institution, the U.S. National Collection of Insects and Mites, a vital resource for insect and mite research and identification services;
4. Develop digital identification tools and databases of taxonomic and biological information.

It is quickly apparent that much of the SEL mission statement overlaps with the duties of insect diagnosticians, making the SEL a very valuable networking resource.

The SEL routinely provides specimen identification assistance as a free service to both governmental and private entities, including federal regulatory agencies, state departments of health and agriculture, university researchers, and private citizens. This service is coordinated by the SEL's Communications & Taxonomic Services Unit (CTSU) and relies on the expertise of SEL scientists and collaborating specialists. The CTSU maintains a relational database allowing for the efficient management of identification assignments. Upon completion, the CTSU is responsible for reporting identifications to the submitter and returning specimens when requested.

The SEL's Insect Identification Service is available to all US citizens to assist with specimen identification. The guidelines that follow provide useful tips to help get the most out of the Insect Identification Service.

SEL SERVICE LIMITATIONS

The SEL provides specimen identifications. The SEL does NOT provide extensive biological information along with its identifications, and lab personnel are not authorized to provide specific information regarding pest control strategies. Questions about managing pest species should be directed to state Cooperative Extension offices.

Not all specimens can be identified. Insects represent the most diverse group of animals on the planet, and existing knowledge of this diversity is still very limited. Additionally, many insects are only known from specific life stages (larva, pupa, or adult) or from a single sex. As a result, it is not uncommon for specialists to be unable to provide a complete species-level determination.

Submit specimens to the CTSU. If sent directly to scientists, there is a greater probability that specimens will be misplaced or lost. The CTSU maintains a database of all pending SEL identifications, and this system allows tracking of submissions within the lab.

SUBMISSION REQUIREMENTS

Due to the large volume of identification requests that are received by the SEL, submission requirements have been instituted to facilitate the processing of ID requests. The requirements that follow fall into two major categories: documentation and specimen preparation. Timely completion of requested identifications is dependent upon compliance with submission standards.

DOCUMENTATION

- All submissions must be accompanied by a completed copy of the SEL's Insect Identification form 748, provided in Figure 9.2.
- Label specimens with the following information and provide a detailed account of the circumstances under which they were collected:
- Provide clear return contact information.

Country: State: County
Town
Host or Specific Location
Date
Collector's Name

IDENTIFICATION REQUEST

SEL

Priority:	Lot Number:
Date Submitted:	Number of Specimens:
Date Needed:	Specimen Disposition: ☐ Return ☐ Keep/Discard
Submitter's Reference Number:	Tentative Identification:

Name:

Address:

Telephone: | FAX:

E-mail:

Affiliation:
- ☐ APHIS/PPQ
- ☐ ARS
- ☐ CICP
- ☐ Commercial Organization
- ☐ US Department of Defense
- ☐ Foreign
- ☐ US Forest Service
- ☐ Private Individual
- ☐ Other Federal (US)
- ☐ Other State Agency
- ☐ Private University
- ☐ State Agriculture Agency
- ☐ State University

Level of Identification Requested: ☐ Family ☐ Genus ☐ Species

Host:

Reason for Identification:
- ☐ A – Biological Control
- ☐ B – Damaging Crop/Plants
- ☐ C – Suspected Pest of Regulatory Concern
- ☐ D – Stored Product Pest
- ☐ E – Livestock, Wildlife, or Domestic Animal Pest
- ☐ F – Danger to Human Health
- ☐ G – Household Pest
- ☐ H – Possible Immigrant
- ☐ I – Reference Collection
- ☐ J – Survey
- ☐ K – Thesis Problem
- ☐ L – Other (elaborate below)

Project Support:
☐ APHIS/PPQ ☐ ARS ☐ DOI ☐ EPA ☐ FAO ☐ FS ☐ Hatch ☐ NIH ☐ NRCS ☐ NSF ☐ USAID ☐ Other

Collecting Permits:
☐ Required ☐ Not Required If required, please submit copies with specimens.

Project Description:

Remarks:

USDA

Communications & Taxonomic Services Unit – Systematic Entomology Laboratory
Building 005 – Room 137 – BARC-West
10300 Baltimore Avenue – Beltsville – Maryland – 20705

ars

OMB 0518-0032-sel-1 (2/2006) -- SEL Identification Request

FIGURE 9.2 SEL's Insect Identification form 748.

Privacy and Paperwork Reduction Act Statements

Privacy Act Information: This information is provided pursuant to Public Law 93-579 (Privacy Act of 1974) for individuals completing Federal records and forms that solicit personal information. The authority is Title 5 of the U.S. Code, sections 1302, 3301, 3304, and 7201.
Purpose and Routine Uses: The information from this form is used solely to respond to you regarding the service you have requested. No other uses will be made of this information.
Effects of Non-Disclosure: Providing this information is voluntary, however, without it we may not be able to respond to you regarding the service you have requested.
Paperwork Reduction Act Statement: A Federal agency may not conduct or sponsor, and a person is not required to respond to a collection of information unless it displays a current valid OMB control number.
Public Burden Statement: Public reporting burden for this collection of information is estimated to vary from two to four minutes with an average to three minutes per response, including time for reviewing instructions, and completing and reviewing the collection of information. Send comments regarding this burden estimate or any other aspect of this collection of information, including suggestions for reducing this burden, to Department of Agriculture, Clearance Officer, OIRM, Room 404-W, Washington, D.C. 20250, and to the Office of Information and Regulatory Affairs, Office of Management and Budget, Washington, D.C. 20503.

FIGURE 9.2 Cont'd

SPECIMEN PREPARATION
The ability of SEL scientists and collaborators to provide complete and accurate identifications depends heavily on the quality of the specimens that are submitted for examination. In short, all hard-bodied insects should be pinned or point-mounted while larvae and other soft-bodied specimens should be preserved in 70% ethanol. All specimens should be appropriately labeled with collection locality data. Additionally, all adult Lepidoptera should be submitted with their wings spread to allow for easy examination of wing venation.

Questions or inquiries regarding SEL's Identification Service should be directed to: e-mail: IDService@ars.usda.gov; phone: 301.504.7041 or FAX: 301.504.6482

GUIDELINES FOR SEL SUBMISSION SEL scientists actively collaborate with members of academia, providing taxonomic support for a broad range of basic and applied biological projects. Every attempt is made to provide timely service; however, SEL scientists' diversity of commitments and shortage of technical support staff limit the speed at which identification requests can be completed. The information that follows includes suggestions of steps that facilitate the processing of specimens.

General Considerations

1. Provide tentative IDs. In most cases, tentative ID to at least the family level is possible. This information is essential for CTSU staff to assign specimens to the appropriate specialist. By providing tentative IDs the amount of time CTSU requires to process each submission is reduced.
2. Sort specimens. If submitting a mixed-lot, please sort specimens by family. Separating taxa and repackaging them for distribution to multiple specialists can be a time-consuming task, especially for very large, diverse lots.
3. Contact specialists prior to submission. SEL scientists have a variety of professional commitments that limit the time they have available to perform routine identifications. Therefore, contact SEL specialists prior to submission and inquire about projected turn-around times.
4. Submit specimens to CTSU. If sent directly to SEL scientists, there is a greater probability that the specimens will be misplaced or lost. CTSU maintains a database of all pending SEL identifications,

and this system allows tracking of submissions within the lab.

Submission Requirements

- Please adhere to SEL's standard submission requirements.
- Include copies of collecting permits when required.

Contact Information

Communications & Taxonomic Services Unit
USDA-ARS-Systematic Entomology
Laboratory 005
Room 137
BARC-West
10300
Baltimore Avenue
MD 20705-2350
e-mail: IDService@ars.usda.gov
Phone: 301.504.7041
FAX: 301.504.6482

Guidelines for General Submissions

Communications & Taxonomic Services Unit
USDA-ARS-Systematic Entomology
Laboratory
Building 005, Room 137
BARC-West 1030
Baltimore Avenue
Beltsville, MD 20705-2350
e-mail: IDService@ars.usda.gov
Phone: 301.504.7041
FAX: 301.504.6482

Table 9.1 provides a quick reference to SEL's preferred methods for preparing adult insect specimens from specific taxonomic groups. Insect diagnosticians quickly learn the value of networking with personnel from the SEL. Because the SEL is a single location of a large number of experts in systematics, they can identify or validate identifications authoritatively. The major drawbacks of their services to diagnosticians and the public at large are:

1. length of turn-around time;
2. special sample preparation requirements; and
3. the single mode of sample submission (physical samples only).

ANIMAL AND PLANT HEALTH INSPECTION SERVICE

The USDA is charged specifically with protecting against the invasion of foreign pests through a program called the Animal and Plant Health Inspection Service (APHIS). Its mission is to protect agricultural health, specifically animal and plant resources, from agricultural pests and diseases. These efforts support the overall mission of the USDA which is to protect and promote food, agriculture, natural resources and related issues.

Port inspections are a vital regulatory effort and are considered the first line of defense against these pests. In the event that a pest of concern is detected, the APHIS implements emergency protocols and partners with affected states to quickly manage or eradicate the outbreak. This aggressive approach has enabled the APHIS to successfully prevent and respond to potential pest threats to U.S. agriculture.

For example, if the Mediterranean fruit fly and Asian long-horned beetle, two serious, exotic, introduced pests, were left unchecked, they would cause a loss of several billions of dollars in agriculture production annually.

A strong domestic agricultural pest detection system is an essential component in providing pest monitoring and surveys in rural and urban sites across the United States. The APHIS and its state cooperators carry out surveys for high-risk pests through a network of cooperators in the Cooperative Agricultural Pest Survey (CAPS) program.

CAPS

The CAPS program supports the APHIS's goal of safeguarding U.S. agricultural and environmental resources by ensuring that new introductions of harmful plant pests are detected before they have a chance to cause significant damage.

TABLE 9.1 Specimen Preparation Guidelines by Taxa

Taxon	Preferred Preparation Method
Acari (mites & ticks)	Ethanol or Slide
Auchenorrhyncha (planthoppers, treehoppers, froghoppers, etc.)	Pin or Point
Blattaria (roaches)	Pin
Coleoptera (beetles)	Pin, Point, or Ethanol (depending on size)
Collembola (springtails)	Ethanol or Slide (depending on size)
Dermaptera (earwigs)	Pin
Diplura	Ethanol or Slide (depending on size)
Diptera (true flies)	Pin, Point, or Ethanol (depending on size)
Ephemeroptera (mayflies)	Ethanol
Heteroptera (true bugs)	Pin or Point
Hymenoptera (bees, wasps, ants, etc.)	Ethanol, Pin, or Point (depending on size)
Isoptera (termites)	Ethanol
Lepidoptera (butterflies & moths)	Pin or Point
Mantodea (mantids)	Pin
Mecoptera (scorpionflies, earwigflies, etc.)	Pin or Point
Megaloptera (dobsonflies, fishflies, etc.)	Ethanol or Slide (depending on size)
Microcoryphia (bristletails)	Ethanol
Neuroptera (lacewings, antlions, owlflies, etc.)	Pin or Point
Odonata (dragonflies & damselflies)	Pin
Orthoptera (grasshoppers, crickets, & katydids)	Pin
Phasmida (walking-sticks)	Pin
Phthiraptera (lice)	Ethanol or Slide (depending on size)
Plecoptera (stoneflies)	Ethanol
Protura	Slide
Psocoptera (plant lice)	Ethanol or Slide (depending on size)
Siphonaptera (fleas)	Ethanol or Slide (depending on size)
Sternorrhyncha (aphids, scales, whiteflies, psyllids)	Ethanol or Slide (depending on size)
Strepsiptera (twisted-winged parasites)	Ethanol or Slide (depending on size)
Thysanoptera (thrips)	Ethanol or Slide (depending on size)
Trichoptera (caddisflies)	Ethanol
Zoraptera (webspinners)	Ethanol or Slide (depending on size)
Zygentoma (silverfish)	Ethanol

The CAPS program conducts science-based national and state surveys targeted at specific exotic plant pests identified as threats to U.S. agriculture and/or the environment. Survey activities are accomplished through cooperative agreements with state departments of agriculture, universities, and other entities. Surveys conducted through the CAPS program represent a second line of defense against the entry of harmful plant pests and weeds. These efforts support inspections of commodities, conveyances, and passenger baggage conducted by the Department of Homeland Security, Customs and Border Protection (CBP) at sea ports, airports, and land border crossings.

Additionally, the CAPS program develops commodity-based and resource-based surveys. These surveys enable the program to target high-risk hosts and commodities, gather data about pests specific to a commodity, and establish better baseline data about pests recently introduced in the United States.

The mission of the CAPS program is to provide a survey profile of exotic plant pests in the United States deemed to be of regulatory significance through early detection and surveillance activities.

Insect diagnosticians support USDA, APHIS and CAPS personnel in their mission of protecting US agriculture and the environment. In turn, diagnosticians are often the first to receive samples of introduced pests. The USDA relies on the expertise of diagnosticians to recognize potentially regulated pests. Close associations between diagnosticians and government agencies can strengthen homeland security by facilitating detection of potential threats and improving the training of first responders. Government regulators and identifiers are important to a diagnostician's professional network.

The National Plant Diagnostic Network

The National Plant Diagnostic Network (NPDN) was constructed to detect, diagnose, and respond to intentional and accidental introductions of plant diseases, insects and other pests that have been deliberately or accidentally introduced into U.S. agricultural and natural ecosystems, identify them, and report them to appropriate state and federal responders and decision-makers.

The NPDN facilitates plant diagnostic laboratory activities for a consortium of land-grant institutions, state departments of agriculture and the U.S. Department of Agriculture. These laboratories support insect identification and provide a common database to facilitate responses and record-keeping.

The NPDN was established in 2002 with support from the USDA and through the collective efforts of many individuals including diagnosticians representing land-grant universities, federal agencies, state departments of agriculture, and other stakeholders. The NPDN has grown into an internationally respected consortium of pest diagnostic laboratories.

Most diagnosticians share in the purpose of the NPDN, which is to provide a cohesive, yet widely distributed system to quickly detect and identify pests and pathogens of concern. NPDN laboratories immediately report critical findings to appropriate responders and decision-makers.

To accomplish its mission the NPDN has:

1. invested in diagnostic laboratory infrastructure and training;
2. developed an extensive network of first detectors through education and outreach; and
3. enhanced communication among public agencies and stakeholders responsible for responding to and mitigating new outbreaks (Table 9.2).

TABLE 9.2 NPDN Diagnostic Laboratories by State

Alabama (SPDN)	Maine (NEPDN)	Oregon (WPDN)
Alaska (WPDN)	Maryland (NEPDN)	Pennsylvania (NEPDN)
Arizona (WPDN)	Massachusetts (NEPDN)	Puerto Rico (SPDN)
Arkansas (SPDN)	Michigan (NCPDN)	Rhode Island (NEPDN)
California (WPDN)	Minnesota (NCPDN)	South Carolina (SPDN)
Colorado (GPDN)	Mississippi (SPDN)	South Dakota (GPDN)
Connecticut (NEPDN)	Missouri (NCPDN)	Tennessee (SPDN)
Delaware (NEPDN)	Montana (GPDN)	Texas (GPDN)
Florida (SPDN)	Nebraska (GPDN)	Texas (SPDN)
Georgia (SPDN)	Nevada (WPDN)	Utah (WPDN)
Hawaii (WPDN)	New Hampshire (NEPDN)	Vermont (NEPDN)
Idaho (WPDN)	New Jersey (NEPDN)	Virgin Islands - US (SPDN)
Illinois (NCPDN)	New Mexico (WPDN)	Virginia (SPDN)
Indiana (NCPDN)	New York (NEPDN)	Washington (WPDN)
Iowa (NCPDN)	North Carolina (SPDN)	West Virginia (NEPDN)
Kansas (GPDN)	North Dakota (GPDN)	Wisconsin (NCPDN)
Kentucky (SPDN)	Ohio (NCPDN)	Wyoming (GPDN)
Louisiana (SPDN)	Oklahoma (GPDN)	
American Samoa (WPDN)		
Guam (WPDN)		

Tier 4: Local Entities

First Responders

Insect diagnosticians are often called upon to conduct training. Diagnosticians in an academic institution, as part of the Cooperative Extension program, are responsible for training clients in a wide range of settings. Professional pest managers in agriculture, horticulture, urban buildings, landscapes, pesticide certification, and public health regularly invite diagnosticians to participate in training conferences, forums, workshops and seminars. In doing so, diagnosticians form valuable networks with many people and organizations that are considered first detectors. Establishing, cultivating and expanding relationships with potential first detectors significantly advances networking.

First detectors are people who may initially come into contact with a potential new or exotic pest situation capable of threatening our agriculture biosecurity. First detectors are trained to recognize possible threats and how to properly alert others. Because the rapid response of first detectors is so critical for managing a threatening invasive insect or other arthropod, the greater the number of trained first detectors that we can possibly have, the better.

First detectors are individuals who are:

- familiar with some national and/or local pests of concern and are constantly monitoring for high-risk pests;
- trained to successfully submit diagnostic samples to their local diagnostic laboratories;
- understand communication protocols.

Any people who are trained by a diagnostician through conferences, workshops or seminars qualify as first detectors. First detectors also may volunteer from any of several vocations, including but not limited to the following:

- Growers
- Cooperative Extension Service personnel
- Crop consultants and pesticide applicators
- Commercial chemical and seed representatives
- Master gardeners
- Amateur entomologists
- Survey entomologists
- Seed industry
- Agriculture industry
- Food industry
- Urban pest managers
- Health care professionals
- Farmers
- Landscapers
- Gardeners.

Some of these have already been discussed as in regards to their important roles in a diagnostician's network, but all can also serve on the front lines by assisting in rapid detection of exotic invasive insects. For example, master gardeners and extension agents are often the first ones consulted about strange or unusual problems. With training, growers, crop consultants, pesticide applicators commercial seed sales reps, landscapers, backyard gardeners, hobbyists or agricultural product representatives also can be very important first detectors, especially if they have a relationship with diagnosticians or identifiers.

Diagnosticians provide training to first detectors on proper techniques for sampling, monitoring, and identifying pests. They must also teach proper procedures for reporting pest problems and submitting samples or photos to be analyzed.

Rapid evaluation and reporting of potential pest threats, quick response time for diagnosis, real-time consultation with experts, web-based communications links among regional and national diagnostic labs, established links to regulatory agencies including the APHIS, high quality and uniform information coding for samples, high quality record-keeping and reporting of pest outbreaks: all these are made possible with networks of trained first detectors.

Media

Professional networks for diagnosticians must also include media specialists. The media is one of the most powerful tools that diagnosticians can have at their disposal. Having a relationship with media specialists, journalists, television reporters and internet specialists, in advance of an insect pest outbreak, can aid in quickly getting the word out to the public and to first responders. In addition, forewarning and reminding the public about predictable events, such as the annual invasion of perimeter pests and what to do to prevent them, can improve the effectiveness of insect diagnosticians.

A relationship with the media that is developed in advance of a crisis can be beneficial in establishing trust. If a journalist has a relationship with a diagnostician they can rely and act upon a diagnostician's judgment to know how important and how time-sensitive a story is.

Having the media as part of a networking team is fundamental.

Legal and Medical Experts

Insect diagnosticians are sometimes asked to assist in medical and legal cases. Criminal and forensic laboratories, as well as associated personnel, can be valuable parts of a network. Medical professionals are important in assisting with cases of human and health threats by exotic pests. In addition, being able to network with medical personnel in cases of human parasitosis or on delusory parasitosis can be of great value.

Medical and legal professionals can become important first detectors if trained by diagnosticians.

Educators

It is important to include private or public educators in a diagnostician's network. Educators may work in any of several entomological education venues including zoos, botanical gardens and butterfly houses. They often have need of a diagnostician's expertise and in turn can be first detectors.

SPECIALTY LABORATORIES AND INFORMATION SERVICES

Contact information for specialized diagnostic laboratories and information services should be kept on file by diagnosticians.

These laboratories and information services form part of a network that diagnosticians can use themselves or in referring clients. Many services can be found by searching the internet. The number of possible networking services available to diagnosticians is too large and varied for an exhaustive list here. We cite the following by way of example only:

Diagnostic Center For Population and Animal Health
4125 Beaumont Road
Lansing, MI 48910-8104
Phone: 517-353-1683

The Diagnostic Center for Population and Animal Health is a full-service veterinary diagnostic laboratory offering more than 800 tests, including Lyme disease testing for ticks.

UMass Extension Tick Assessment
Agricultural Engineering Building
250 Natural Resources Way
University of Massachusetts
Amherst, MA 01003
Phone: 413-545-1055

The UMass Laboratory for Medical Zoology, in cooperation with UMass Extension, identifies submitted ticks and tests them to determine whether or not they carry the bacterium that causes Lyme disease and other tick-borne pathogens.

Clongen Laboratories, LLC
12321 Middlebrook Road
Suite 120, Germantown, MD 20874
Phone: 301-916-0173

Clongen Laboratories provides services related to molecular diagnostics. They focus on using rapid detection methods such as PCR, real time PCR and DNA sequencing for identification of human pathogens with a specialty in tick-borne illness diagnostics. They also offer a highly sensitive quantitative Western Blot for Lyme disease.

Analytical Services, Inc.
130 Allen Brook Lane
Williston, VT 05495
Phone: 800-723-4432

Analytical Services, Inc. provides comprehensive environmental microbiological services for microbiological issues associated with humans and the environment. Part of their services includes testing of ticks for diseases.

National Poison Center
Phone: 1-800-222-1222

For a POISON EMERGENCY or information regarding poisoning; covers the entire United States. The center is manned 24 hours a day,

7 days a week by experts in poisoning. This is a free and confidential service. All local poison control centers in the United States use this national number. You should call if you have any questions about poisoning or poison prevention.

National Pesticide Information Retrieval System
1435 Win Hentschel Blvd, Ste. 207
West Lafayette, IN 47906-4161
FAX:765-494-9727
Website: http://state.ceris.purdue.edu

The National Pesticide Information Retrieval System public website contains information pertaining to pesticides either currently or previously licensed for distribution and sale in the United States and is provided for informational purposes only.

The National Pesticide Information Center
Phone: 1-800-858-7378
Web site: npic@ace.orst.edu

This center is a collaboration between Oregon State University and the United States Environmental Protection Agency created to provide objective, science-based information about a wide variety of pesticide-related topics, including pesticide product information, information on the recognition and management of pesticide poisonings, toxicology and environmental chemistry.

Highly trained specialists can provide referrals for the following: laboratory analyses, investigation of pesticide incidents, emergency treatment information, safety information, health and environmental effects, and clean-up and disposal procedures. NPIC also produces pesticide fact sheets, frequently-asked questions and podcasts.

Plant Protection and Quarantine (PPQ) Containment Facilities

PPQ offers the following information on their website (www.aphis.usda.gov/plant_health/permits/organism/containment_facility_inspections.shtml) that relates to constructing and using a containment facility:

CONSTRUCTING AND USING A CONTAINMENT FACILITY

One of the purposes of PPQ permits is to prevent the dissemination of plant pests into or through the United States. As a consequence, PPQ may only issue permits for certain organisms when the receiving facility can adequately contain the organisms so as to prevent dissemination of the organisms.

Containment of regulated organisms may be accomplished by a combination of proper handling of the regulated organisms and by physical and security attributes of the premises where the regulated organisms will be held. PPQ containment specialists can determine the adequacy of a facility either by a computer assisted facility evaluation (CAFÉ) or by a full inspection which includes an on-site inspection of the facility. The criteria used to determine the kind of evaluation (CAFÉ or full) needed before permit issuance is described below.

All facilities with PPQ-regulated organisms may be inspected at any time by agency officials during normal business hours.

Criteria used to Determine the Kind of Evaluation (CAFÉ or Full) Needed Before Permit Issuance

- Guidance for Containment Evaluation of Arthropod/Snail Permit Applications When Containment is Required

CONSTRUCTING AND USING A CONTAINMENT FACILITY *(cont'd)*

- Guidance for Containment Evaluation of Plant Pathogens and Noxious Weeds Permit Applications When Containment is Required
- Guidance for Containment Evaluation of Diagnostic Permit Applications

Related Information

If you are applying for a PPQ permit that meets the criteria for an inspection and your facility or laboratory has not been previously inspected, then you should anticipate a one to three month delay in processing your application.

When PPQ evaluates the containment capabilities of a facility, physical and operational characteristics are examined relative to the risks of the organisms and their methods of actual or potential dispersal from the facility. The operational characteristics of the facility are developed by the applicant and the attached Outline for Standard Operating Procedures (SOPs) may be used to prepare this document. For butterflies, use the attached Outline for Standard Operating Procedures. These documented operational features will be confirmed during the inspection.

APHIS has developed guidelines for the containment of organisms. These guidelines vary depending upon the type of organism to be contained and the risk posed by those organisms.

Not all elements in the following guidelines apply to all organisms to be contained and the guidelines are guidelines not regulations:

- Containment Guidelines for Educational Displays of Adult, Butterflies and Moths (Lepidoptera)
- Containment Facility Guidelines for Noxious Weeds and Parasitic Plants
- Containment Guidelines for Non-Indigenous, Phytophagous Arthropods and Their Parasitoids and Predators
- Containment Guidelines for the Receipt, Rearing and Display of Non-Indigenous Arthropods in Zoos, Museums, and Other Public Displays
- Containment Guidelines for Plant Pathogenic Nematodes
- Containment Guidelines for Non-Indigenous Snails
- Containment Guidelines for Plant Pathogenic Bacteria
- Containment Facility Guidelines for Viral Plant Pathogens and Their Vectors
- Containment Facility Guidelines for Fungal Plant Pathogens
- Before beginning construction of a containment facility, you should contact Pest Permit Evaluations staff and discuss our containment requirements with a Containment Facility Evaluation Specialist.

Frequently Asked Questions

View a list of the commonly asked questions and concerns associated with containment facility inspections.

Contact Information

USDA/APHIS/PPQ/PHP/PPB
4700 River Road, Unit 133
Riverdale, MD 20737
Phone: 866.524.5421
Fax: 301.734.5392
Email: Pest.Permits@aphis.usda.gov

References

Arnett, R.H., Jacques, R.L., 1981. Simon & Schuster's Guide to Insects. Simon & Schuster, New York, 512pp.

Blum, M.S., Woodring, J.P., 1963. Preservation of insect larvae by vacuum dehydration. J. Kans. Entomol. Soc. 36, 96–101.

Bennett, G., Owens, J.M., Corrigan, R.M., 1997. Truman's Scientific Guide to Pest Control Operations, fifth edn. Advanstar Communications, Inc., Santa Monica, 520pp.

Borror, D.J., Triplehorn, C.A., Johnson, N.F., 1989. An Introduction to the Study of Insects, sixth edn. Sanders College Publishing, Philadelphia, 875pp.

Borror, D.J., White, R.E., 1998. A Field Guide to Insects: America North of Mexico. Houghton Mifflin Co, Boston, 404pp.

Bosik, J.J., 1997. Common Names of Insects and Related Organisms. Entomol. Soc. of America, College Park, MD, 232pp.

Chapin, J.B., 1989. Common names of insects. Bull. Entomol. Soc. Am. 35 (3), 177–180.

Cranshaw, W., 2004. Garden Insects of North America. Princeton University Press, Princeton and Oxford, 656 pp.

Dallwitz, M.J., 2005. A comparison of interactive identification programs. http://delta-intkey.com/www/comparison.htm.

Dallwitz, M.J., 2006. Programs for interactive identification and information retrieval. http://deltaintkey.com/www/idprogs.htm.

Ekbom, K.A., 1938. Der praesenile dermatozoenwahn [presenile delusional parasitosis]. Acta. Psychiatr. Neurol. Scand. 13, 227–259.

Fessenden, G.R., 1949. Preservation of agricultural specimens in plastics. U.S. Dept. Agric. Misc. Publ. 679.

Fichter, G.S., Zim, H.S., 1966. Insect Pests. Golden Press, New York, 160pp.

Foote, R.H., 1948. A synthetic resin imbedding technique. U.S. Public Health Serv. CDC Bull. 1948 (July–Sept.), 58–59.

Gibb, T.J., Oseto, C.Y., 2006. Arthropod collection and identification-laboratory and field techniques. Elsevier/Academic Press, New York, 311pp.

Gordh, G., Hall, J.C., 1979. A critical point drier used as a method of mounting insects from alcohol. Entomol. News 90, 57–59.

Gotelli, N.J., 2004. A taxonomic wish-list for community ecology. Philos. Trans. R. Soc. London, B. 359, 585–597.

Gurney, A.B., 1953. An appeal for clearer understanding of the principles concerning the use of common names. J. Econ. Entomol. 46, 201–211.

Hebert, P.D.N., Cywinska, A., Ball, S.L., deWaard, J.R., 2003. Biological identifications through DNA barcodes. Proc. R. Soc. London B. 270, 313–321.

Hinkle, N.C., 2000. Delusory parasitosis. Am. Entomol. 46, 17–25.

Hinkle, N.C., 2010. Ekbom's Syndrome: The challenge of 'invisible bug' infestations. Ann. Rev. Entomol. 55, 77–94.

Hocking, B., 1953. Plastic embedding of insects – a simplified technique. Can. Entomol. 85, 14–18.

Inouye, D.W., 1991. Quick and easy insect labels. J. Kans. Entomol. Soc. 64, 242–243.

Jacques, H.E., 1987. How to Know the Insects. William C. Brown Co, Dubuque, Iowa, 205pp.

Johnson, W.T., Lyon, H.H., 1976. Insects that Feed on Trees and Shrubs. Cornell University Press, Ithaca, NY, 556pp.

Kaston, B.J., 1978. How to Know the Spiders. William C. Brown Co, Dubuque, Iowa, 272pp.

Koblenzer, C.S., 1993. The clinical presentation, diagnosis and treatment of delusions of parasitosis: a dermatologic perspective. Bull. Soc. Vector Ecol. 18, 6–10.

Kushon, D.J., Helz, J.W., Williams, J.M., Lau, K.M.K., Pinto, L., St. Aubin, F.E., 1993. Delusions of parasitosis: a survey of entomologists from a psychiatric perspective. Bull. Soc. Vector Ecol. 18, 11–15.

Lim, C.S.H., S.L. Lim, S.L., Chew, F.T., Ong, T.C., Deharveng, L., 2009. Collembola are unlikely to cause human dermatitis. J. Insect. Sci. 9, 3.

Levi, H.W., Levi, L.R., 1990. Spiders and Their Kin. Golden Press, New York, 160pp.

Mallis, A., 2004. Handbook of Pest Control. GEI Media, Inc., Toronto, 1400pp.

Marshall, S.A., 2006. Insects: their Natural History and Diversity: With a Photographic Guide to Insects of Eastern North America. Firefly Books Richmond Hill, Can., 704pp.

Mayr, E., 1969. Principles of Systematic Zoology. McGraw-Hill, New York, 428pp.

McAleer, P., 2005. A national survey of master gardener volunteer programs. CSREES April 2005. www.csrees.usda.gov/nea/plants/pdfs/survey_master_gardener_programs.pdf.

Metcalf, C.L., Flint, W.P., 1962. Destructive and Useful Insects. McGraw-Hill Book Company Inc., New York, 1087pp.

Mitchell, R.T., Zim, H.S., 1964. Butterflies and Moths. Golden Press, New York, 160pp.

Peterson, A., 1960. Larvae of Insects. Edwards Brothers, Inc, Lillington, NC. Vol. 1, 416pp. and Vol 2, 315pp.

Poorbaugh, J.H., 1993. Cryptic arthropod infestations: separating fact from fiction. Bull. Soc. Vector Ecol. 18, 3–5.

Pankhurst, R.J., 1991. Practical Taxonomic Computing. Cambridge University Press, Cambridge, UK, 214pp.

Quicke, D.L.J., 1993. Principles and Techniques of Contemporary Taxonomy. Blackie Acad. & Prof, London, 311pp.

Shull, E.M., 1972. Butterflies of Indiana. Indiana Academy of Science. 262pp.

Stehr, F.W., 1987. Immature Insects. Kendall/Hunt Publishing Company, Dubuque. IA, 754pp.

Stoetzel, M.B., 1989. Common Names of Insects and Related Organisms. Entomol. Soc. of America. College Park, MD, 199pp.

Triplehorn, C.A., Johnson, N.F., 2005. An Introduction to the Study of Insects, seventh edn. Thomson, New York, 864pp.

Voss, E.G., 1952. The History of Keys and Phylogenetic Trees in Systematic Biology. J. Sci. Lab. Denison. Univ. 43, 1–25.

Walter, D.E., Winterton, S., 2007. Keys and the crisis in taxonomy: extinction or reinvention? Ann. Rev. Entomol. 52, 193–208.

Waltz, R.D., McCafferty, W.P., 1984. Indication of mounting media information. Entomol. News 95, 31–32.

Weaver III, J.S., White, T.R., 1980. A rapid, steam bath method for relaxing dry insects. Entomol. News 91, 122–124.

Webb, J.P., 1993. Case histories of individuals with delusions of parasitosis in southern California and a proposed protocol for initiating effective medical assistance. Bull. Soc. Vector Ecol. 18, 16–25.

Westcott, C., 1973. The Gardener's Bug Book. Doubleday and Company, New York, 689pp.

Wilson, J.W., Miller, H.E., 1946. Delusion of Parasitosis. Arch. Dermatol. Syphilol. 54, 39–56. Chicago.

Zim, H.S., Cottam, C., 1956. Insects. Golden Press, New York, 160pp.

Index

Note: Page numbers followed by "f" indicate figures; "t" tables; "b" boxes.

Q

R

S

T

U

V

W

Printed in the United States
By Bookmasters